GERMAN WARSHIPS
OF WORLD WAR I

GERMAN WARSHIPS OF WORLD WAR I

The Royal Navy's Official Guide to the Capital Ships, Cruisers, Destroyers, Submarines and Small Craft, 1914–1918

INTRODUCTION BY NORMAN FRIEDMAN

Greenhill Books, London

This edition of *German Warships of World War One* first published 1992
by Greenhill Books, Lionel Leventhal Limited, Park House,
1 Russell Gardens, London NW11 9NN

British Library Cataloguing in Publication Data

Royal Navy
German Warships of World War One: Royal Navy's Official Guide to Capital
Ships, Cruisers, Destroyers, Submarines and Small Craft
I. Title
623.8

ISBN 1–85367–123–1

Publishing history

The Confidential Books which form the basis of this work were originally printed and bound in individual parts, each with its own sequence of page, figure and plate numbers. These have been retained to make sense of the text references to the illustrations, but a new continuous sequence of folios has been added to the foot of each page; the Contents page and Index refer to these new page numbers. All the original illustrations have been included, although in some cases the positions and sizes have been modified to suit modern book production methods. Individual parts, corresponding to the Section divisions of this book, were revised from time to time and those issued nearest the end of the First World War – the most complete – have been chosen for reproduction, but they are not all of a single date. In reprinting in facsimile from the wartime originals any imperfections in the typesetting are inevitably reproduced and the quality may fall short of the high standards normally to be expected in a modern book.

Printed by Butler & Tanner Limited, Frome, Somerset

CONTENTS

INTRODUCTION

This reference work, issued for the first time since 1918, is drawn from the series of Confidential Books compiled and regularly updated by British Naval Intelligence during the First World War, which were carried to sea by British warships when they faced the German fleet. The information was obtained painfully, by long observation, by interrogation of captured officers and by capturing German handbooks, and was then distilled into these Confidential Books. They were used by the Navy for reference partly on German naval practice and matériel, and partly for actual combat. They were issued in parts, each devoted to a particular type of warship, and summarised under standard headings the essential features of each ship or class. Thus in combat, for example, submariners who often estimated range by measuring the angle subtended by a known height, could check the tables of heights for the various German capital ships in these books.

It was not only the British navy to whom this information was invaluable. When the United States entered the war in April 1917, it had only a rudimentary intelligence service. The British supplied the printing plates of the books to the US, and in turn they became the standard US publications. Thus the Confidential Books, in the late 1918 editions from which *German Warships of World War One* is compiled, represented what both major navies knew on the eve of the Armistice, a knowledge that shaped their policies regarding, for example, the disposition of the German fleet.

The situation in 1918 was quite unlike that of 1945. Germany had not surrendered at the end of World War One; she had agreed to an armistice during which peace terms would be worked out. Germany was not occupied; former members of the Imperial Navy were not obliged to answer whatever operational or technical questions the winners might think to ask. Indeed, a German government survived the war. Allied officers could and did visit Germany to inspect and to verify disarmament (much as UN officials have visited modern Iraq), but the core of the German military and naval establishments resisted. The picture of the German wartime navy given in *German Warships of World War One* was the only one available outside Germany until partial details of the Kaiser's navy were published in the 1930s, in such works as *German Warships, 1815–1945* by Erich Groener. Even then, the data that were released were not altogether consistent.

The Allies demanded at the Armistice that the most modern German warships be interned in a neutral port, but shortly changed this requirement to internment at Scapa Flow. This fleet yielded relatively little information, however. While the Allies argued as to who would get which ships, the Germans still on board scuttled the fleet. A few ships were saved by grounding, but most, including all but one of the capital ships, ended up on the bottom of Scapa Flow. The battleship SMS *Baden* was saved for British tests, and detailed reports were made of her equipment, general design, and protection. Even then the Allies had to guess at the rationale for her fittings, often without the sort of detailed prior knowledge that could have proved illuminating. For example, the Germans never explained their concepts of armour or of underwater protection. Other capital ships formed the basis of an epic salvage story in the 1930s.

As *German Warships of World War One* reveals, by the very absence of information, the British (and by extension the other Allied navies) were unaware until after the war that the Germans had used stereo rather than coincidence rangefinders. That was why British anti-rangefinder measures (the optical equivalents of jammers) had been totally unsuccessful. In the absence of reliable German operational accounts, the British decided postwar that the stereo instruments they recovered from the sunken fleet had been ineffective, because they had required too high a degree of concentration. Other navies thought otherwise: the US Navy, for example, did adopt stereo rangefinders. In this, as in other ways, *German Warships of World War One* is an invaluable guide to what the Allies did, or did not, know at the end of hostilities.

U-boats, on the other hand, provide an example of the positive information conveyed by the Confidential Books. They were the most interesting examples of German naval technology, the only area in which the Allies would admit great German superiority. As in 1945, submarine technology was then the greatest naval prize. The main German advance in this field was in powerplants, particularly in submarine diesels, work on which had pressed ahead very vigorously in war-time. By 1918 U-boats were substantially more reliable than anything the Allies possessed. For example, the U-cruisers could operate for four months at a time without returning to base. By way of contrast, typical British submarines went to sea for a week or two at a time and US boats were worse.

During the war the Allies obtained many U-boats, and enormous effort went into analysing them. After the Armistice many were surrendered (not interned with their crews aboard) and the Allies ran extensive postwar trials despite German attempts to sabotage them. U-boat technology, particularly in diesel engine design, was the most thoroughly exploited by all the Allies in the naval field. However the evidence of that technology was only the mute surrendered U-boats, not their designers and builders, or even their blueprints. Thus exploitation of German technology often resulted merely in blind copying of what had been seized. For example, for over a decade after World War One, the standard US Navy submarine diesel was an unlicensed (and unassisted) copy of a standard German MAN submarine engine. The Versailles Treaty after all prohibited the Germans from further development of submarine technology (the US Navy complained that the Japanese, oblivious to the treaty, were quite happy to hire German engine designers and to buy new diesels from Germany). By way of contrast, after 1945 both the West and the Soviets managed to acquire not merely German submarine parts but also those responsible for their design.

Former U-boat commanders had little interest in revealing their methods of operations and the lessons they had learned. The German official histories that did appear, and which were obtained and translated by other navies, concentrated on overall operational issues, not on individual submarine tactics and tactical lessons, or on the efficacy of wartime ASW measures. Thus the notes on U-boat operational procedure and tactics published in *German Warships of World War One* are particularly important. They were virtually all the Allies ever discovered about just how the U-boats had operated against them, for they learned almost nothing further after the Armistice (they got hardware but not publications or evaluations). Furthermore, World War One U-boat tactics became the basis of World War II operations, making this information even more valuable to the Allies.

Data described in *German Warships of World War One* as 'particularly reliable' are significant for they reflect major British intelligence successes, often prewar. For example, the British bought the plans of the battlecruiser *Seydlitz* in 1913, and analysed her design, a process we might call reverse naval architecture, as was in the case, for example, of Soviet warships. Admiral Jellicoe was much impressed by such data, which convinced him that the Germans built true fast battleships, unlike his relatively flimsy battle-cruisers. Jellicoe published some of the results of this analysis in his postwar memoirs, as evidence that his escape at Jutland had been narrow indeed.

This reference work paints a more rounded picture of the German navy than we get from later descriptions of German warships in other respects than the purely technical. For example, destroyer and submarine tactics are described in some detail. The former section includes a generally neglected topic, German counter-measures to Allied submarines. After all, the postwar US Navy was quite impressed by the use of British submarines to warn of wartime sorties by the German fleet. One submarine even torpedoed the battle-cruiser *Moltke* during an abortive 1918 anti-convoy operation, and the Germans were forced to arm their own destroyers to counter the British submarine force.

Generally, comparison with data now available shows that British naval intelligence was quite accurate, although it erred somewhat on the side of expecting the Germans to complete programmes more quickly than was actually the case, and apparently in missing the general abandonment of capital ship construction after 1917. For example, the British did not realise that the dreadnought SMS *Rheinland* was never repaired after severe grounding in the Baltic. Overall, the internal evidence of the 1918 Confidential Books suggests that British naval intelligence was much better in the German naval bases than in the surface ship-building yards. In some cases, for example, as in the list of war-built destroyers, signals intelligence would have indicated the existence of ships without giving much information on their characteristics.

As British intelligence did not register the decision to cease building capital ships, the last two ships of the *Bayern* class were treated as completed (they were launched but stopped well short of completion). Similarly, the battle-cruiser *Mackensen* is assumed to have been completed (and was thus demanded as part of the interned fleet at Scapa Flow), although in fact work on her had ceased in 1917; she was also a new design not a sister to the *Hindenburg*). At that time British fears of new German battle-cruiser construction were such that the Government tried to buy or lease the *Kongo* class battle-cruisers from Japan. Although that country was allied to Britain, apparently the Royal Navy did not want Japanese participation in the Grand Fleet. We know now that *Mackensen* was the first of a class of four, of which both she and *Graf Spee* were launched in 1917, armed with 14-in guns (rather than the 12-in of the *Derfflingers*). A further class of three was planned but never brought near completion (work continued after 1917 only to keep the yard work forces employed).

British intelligence also somewhat underestimated the last German light cruiser programme. Although it correctly stated that only two ships had been completed, it listed only four others, whereas eight had been laid down. Of these, five had actually been launched, and some were quite close to completion. On the other hand, the British seem to have believed that the Germans were building rather faster than they were; the first ship of the class, SMS *Cöln*, was listed as completed for trials in November 1917 (she was actually commissioned in January 1918).

There is one interesting omission. In 1916 the Germans considered whether to build either small cruisers ('fleet cruisers') vaguely similar to the contemporary British C-class, or else large destroyer leaders. They chose the big destroyers, displacing about 2,300 tons, and armed with 5.9-in guns and with a new 60cm (23.6-in) torpedo tube. Ships were ordered from all four of the main yards, and two of them, *S 113* and *V 116*, were completed post-war, and taken over by the French and Italian navies. These ships were considered revolutionary for their time, and they inspired both navies' interest in what became known as super-destroyers. They were launched in January and March 1918, at Elbing (Schichau) and at Stettin (Vulkan).

There is considerable evidence beyond these pages that the Allies never suspected that the Germans were building destroyers with cruiser guns. For example, a US programme to mount 5-in guns on board flush-deck destroyers was justified entirely on the basis that German U-boats were appearing with 5.9-in guns, with no reference whatsoever to expected surface opposition.

The Confidential Books were issued as a series of pamphlets, each covering a different range of classes. Thus the capital ship section of *German Warships of World War One* which is dated October 1918, is correct as of late November. The cruiser section reflects information up to about July 1918; the destroyer section up to August. The submarine section is dated April 1918. Each section was separately paginated, but a consistent series of page numbers has been applied in this compiled volume.

The many paintings of German warships accompanying the text are by Dr Oscar Parkes, the noted British naval expert who was later long-time editor of *Jane's Fighting Ships* and who wrote the first of the modern run of design histories, *British Battleships* (first published by Seeley Service in 1957, shortly after Parkes died). Presumably paintings were used to avoid reproducing the backgrounds of the photographs used as source materials, which would disclose where and when the photographs had been taken. Since the German fleet spent very little of its time at sea, virtually all the source material must have been collected at the main German bases. On the other hand, photographs of U-boats could easily be published, since so many of them had surrendered at sea or had been salvaged after sinking in shallow coastal waters. The reader will notice that some paintings are marked 'from a sketch' implying that most were taken from photographs. This practice was paralleled in recent US publications on Soviet military power, in which paintings were often substituted for unpublished satellite photographs.

More than seventy years later, the information contained in *German Warships of World War One* remains of value, for it is our best guide to what the British did, and did not, know of their enemy during the war, and therefore is a guide to the logic of much of British wartime naval policy. In its wealth of detail, a rich picture of the German navy emerges – one in which the Navy trusted as it faced the German warships at the Falklands, Dogger Bank, in the U-boat war and at Jutland.

1992 **Norman Friedman**

SECTION I
GENERAL REMARKS

GENERAL REMARKS CONCERNING SHIPS, ETC.

Accommodation.—The following detailed information is derived from a German official publication, and is believed to apply to the earlier German Dreadnoughts. In general it may be stated that the accommodation in German vessels appears, as a rule, more crowded than in British vessels of a similar class. This is due mainly to the larger complements carried, and is partly counteracted by the neat and convenient arrangements for stowage of clothes, bedding, etc., and by the manner in which every available space is utilised.

Description of Accommodation.	Accommodation provided.	Remarks.
CABINS.*		
Admiral commanding Squadron or Division.	1 reception and dining room - 1 day cabin. 1 sleeping cabin. 1 bathroom with closet.	Cabins leading as far as possible from one to another; preferably on starboard side of the ship where the cabins do not occupy the whole breadth of the ship
Captain - - -	Do. do. - -	In flagships a special captain's mess is not fitted.
Chief of Staff - -	1 day cabin - - - - - 1 sleeping cabin. 1 bathroom with closet.	In flagships with squadron staff only.
Executive Officer -	1 day cabin - - - - 1 sleeping cabin. (In small cruisers, a single larger cabin only.)	In as quiet a place as possible on the upper deck, and well lighted.
Officers: Navigating, Gunnery, Torpedo Officers, Watchkeepers, Adjutant, Surgeons, Engineers, Fleet Constructor, Paymaster, Chaplain.	1 cabin each - - - -	The cabins on the upper deck are primarily for the Navigating Officer, Gunnery Officer, and Staff Surgeon. The bunks are placed amidships.
Sub-Lieutenants: (For every 2 Sub-Lieutenants).	1 cabin as above, or an enclosed living space (berth) and wash-place combined.	With double bunk or sofa bunk for 2.
Disposable: (For the use of supernumeraries: Constructor, Assistant Constructor, Reserve Officers, Umpires at manœuvres, &c.).	1 cabin with 2 bunks; in small cruisers with 1 bunk.	All officers' cabins are to be adequately lighted by natural light.
Squadron or Division Staff: Flag-Lieutenant, Engineer, Paymaster, Chaplain, Signal Officer, Admiralty Staff Officer, Surgeon, Judge-Advocate, and Secretary.	1 cabin each - -	Flag-Lieutenants' and Signal Officers' cabins are to be as close as possible to the fore bridge. Admiral's Secretary's cabin next to the Admiral's office.
Warrant Officers: Senior Chief E.R. Warrant Officer, Gunner, Navigating Warrant Officer, Boatswain, Storekeeper, Pumpmaster, Master-at-Arms, Carpenter.	1 cabin each - - - -	In flagships a cabin is provided for the bandmaster as well if there is sufficient space. If not, he is accommodated with the carpenter.

* Officers' cabins are fitted for steam-heating. The bunks, wherever possible, are placed away from the ship's side against the inner fore-and-aft bulkhead. Floor space in general 8' 4" long and 8' 3" wide. Height 6¾' to 7½'. The walls and partitions are of steel and are covered with an insulating material.

C 4

Description of Accommodation.	Accommodation provided.	Remarks.
Warrant Officers—*cont.* The rest of the warrant officers borne as part complement: for every 2 W.O.s.	1 cabin - - - -	If necessary 3 Warrant Officers may occupy 1 cabin.
Cooks and stewards -	1 cabin.	
If more than 4 are allowed in complement.	2 cabins.	
BATHROOMS AND WASHPLACES.		
Officer commanding Squadron, Division, or Flotilla, Chief of Staff, Captain.	1 bathroom each, with fixed bath and shower-bath, adjoining the sleeping cabin.	When there is lack of space, a common bathroom is provided for the Captain and the Chief of Staff.
Members of the Officers' Mess.	2 in battleships and large cruisers (including 1 for the engineers); —in other ships 1 bathroom as above.	
Midshipmen and Cadets.	In training ships, as large a room as possible, with shower-baths and hand basins (1 for every 2 cadets, and 1 bath partitioned off).	If there is not enough space for the provision of a separate bathroom with fixed bath, the midshipmen are to make use of the bath accommodation provided for members of the Officers' Mess.
Members of Warrant Officers' Mess.	Bathroom, with separate shower-baths for one-third of the complement of W.O.s of the engine-room personnel.	With partition—the height of a man—seat, 3 cupboards, and waterproof curtains. In the larger ships, if there is space, a second bathroom with bath may be provided for the other Warrant Officers also.
Other Engine-room ratings.	Bathroom with washing and shower-bath arrangements for at least one-sixth to one-third of the ratings of the engineering branch, 1 shower for every 2–3 men, and 1 hand basin for each man.	These rooms have two entrances. Half the showers and all hand basins have fresh-water supply, two-thirds of the showers salt-water supply. Partition for P.O.s, 5 ft. high.
Remainder of the ship's company.	Joint use of the bathrooms for ratings of the engineering branch.	In addition, in suitable places 2 to 10 hand basins, each with a tap from fresh-water supply, and a few shower-baths with fresh-water supply and drainage, are provided.
SICK BAY - -	- - - - - - -	Situation must be light and airy. Entrance at least $3\frac{1}{4}$ ft. broad. Battleships and large cruisers are provided with swinging cots for 1·5 per cent., small cruisers for 2 per cent. of complement. Floor space for each swinging cot is at least $53\frac{3}{4}$ sq. ft.; and there is an interval of at least $17\frac{3}{4}$ in. between 2 cots, so that they can swing 20 degrees either side. Bathroom close to sick bay with bath and shower-bath, fresh-water supply and heating apparatus.
DISPENSARY - -	- - - -	As near as possible to the sick bay. In ships with complement of 100 and over, at least $8\frac{1}{4}$ ft. deep, $8\frac{1}{2}$ ft. broad, with regulation fittings. Must be easily accessible at night, light and airy. Small ships are fitted with a cupboard for medicines and surgical dressings, or carry a medicine chest only.

Description of Accommodation.	Accommodation provided.	Remarks.
DRESSING STATION -	- - - - - - -	In battleships and large cruisers, 21¼ ft. long, 14 ft. broad, without hatchways, and in addition a gangway at least 5 ft. broad. Behind armour, cool, easily accessible, and as far as possible, enclosed. Near well-protected and airy spaces which are suitable for accommodating wounded. Space for depositing 10 per cent. of the complement. Hatchways and doors for transport, 4¾ ft. × 2¼ ft. Other ships, in so far as they possess dispensaries, are provided with a surgical dressing cupboard in a protected position. 2 operating tables 6¼ ft. × 2¼ ft., with a 7½ ft. gangway between. At the free ends and sides a space at least 2 ft. wide. Arrangements are fitted for lighting, ventilation, water supply and drainage.
CELLS - - -	Battleships, 3–4. Large cruisers, 3. Small cruisers, 1.	Contain plank bed with hinged table. Must not be near the engine rooms, and must be well ventilated.
DRYING ROOMS -	43 sq. ft. floor space in all ships for every 50 men. Existing sources of heat to be utilised if possible. If the room is used for other purposes as well, *e.g.*, as a sleeping place, double the floor space must be provided.	Ceilings and free walls at sides to be insulated with cork. With a temperature on deck of 14° F., the temperature in the drying room can be raised to from 63° to 72° F. Artificial ventilation.
WIRELESS ROOM -	- - - - - -	In battleships and large cruisers, behind armour; in small cruisers, on the upper deck aft. Floor space at least 8¼ × 8¼ ft. Height 7¼ ft. Door on after side and to open outwards. Position to be chosen with reference to the most favourable lead for the wires using both masts. Ventilation, heating, and insulation of the walls to be provided. A table against 2 or 3 walls 31½ ins. in breadth and the same in height.
CUPBOARDS AND LOCKERS:		
Cadets' cupboards -	A number corresponding to the midshipmen, naval cadets, or engineer aspirants in complement.	Of galvanised iron. In or outside the mess. The cupboards are about 6 ft. high, 2 ft. 6 ins. wide and 1 ft. 6 ins. deep and are built into the ship. They are neat and serviceable, and there is sufficient room for kit and all effects.
Clothes lockers - -	Corresponding to the numbers sleeping in hammocks, together with a small number as spare. Each with 2 compartments side by side— 17·7″ high × 12·7″ wide × 19·7″ deep for clothes. 17·7″ high × 7·3″ wide × 19·7″ deep for boots. A movable shelf divides each of the compartments into 2 parts, 8″ high (above) and 9¾″ (below).	Of galvanised iron. In the same compartment as the mess tables if possible, but in any case on the same deck. Each locker has a thin iron door which is padlocked. The lockers are numbered with brass letters. They are built into the ship, and extend from the deck to a height of about 5 to 6 ft. They are fitted into any available space, there being a great saving of space as compared with bags and bag racks.
Stokers' lockers for dirty clothes. (These are in addition to the clothes lockers previously mentioned.)	For the established number of engine-room personnel, plus 20 spare for battleships, 14 for large cruisers, 10 for small cruisers. Divided into 2 compartments by a horizontal shelf of thin iron sheeting— 17·7 × 13·8 × 7·9 ins. (below), 17·7 × 13·8 × 5·9 ins. (above).	Of galvanised iron. Outside the bathroom or in the bathroom (above the hand basins).

O AS 2016—3

D

Alarm Bells.—In all modern vessels, alarm bells for use in night defence are fitted outside officers' cabins and on the mess decks.

Anchors, Capstans, and Chain Cables.—ANCHORS.—Hall's stockless anchors are carried almost exclusively by all larger vessels.

Admiralty pattern anchors are used in river gunboats, torpedo boats, and generally in destroyers.

Bower Anchors.—Two are carried by all larger vessels—destroyers carry one only.

Sheet Anchor.—A sheet anchor is supplied, as a rule, to battleships and battle cruisers only.

Stern Anchor.—Almost all ships carry one—destroyers and torpedo boats excepted. Whenever possible, the stern anchor (stockless) is carried in a stern hawse-pipe in the centre line of the ship ready for letting go instantly.

Kedge Anchors.—Are supplied to cruisers and light cruisers only.

CAPSTANS.—Baxter capstans are usually fitted. In large ships there are two on the forecastle before the turret and one on the quarterdeck. In addition in later ships one warping capstan, worked by the capstan engine, is fitted on the forecastle and another on the quarterdeck.

CHAIN CABLES.—Each cable of 1·53 ins. thickness and over consists of 9 shackles; if under 1·53 ins. of 7 shackles. A shackle is 82 ft. long. 18 shackles are carried for the bower anchors and 3 as spare. Where no chain cable is supplied for the stern anchor, the heaviest steel hawser carried is utilised instead.

Full Load Displacement of Ship. Tons (British).	Anchor.			Chain Cable for Bower Anchor.
	Bower.	Stern.	Kedge.	
	Weight in Tons.			Thickness in inches.
1,480— 2,460	2	·7	·25	1·65
2,460— 3,440	2·5	·9	·3	1·77
3,440— 4,430	3	1·1	·3	1·89
4,430— 5,900	3·5	1·3	·4	2·00
5,900— 7,380	4	1·6	·5	2·13
7,380— 8,860	4·5	1·85	·6	2·24
8,860—10,330	5	2·1	·7	2·36
10,330—12,300	5·5	2·4	·7	2·36
12,300—14,760	6	2·7	·8	2·48
14,760—18,210	6·5	3·1	·8	2·60
18,210—22,150	7	3·5	·9	2·71
22,150—26,570	7·5	3·75	·9	2·83
26,570—32,480	8	4	1·0	2·95

Armour, &c.—General disposition of side armour and under-water protection in Dreadnought battleships and battle cruisers:

The system of armouring is that of a closed caisson extending between the extreme turrets, the main side armour being closed transversely at the ends by vertical armoured bulkheads. Beyond the caisson, the armour is continued to bow and stern by armour of lesser thickness.

The under-water protection against torpedoes and mines consists of (*a*) protective coal bunkers; and (*b*) a closed caisson, of the same length as the principal belt, formed by a torpedo protection bulkhead on each side, and a transverse bulkhead connecting the ends forward and aft.

The best description available of the armour and other protection in a German capital ship is that given for the battle-cruiser *Seydlitz*. The details there given may be accepted as absolutely authentic.

An equally reliable, though less detailed, description is that given for the *Derfflinger* (*see* Section 2, Battleships, &c.). It should be noted that in this vessel the under-water protection has been augmented by (*c*) supply bunkers on the inner side of the torpedo protection bulkhead.

Boats.

	Dimensions.			Weights.					Engines or *Motors.*	Boilers.	Speed.	Cost, excluding Equipment.	Numbers that can be carried in moderate weather with all Stores on board.
	Length.	Breadth.	Depth.	Hull.	Machinery.	Equipment.	Fuel.	Total.					
				Cwt.	Cwt.	Cwt.	Cwt.	Cwt.				£	
Motor Boats, Class A. -	47′ 7″	9′ 2″	–	–	60	–	–	180	Daimler - -	—	13	1,960	–
A. -	49′ 10″	9′ 3″	–	–	85	–	–	195	Körting - -	—	14½	1,910	–
C.* -	39′ 4″	8′ 4″	–	–	40	–	–	120	Daimler or Körting	—	11	1,560	–
I. -	32′ 10″	8′ 2″ to 9′ 0″	–	–	17	-	-	80	Daimler or Körting	—	8	880	–
II. -	27′ 10″	6′ 7″	–	–	14	–	–	40	Daimler - -	—	8	440	–
III. -	24′ 7″	6′ 3″	–	–	10	–	–	20	Naval type, Daimler or Körting.	—	7	300	–
Pulling Launch, with Auxiliary Motor, No. O.	45′ 11″	11′ 10″	–	-	17	-	-	90	Daimler or Körting	—	7½	610	-
Steamboat, Class A. -	52′ 6″	10′ 3″	4′ 6″	120	150	30	40	340	3-cylinder compound.	Water-tube -	12	2,310	30
A. -	52′ 6″	10′ 3″	4′ 6″	110	120	35	60	325		Locomotive -	12	2,130	30
I. -	32′ 9″	8′ 9″	4′ 3″	45	80	35	10	170	2-cylinder compound.	Locomotive -	8	760	25
II. -	29′ 6″	8′ 0″	4′ 0″	35	60	30	10	135			7½	660	20
III. -	26′ 3″	7′ 3″	3′ 6″	30	50	25	5	110			7	595	14
Naphtha Boat, Class II. -	28′ 0″	6′ 10″	2′ 9″	20	10	25	5	60	3-cylinder single acting.	Coiled tubes -	5¾	415	28
III. -	26′ 3″	6′ 10″	2′ 9″	20	10	20	5	55			6	410	24
Pulling Boats:—													
Launch, Class I. - -	42′ 6″	11′ 0″	4′ 0″	65	–	21	–	86	—	—	–	245	100
II. - -	39′ 0″	10′ 3″	3′ 9″	50	–	20	–	70	—	—	–	220	90
III. - -	36′ 0″	9′ 6″	3′ 6″	38	-	19	–	57	—	—	–	205	80
IV. - -	33′ 0″	9′ 0″	3′ 6″	42	–	17	-	59	—	—	-	185	70
Pinnace, Class O. -	36′ 0″	9′ 6″	3′ 6″	49	–	14	–	63	—	—	-	175	80
I. - -	33′ 0″	9′ 0″	3′ 6″	35 / 38	–	14	–	49 / 52	—	—	–	160	62
II. - -	31′ 0″	8′ 9″	3′ 3″	31	-	13	–	44	—	—	-	150	56
Cutter, Class O. - -	33′ 0″	8′ 3″	3′ 0″	26	-	11	–	37	—	—	-	125 †105	50
I. - -	29′ 6″	7′ 3″	2′ 9″	24	–	11	–	35	—	—	–	115 †90	42
II. - -	28′ 0″	7′ 0″	2′ 9″	22	–	11	–	33	—	—	-	95 †85	35
III. - -	26′ 0″	7′ 0″	2′ 9″	18	-	10	-	28	—	—	-	90 †80	30
IV. - -	24′ 6″	6′ 6″	2′ 6″	17	–	9	–	26	—	—	-	80 †70	28
Gig, Class I. - -	33′ 0″	6′ 3″	2′ 6″	12	-	8	–	20	—	—	-	90 †75	16
II. - -	30′ 6″	6′ 0″	2′ 6″	11	-	7	-	18	—	—	–	85 †65	14
III. - -	26′ 3″	6′ 0″	2′ 3″	9	-	6	-	15	—	—	–		11
Jolly Boat, Class I. -	19′ 6″	6′ 3″	2′ 6″	11	–	5	–	16	—	—	-	60 †45	20
II. -	18′ 0″	6′ 0″	2′ 3″	10	–	5	–	15	—	—	-	55 †40	16
Skiff-dinghy - -	12′ 0″	4′ 3″	1′ 9″	2½	-	0¾	–	3¼	—	—	-	25 †20	3
Whaler - - -	25′ 0″	6′ 3″	2′ 6″	9	-	6	-	15	—	—	-	†60	20
Boat for T.B.D.'s and T.B.'s.	12′ 6″	4′ 3″	1′ 6″	2½	-	0¼	-	2¾	—	—	-	20	3

* Admiral's boat. † Carvel built.

Remarks.

Boats during Action.—When going to sea in expectation of an action, modern battleships and battle cruisers leave all their boats in harbour.

Motor Boats.—These have very largely replaced steam boats, but there appears to be no present intention of doing away with the latter entirely, more especially on foreign stations where there may be difficulty in obtaining supplies of fuel. In some cases (*e.g.*, light cruiser *Bremen*) one gig has been replaced by a small motor boat for the Captain's use, run by two men in order to economise in men and thus not interfere with gun drill. They are very lightly built and would probably not stand rough work, such as towing targets, sweeping, &c. The class "A" boats are fast in smooth water, but are reported soon to lose their speed in a seaway.

In most of the boats either benzine or benzol is used as fuel, but it is stated alcohol may be used alternatively.

Scale of Boats allowed.

Class and Type of Ship.	Ship's Displacement.	Motor Boats.		Steam Boats.		Pulling Boats. Launch.		Pulling Boats. Pinnace.		Pulling Boats. Cutter.		Pulling Boats. Gig.		Pulling Boats. Jolly Boat.		Pulling Boats. Skiff Dinghy.	Total No. of Boats carried
		No.	Class.	No.	Class.	No.	Class.	No.	Class.	No.	Class.	No.	Class.	No	Class.	No.	
	Tons.																
Battleships :																	
König - - -	25,390	2 1 1	A. C. III.	1	I.	2	0	–	–	2	0	–	–	2	I.	–	11
Nassau - - -	18,600	1 1	I. III.	2	I.	2	0	–	–	2	0	–	–	2	I.	–	10
Deutschland - -	13,040	1 1	A. C.	2 1	I. II.	2	0	–	–	2	0	2	I.	2	I.	–	13
Braunschweig -	12,988	–	–	1 1	A. I.	2	0	–	–	2	0	1 1	I. II.	2	I.	–	10
Wittelsbach - -	11,611	–	–	1 1	A. I.	2	0	–	–	2	0	1 1	I. III.	2	I.	–	[illegible]
Kaiser Friedrich -	10,474	–	–	1 1	A. I.	2	0	–	–	2	0	1 i	I. II.	2	I.	–	10
Brandenburg -	9,901	–	–	1 1	T. I.	2	I.	–	–	2	I.	1 1	I. III.	2	I.	–	10
Battle Cruisers :																	
Derfflinger - -	26,180	1 1 1	A. C. III.	1	A.	1	0	1	I.	2	0	1 1	I. III.	–	–	–	10
von der Tann -	19,100	1 1 1	A. C. III.	1	A.	1	0	1	I.	2	0	–	–	–	–	–	8
Cruisers :																	
Roon - - -	9,348	1	C.	2	I.	1	I.	1	I.	2	0	2	I.	3	I.	1	13
Prinz Heinrich -	8,756	–	–	1 1 1	A. I. III.	1	I.	1	I.	2	0	2 1	I. III.	2	I.	1	13
Fürst Bismarck -	10,520	–	–	1 1 1	A. I. III.	1	0	2	0	2	0	2 1	I. III.	2	I.	1	14
Light Cruisers :																	
Stettin - - -	3,494	1	II.	1	II.	–	–	1	II.	2	II.	1	II.	3	II.	1	10
Berlin - - -	3,200	–	–	1	III.	–	–	1	II.	2	II.	2	II.	2	II.	1	9
Arcona - -	2,656	–	–	1	III.	–	–	1	II.	2	II.	2	II.	2	II.	1	9
Gazelle - -	2,558	–	–	1	III.	–	–	1	II.	2	II.	2	III.	1	II.	1	8
Hertha - -	5,575	1	I.	1	I.	2	II.	1	II.	2 4	I. II.	1	II.	2	II.	2	16
Gefion - - -	3,705	–	–	1	I.	–	–	1	II.	2	II.	1 1	II. III.	1	II.	1	8
Kaiserin Augusta -	5,960	–	–	1 1	T. I.	1	II.	1	II.	2	II.	1 1	II. III.	2	II.	–	10
Special Vessels :																	
Blitz - - -	1,366	–	–	1	III.	–	–	–	–	2	II.	2	II.	2	II.	1	8
Schwalbe - -	1,100	–	–	1	III.	–	–	–	–	1	II.	1	III.	1	II.	1	5
Gunboats :																	
Eber - - -	984	–	–	1	III.	–	–	–	–	1	II.	1	II.	2	I.	1	6
Geier - - -	1,590	–	–	1	III.	–	–	–	–	1	II.	2	III.	2	II.	1	7

Remarks—continued.

This fuel appears formerly to have been stowed below, but since the occurrence of a severe explosion on board the cruiser *Yorck*, where the receptacles were in a wing or passage alongside the engine-room, it is stowed in safety tanks on the upper deck.

The new fast boats ("Schnellboote") are direct-driven by Diesel motors with mechanical reversing gear, and exhaust by a short funnel. They are 41 feet in length, weigh about 10 tons, and have a speed of 14 knots.

Naphtha Boats.—Are obsolescent.

D 3

Part III. Section 1.

General Remarks.

Classification.—German war vessels are classified as follows:—

German classification.	Translation.	British classification.
Linienschiffe	Battleships	Battleships.
Küstenpanzerschiffe	Armoured Coast Defence Ships.	Coast defence ships.
Grosse Kreuzer	Large cruisers	Battle cruisers and cruisers.
Kleine Kreuzer	Small cruisers	Light cruisers.
Kanonenboote	Gunboats	Gunboats.
Flusskanonenboote	River gunboats	River gunboats.
Schulschiffe	School ships	Special vessels.
Spezialschiffe	Special ships	Special vessels.
Grosse Torpedoboote	Large torpedo boats	Destroyers and division boats.
Kleine Torpedoboote	Small torpedo boats	Torpedo boats.
Unterseeboote	Submarines	Submarines.

The classification in the last column has been followed as nearly as possible in this publication. The following exceptions are noted.

The five training ships for cadets and boys—*Freya*, *Hansa*, *Hertha*, *Victoria Louise*, and *Vineta*, as well as the *Kaiserin Augusta*—are classified in the German list as large cruisers whereas in this work they are classified as light cruisers.

The British classification of cruisers (except battle cruisers) was changed on 31st January 1913. Prior to that date:—

Cruisers were termed - - - Armoured cruisers. / Protected cruisers 1st class.

Light cruisers were termed - - Protected cruisers 2nd class. / Protected cruisers 3rd class. / Unprotected cruisers. / Scouts.

All German ships now classified as cruisers were previously termed armoured cruisers. The present light cruiser list, in addition to vessels previously classified as protected cruisers 2nd class, includes:—

Protected Cruisers 3rd class.

Amazone.
Arcona.
Gazelle.
Medusa.
Niobe.
Nymphe.
Thetis.

Coaling.—German ships in home waters almost invariably coal from lighters and not from colliers. The best pre-war coaling performance was that of the battleship *Helgoland*, which, in 1913, was reported to have taken in 744 tons in one hour. Under war conditions battleships usually coal about once a week, taking in, in the case of the *König* class, 700–800 tons at the rate of about 300 tons an hour.

Colour.—With the following exceptions, all German war vessels are painted light grey. Destroyers and torpedo boats are usually painted black, but some modern destroyers are painted grey. Submarines are usually painted grey.

For funnels, *see* p. 32.

Compasses.—The Anschütz gyro-compass is in extensive use supplemented by liquid compasses.

Cost.—Particulars of the cost of each type of ship will be found in the detailed description of the type.

The cost of ships is shewn in the German Navy Estimates under three broad headings—Shipbuilding, Gun Armament, and Torpedo Armament; the first of these also includes cost of machinery and of trials. It is not clear whether torpedo nets and mines are included under these heads.

For more detailed information regarding cost of Gun Armament, *see Ostfriesland* (Section 2, Battleships, &c.).

The cost of British and German ships is not directly comparable, as in Germany. the cost of the reserve ammunition as well as of the first outfit of ammunition is included under the head of Gun Armament. It is not clear whether reserve guns and mountings are also included, but this is probable.

Part III. Section 1.

General Remarks.

Before the war money for construction was voted in annual instalments as follows :—

Battleships - - - - -	4 instalments.
Large cruisers - - - - -	4 instalments.
Small cruisers - - - - -	3 ,,
Flotillas (of 12 destroyers) - - -	2 ,,
Submarines - - - - -	A lump sum was voted annually, but the number to be built was not disclosed.
Other vessels - - - - -	The number of instalments varied.

The instalments *voted*—the actual expenditure was not published—for the last ships for which complete figures are available, are :—

Instalment.	Battleship "König."				Battle-Cruiser "Derfflinger."			
	Hull, Machinery, &c.	Gun Armament.	Torpedo Armament.	Total.	Hull, Machinery, &c.	Gun Armament.	Torpedo Armament.	Total.
	£	£	£	£	£	£	£	£
1st - -	269,080	195,695	14,677	479,452	244,618	146,771	9,785	401,174
2nd - -	513,699	232,387	21,526	767,612	538,160	134,540	13,699	686,399
3rd - -	440,313	256,849	14,188	711,350	415,851	256,849	10,763	683,463
4th - -	244,619	196,184	16,634	457,437	251,957	197,162	13,209	462,328
Totals -	1,467,711	881,115	67,025	2,415,851	1,450,586	735,322	47,456	2,233,364

Instalment.	Light Cruiser "Graudenz."				Total for Twelve Destroyers, V 25—S 36.			
	Hull, Machinery, &c.	Gun Armament.	Torpedo Armament.	Total.	Hull, Machinery, &c.	Gun Armament.	Torpedo Armament.	Total.
	£	£	£	£	£	£	£	£
1st - -	122,309	24,462	4,892	151,663	489,237	48,924	90,509	628,670
2nd - -	122,309	24,462	10,763	157,534	479,452	44,031	119,961	643,444
3rd - -	73,385	24,462	10,763	108,610	—	—	—	—
Totals -	318,003	73,386	26,418	417,807	968,689	92,955	210,470	1,272,114

Displacement.—The official displacements of German war vessels—excepting destroyers, torpedo boats and submarines—are published in the German Navy List (Rangliste der Kaiserlichen Marine) and are quoted in metric tons. This Navy List displacement is the designed load displacement, calculated on the basis of what is called the "normal" stowage of fuel and the "authorised" equipment. The "normal" stowage of fuel is always less than half the maximum stowage—in the case of large vessels very considerably less. In practice, therefore, the designed displacement will almost always be exceeded except when the vessel is very light.

Docking.—Ships in commission are usually docked once a year during their annual refit; but where considerable loss of speed arises through fouling of bottom, they may be docked a second time.

Destroyers and torpedo boats which are regularly, though not necessarily continuously, in commission throughout the year, are docked twice a year and have their bottoms coated. The first docking takes place prior to commissioning, the second at some suitable period whilst in commission.

Powder and filled shell are discharged if a stay of more than eight days in dock is anticipated.

Battleships of *Deutschland* type and earlier, as a rule, are docked with empty upper bunkers, and with at most 600 tons in their lower bunkers. The same applies to cruisers of *Roon* type and earlier, except that they may have 750 tons in their lower bunkers.

D 4

Part III.
Section 1.
General Remarks.

Draught.—The draught quoted in this publication, unless otherwise stated, is the designed draught, *i.e.*, extreme draught at designed load displacement. In practice this draught is almost always exceeded except when the vessel is in a very light condition.

Draught Marks.—Draught marks are applied on both sides of the ship—as a rule forward and aft only; but in ships whose keel takes a convex form, the marks are also applied on both sides above the lowest point of the keel, from which in this case the depth is reckoned. Arabic numerals are used, the lower edge of which marks the depth indicated.

The zero of the after draught marks is reckoned from the horizontal plane passing through the deepest point of the ship aft—keel, sternpost, bottom of rudder, deepest point reached by the propeller, &c.

The zero of the forward draught marks is reckoned from the horizontal plane passing through the deepest part of the keel forward.

In ships of over 100 tons displacement, the figures are 100 mm. (3·9 in.) high and represent decimetres. Even numbers only are affixed, and are carried at least 19·7 in. above the load waterline.

In smaller vessels and torpedo boats the figures are 50 mm. (2 in.) high and are carried at least 9·8 in. above load waterline.

The draught marks in both cases are carried downwards to the draught the vessel will have when placed in II. Reserve, *i.e.*, when ready for commissioning at any time with stores partially on board.

Electric Leads.—In large vessels the lighting circuits are on the ring system, main leads below the armoured deck. Single twin-cored cable is used. The internal lighting is on five main circuits—

(1)—Forward and above armoured deck;
(2) and (3)—Space below armoured decks;
(4)—After section and above armoured deck;
(5)—Upper deck.

Lighting in charthouse and all fighting positions, *e.g.*, magazines, &c., is fed from two circuits.

Motor circuits are not on the ring system.

The voltage is 220 in modern large vessels, 110 in destroyers and T.B.'s.

Flooding Arrangements.—In modern vessels the arrangements fitted for flooding the magazines are:—

(a) For quick flooding, pipes in direct communication with the sea.
(b) For drenching charges and keeping down the temperature of the magazine, pipes are fitted just below the crown of the magazine, perforated with small holes, and supplied with water by means of a small pump.

The magazines can be flooded from within or from a small compartment outside.

Fuel.—*Westphalian Coal* is used for the steam trials of newly built ships, except, perhaps, torpedo craft. It is in general use in all ships in Home Waters, including torpedo craft, after their entry into the fleet.

Tar Oil derived from pit (black) coal is used in the larger German ships fitted to burn oil fuel, except when residual petroleum can be obtained more cheaply. This tar oil is of low viscosity, similar to shale oil, and has a specific gravity of 1·0 to 1·1; the heavier varieties can consequently be flooded in case of fire.

Heavy Oil, as a rule a first distillate of tar oil, specific gravity about 0·9, is used in all but the earliest submarines.

Benzine and *Benzol*, or alternatively *Alcohol*, are used for motor boats.

Funnels.—During the war the peace-time system of painting bands on the funnels appears to have been entirely discarded. On the High Sea Fleet proceeding to sea, the funnels of all the larger vessels and their attendant destroyers are sometimes repainted a special colour. This is done after leaving harbour. Yellow, brown, and blue are the most usual colours.

Furniture.—Almost all furniture of cabins and messes is made of thin steel, and is manufactured, as a rule, in the Imperial Dockyards. In use, this furniture is found to be only slightly liable to rust, but it is not possible to prevent sweating entirely, and the furniture is therefore lined with pegamoid.

Gas Masks.—Gas masks were just being supplied to the fleet at the time of the battle of Jutland. Sufficient are now carried for every man in the complement.

Magazines.—In all modern battleships, battle cruisers, and light cruisers, the magazines are below the shell rooms, the latter being immediately under the armoured deck.

For magazine cooling arrangements, *see under* "Refrigerating Plant."

Masts.—The heavy fighting masts fitted in pre-Dreadnought battleships and cruisers were abandoned in Dreadnought battleships and battle cruisers, light pole masts being fitted in lieu. The battleships *Kronprinz* and *Grosser Kurfürst* have been given a heavier foremast since the battle of Jutland.

In the cruiser *Blücher* a tripod mast was put in forward with great secrecy in February 1914, but this type of mast was not adopted for any other ships at that time. The new *Bayern* class of battleships and the battle cruiser *Hindenburg*, however, have a tripod foremast, and the battle cruiser *Derfflinger* has also received a tripod foremast since the battle of Jutland.

Mines.—All classes of vessels at times carry mines, but there is no permanent stowage for them in capital ships or light cruisers, except those specially appropriated for minelaying duties, or in surface torpedo craft. Consequently light cruisers and destroyers, when employed as minelayers, cannot efficiently perform their ordinary duties. The number of mines carried by light cruisers varies from 70 to 120.

Net Defence.—Torpedo net defence, fitted in the earliest battleships, but discarded since about 1896, was re-introduced in the form of beam defence in all Dreadnought battleships and battle cruisers. The booms were fitted about 33 ft. apart and the nets were hung double. For further details, *see* Part IV., Section 3. But since the battle of Jutland net defence has again been discarded.

Propellers, &c.—In ships with three propellers, the centre propeller is left-handed whilst the others turn outwards. In ships of the *Kaiser Friedrich* and *Wittelsbach* classes, the outer propeller shafts converge slightly towards the engines, but in the *Braunschweig* and later classes they are parallel.

Refrigerating Plant.—CO_2 plant is used, and is generally placed behind armour. In large ships two sets are sometimes fitted. For magazines and refrigerating rooms a 25 per cent. brine-solution, usually cooled to a temperature of between 28° and 32° F., is used and is pumped through circulating pipes in the compartments in question.

Besides cooling the magazines, the refrigerating plant is also employed for cooling drinking water to 46° F., and the refrigerating room to 28°—32° F.

A motor-driven refrigerating apparatus has recently been introduced.

Searchlights.—In all modern capital ships up to and including the *König* class, eight searchlights are carried, viz., four forward and four aft. In the *Bayern* class, and also in the *Derfflinger*, which is now fitted with a tripod foremast, the searchlights forward are somewhat differently arranged, but apparently the number is unaltered.

In recent ships stowage is provided for all searchlights behind armour, and two spare searchlights are carried.

Under war conditions searchlights are stowed below during the day.

For further details, *see* Part IV., Section 3.

Special smaller searchlights are carried for signalling purposes.

Shades for Electric Lights.—In some of the latest ships blue glass shades are fitted to the lights on the mess decks for night defence purposes. The shades are permanently secured to the light fitting and are made in two halves, which are ordinarily kept clear of the lights, but can be lowered and clamped round them when required.

Signalling.—Great importance was attached before the war to having efficient arrangements for signalling in action. In all modern capital ships, good battle signal stations are fitted on the battery deck. For a description of these stations, *see Kaiser* class.

Considerable use is made of searchlights by day and of the short-range wireless set. Masthead flashing lamps are fitted but are not often used under war conditions. The use of the night signalling apparatus, consisting of 3 or 6 lights, each red, white or green, is also probably very limited in war time, but these lights are used for recognition signals. Capital ships, and probably light cruisers also, in addition

General Remarks.

have two rings of lights fitted round the foremost funnel, these lights likewise being red, white, and green, and being used for recognition signals.

Smoke-producing Devices.—During the war, auxiliary cruisers, destroyers, submarines and armed trawlers have frequently made use of smoke clouds and smoke screens. In the case of destroyers the smoke is produced from the funnels or from metal cylinders, carried one on either quarter; these cylinders contain some chemical composition and are in connection with the compressed air service. Submarines frequently rig up a dummy funnel and emit heavy black smoke from this. In the case of trawlers, tin boxes, about 18-ins. square, are carried, containing some substance which emits a thick white smoke on a hole being made in the lid. The box can be placed anywhere on board, according to the direction of the wind, or may be thrown overboard, when it will float. Auxiliary cruisers carry cylinders, about 4 ft. high, fitted with a cock, and probably containing a similar chemical composition.

Battle cruisers and light cruisers have recently (April 1917) been fitted with smoke-producing cylinders of a similar type to those carried by destroyers and likewise placed one on either quarter.

Steam, Raising.—The time allowed for raising steam under peace conditions is :—

		Normal.	Accelerated.
Water tube boilers	cold water	$2\frac{1}{2}$ hrs.	$1\frac{1}{2}$ hrs.
	warm water	$1\frac{1}{2}$,,	1 hr.
Cylindrical boilers	cold water	12 ,,	6 hrs.
	warm water	4 ,,	2 ,,

Steam Heating.—Steam heating is fitted in practically all ships and torpedo craft, the steam being derived from the main boilers. It is so arranged that the temperature of the cabins of flag officers, captains, and heads of departments, messes, offices, sick-bay, and bathrooms can be raised to 59° F.; the cabins of other officers, mess decks, charthouse, and pantries to 50° F.; engine-rooms, boiler-rooms, torpedo flats, w.c.'s, and passages to 41° F.

Steering Engines and Steering Positions.—In the large ships there are two steering engines placed side by side in separate rooms aft: (*a*) the main steering engine on the centre line, and (*b*) the reserve steering engine on the starboard side.

In battleships, battle cruisers, and cruisers* there are, as a rule, five steering positions, viz.:—

(1) On the upper bridge—for use during fog, in the Kaiser Wilhelm Canal, when docking &c.;

(2) In the fore conning tower, for use during ordinary navigation and in action;

(3) In the central conning station (below the fore conning tower, at the foot of the communication tube and under the armoured deck), which is in communication with the fore conning tower by voice-pipe, and is used in case of derangement of the gearing situated between conning tower and central conning station;

(4) In the two steering engine-rooms aft, for use in case of derangement of the gearing situated between them and the central conning station;

(5) The auxiliary hand-steering gear in the tiller flat.

Except as regards (1) and (2), all steering gear is below the armoured deck.

Voice-pipes lead from the fore conning tower, as well as from the central conning station, to each of the steering engine-rooms, and to the tiller flat, and in addition there are electric helm telegraphs.

Stores.—

In peace time, *Ships in commission in Home Waters* were supposed always to have on board—

Ammunition	Full fighting establishment.
Coal	One-half bunker storage.
Lubricating oil	One-third establishment.
Reserve feed water	Sufficient to fill up one-half the main boilers to working height.
Provisions	Four weeks' supply.

* In light cruisers the arrangements are similar, but the hand steering wheel is on the quarter deck.

Trials (peace routine).—

Newly built Ships.

The following are some of the steam trials carried out by large ships in peace time.

Prior to commissioning:—

(1) Basin trial—to test main engines, boilers, shafts, propellers, auxiliary machinery, &c.
(2) Contractors' trial or trials—carried out at sea.

After commissioning:—

(3) Acceptance trial.
(4) Six hours' full power trial.
(5) Measured mile trials.
(6) 24 hours' coal consumption trial.

Since the introduction of turbines in capital ships, it has become customary to test them on an ocean passage also.

Acceptance Trial.—This is the first *official* trial at sea which is carried out by newly built ships, and is attended by the Ships' Trials' Committee. Good Westphalian coal is used.

The trial lasts at least eight hours, during which the engines are worked up to the prescribed maximum power; this has to be maintained for at least two hours.

Whilst running at the highest power, the speed is determined on the measured mile or by bearings.

After the trial, the more important parts of the main engines are opened up for inspection and the boilers are examined externally. The same applies as regards the auxiliary machinery which may have been in use.

After thorough inspection by the Committee, the ship is taken over provisionally. This inspection may be begun before the Acceptance trial, but is chiefly carried out afterwards.

Ships built in private yards are under one year's warranty from the contractors from the date of termination of the official trials.

Measured Mile Trials.—The trials on the measured mile, it is stated, are carried out entirely with naval ratings, and with the full equipment of stores, ammunition and water, and the normal supply of coal on board, the ship being otherwise complete as regards armour, guns, &c. Any deficiency in numbers of crew or in stores is made up with ballast.

The trials are carried out in the Baltic, and as regards large ships, the measured mile trials are all carried out on the two-mile course off Neukrug in the Gulf of Danzig.

Other Ships in Commission.

On commissioning, or after undergoing considerable repairs, in peace time, a six hours' trial is carried out, three consecutive hours being at full power without forced draught. In the case of ships with water-tube boilers, three hours of the trial may be carried out at $\frac{7}{10}$ths power, all boilers being alight.

Once a year, as soon as possible after the annual refit, a three hours' full speed trial is to be carried out, using forced draught, and with the proper watches. The boilers and ship's bottom are, if possible, to be clean, and the trial is to be run as nearly as possible at the draught corresponding to that with full equipment. This trial is not carried out by Training or Experimental Ships (unless modern), by Surveying Ships, ships doing duty as Tenders or for Fishery Protection, or by those for acceptance, or commissioned for the period of exercises of a Reserve formation. It is carried out, however, by newly commissioned ships as soon as the officers and men are thoroughly acquainted with the machinery.

Smoke Prevention Trials.—At suitable opportunities runs of four hours' duration are to be made with the least possible development of smoke. The number of boilers in use is to correspond with those required for $\frac{3}{4}$ speed (*Grosse Fahrt*). If mixed firing is provided, oil is not to be used.

Under-water Protection.—*See under* "Armour," p. 26.

SECTION II

BATTLESHIPS AND BATTLE CRUISERS

GERMAN NAVY.

PART III.

SECTION 2.

BATTLESHIPS AND BATTLE CRUISERS.

Name.	Classification.	Page.	Plate No.	Name.	Classification.	Page.	Plate No.
Baden - - -	Battleship	3	1 & 11	Moltke - - -	Battle Cruiser	30	9 & 20
Bayern - - -	,,	3	1 & 11	Nassau - - -	Battleship	15	5 & 15
Braunschweig - -	Obsolete Battleship	39	23	Oldenburg - -	,,	12	4 & 14
Derfflinger - -	Battle Cruiser	21	7 & 18	Ostfriesland - -	,,	12	4 & 14
Deutschland - -	Obsolete Battleship	37	22	Posen - - -	,,	15	5 & 15
Elsass - - -	,,	39	23	Preussen - -	Obsolete Battleship.	39	23
Friedrich der Grosse -	Battleship	8	3 & 13	Prinz Eitel Friedrich	Battle Cruiser	20	16
Graf Spee - -	Battle Cruiser	20	16	Prinzregent Luitpold	Battleship	8	3 & 13
Grosser Kurfürst -	Battleship	5	2 & 12	Rheinland - -	,,	15	5 & 15
Hannover - - -	Obsolete Battleship	37	22	Sachsen - - -	,,	3	1 & 11
Helgoland - -	Battleship	12	4 & 14	Schlesien - -	Obsolete Battleship.	37	22
Hessen - - -	Obsolete Battleship	39	23	Schleswig-Holstein -	,,	37	22
Hindenburg - -	Battle Cruiser	20	6 & 17	Schwaben - -	,,	41	24
Kaiser - - -	Battleship	8	3 & 13	Seydlitz - - -	Battle Cruiser	25	8 & 19
Kaiserin - - -	,,	8	3 & 13	Thüringen - -	Battleship	12	4 & 14
König - - -	,,	5	2 & 12	von der Tann - -	Battle Cruiser	33	10 & 21
König Albert - -	,,	8	3 & 13	Westfalen - -	Battleship	15	5 & 15
Kronprinz Wilhelm -	,,	5	2 & 12	Wettin - - -	Obsolete Battleship.	41	24
Lothringen - -	Obsolete Battleship	39	23	Wittelsbach - -	,,	41	24
Mackensen - -	Battle Cruiser	20	6 & 17	Württemberg - -	Battleship	3	1 & 11
Markgraf - -	Battleship	5	2 & 12	Zähringen - -	Obsolete Battleship.	41	24
Mecklenburg - -	Obsolete Battleship.	41	24				

o (33) AS 4924—23 Pk 2972 1000 11/18 E & S

A

GERMANY.
C.B. 1182 P.
I.D. 1159.
October 1918.

BATTLESHIPS AND BATTLE CRUISERS.

Scale $\frac{1}{2000}$

BATTLESHIPS.

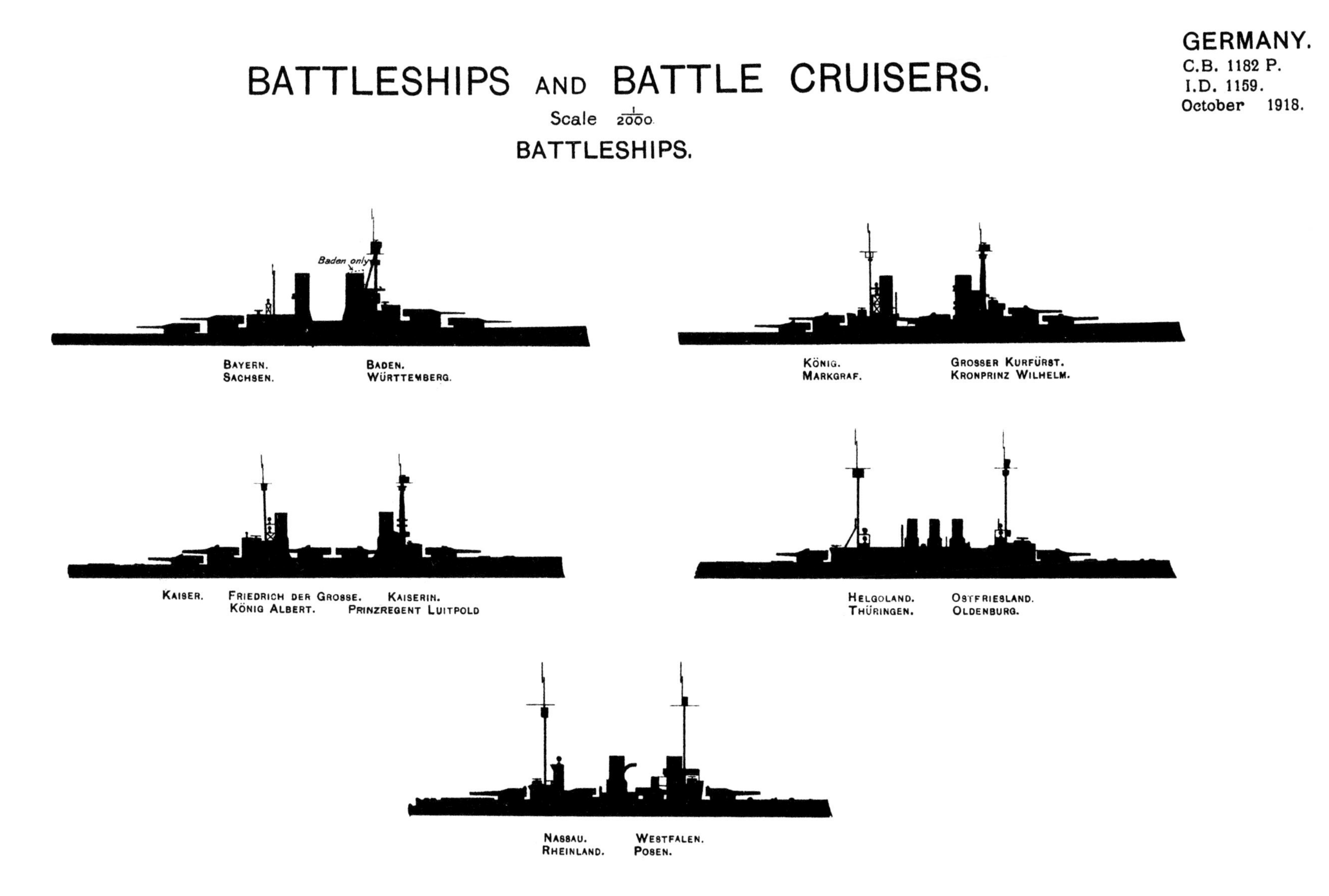

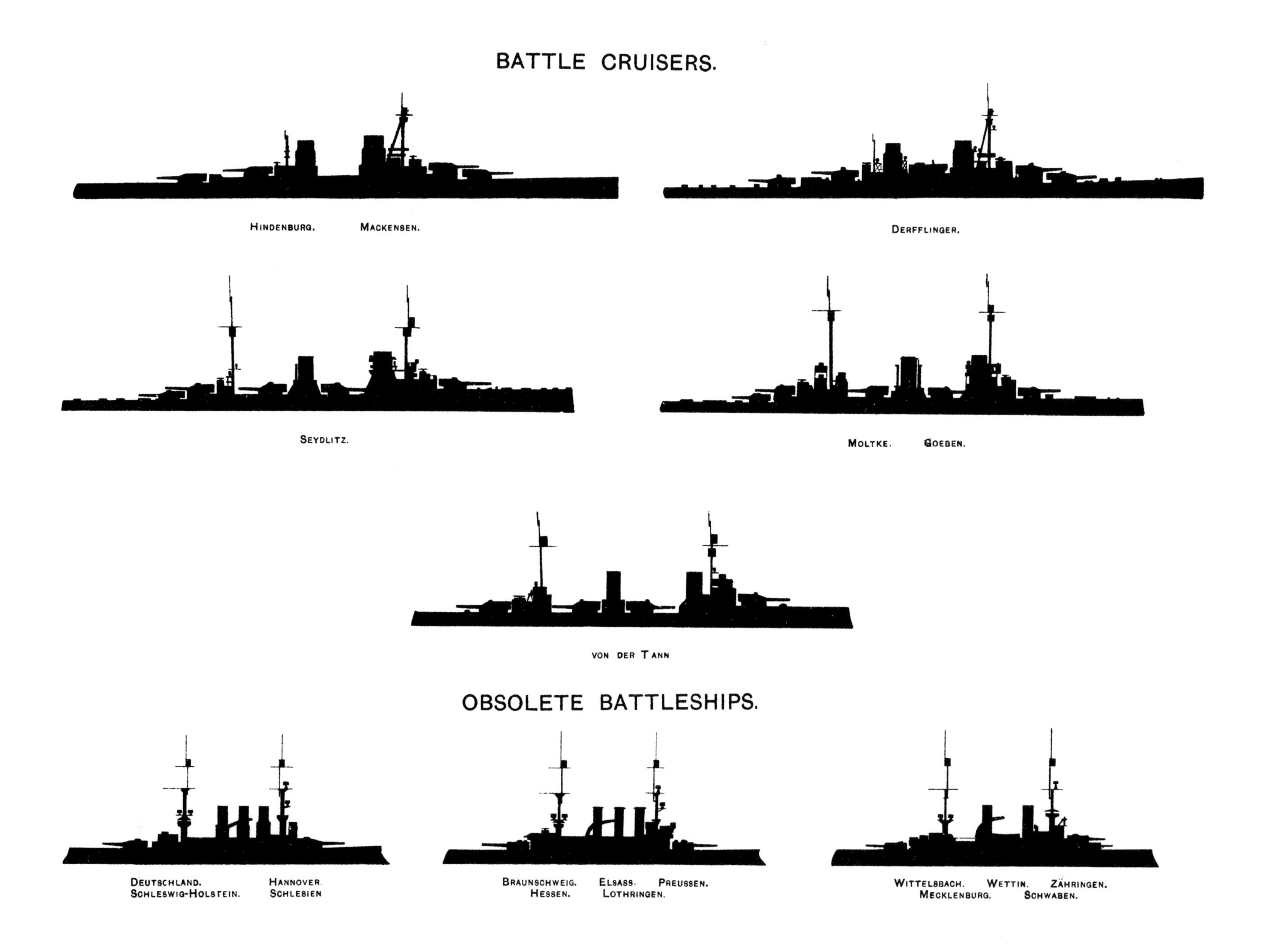
BATTLE CRUISERS.
HINDENBURG. MACKENSEN.
DERFFLINGER.
SEYDLITZ.
MOLTKE. GOEBEN.
VON DER TANN
OBSOLETE BATTLESHIPS.
DEUTSCHLAND. HANNOVER.
SCHLESWIG-HOLSTEIN. SCHLESIEN
BRAUNSCHWEIG. ELSASS. PREUSSEN.
HESSEN. LOTHRINGEN.
WITTELSBACH. WETTIN. ZÄHRINGEN.
MECKLENBURG. SCHWABEN.
Ordnance Survey, November, 1918.

List of Plates.

BATTLESHIPS.

Bayern, Baden, Sachsen, *and* Württemberg.

(*See Plates* 1 *and* 11.)

Ship.	Designation before Launch.	Programme Year.	Built or Building at	Ordered.	Laid Down.	Launched.	Commissioned for Trials.	Completed Trials.
Bayern	"*T*" - -	1913–14	Howaldt's Works, Kiel.	*20.4.13	*–.9.13	18.2.15	17.4.16	*–.9.16
Baden -	Ersatz *Wörth* -	1913–14	Schichau Works, Danzig.	*–.5.13	*19.2.14	*30.10.15	*–.11.16	*–.3.17
Sachsen	Ersatz *Kaiser Friedrich III.*	1914–15	Germania Yard, Kiel.	1914	*23.5.14'	*21.11.16	—	—
Würt-temberg.	Ersatz *Kaiser Wilhelm II.*	1915–16	Vulcan Works, Hamburg.	—	1915	*–.9.17	—	—

* Approximate dates.

General Remarks.

These battleships represent a very great advance on their predecessors, the *König* class, in calibre of main armament, and in length and displacement, and also an advance in speed.

Strictly speaking, the following details apply to the *Bayern* and *Baden* only, but it is believed that in all essentials the *Sachsen* and *Württemberg* are sister ships.

The *Baden* is fitted as fleet flagship.

Complement.—Peace complement, about 1,200; war complement, about 1,400.

General Appearance.—*See* Plate 1 and Silhouette.

General Dimensions.—

Length, L.W.L. - - - - - - - -		623 ft. 4 ins.
Breadth, extreme - - - - - - - -		99 " 9 "
Draught, designed load - - - - - - -		28 " 0 "
Displacement, designed load - - - -	*about*	28,000 tons.

Armour.—(The thicknesses are doubtful.)

Main Belt.—15″ amidships.

Turrets.—The turrets are very noticeably longer than those of the 12-in. guns in previous ships. The lower part of the front armour of the turret is approximately vertical instead of the whole of the front armour being sloping as in previous types.

The roofs of the turrets have sloping sides. This design may have been adopted partly in order to reduce the amount of vertical armour necessary and to permit the thickness of the top of the gun-house to be increased for protection against plunging fire without unduly increasing the weight of the revolving portion of the turret; but no definite information regarding the thickness of the armour is yet available.

The apertures for the range-finder are at the sides of the turret, well towards the front.

Conning and Control Towers.—Fore, 17¾″; the general arrangement appears to be similar to that in *Nassau* (*see* p. 17).

Horizontal.—Two armoured decks are fitted, the upper one being designed to give protection against aircraft attack.

Armament.—(Authentic only as regards *Bayern* and *Baden.*)

Guns.—

(For approximate arcs of training, *see* Plate 11.)

8—15-in. (38 cm.), L/45, mounted in four turrets on the centre line.

16—5·9-in. (15 cm.) Q.F., L/50, in battery.

4—22-pr. (8·8 cm.) semi-automatic, anti-aircraft guns, fitted with curved shields.

Torpedo Tubes.—(Doubtful.)

5—23·6 in. (60 cm.) submerged.

Searchlights.—Probably **8**—43·3-in. (110 cm.) are carried, four on legs of tripod mast, two on after funnel and two on a platform abaft mainmast.

Fire Control.—As in previous ships, the gun control tower and fore conning tower are practically in one, being merely separated by a bulkhead. The control tower is abaft the conning tower and projects above it.

Bayern, Baden, Sachsen, *and* Württemberg—*cont.*

There appear to be two roomy spotting and control tops on the tripod mast; above the upper tops is a platform fitted with weather screens.

Range-finder Positions.—As in previous ships, a range-finder is mounted on each of the two gun control towers and in each turret; and probably there are also the usual armoured hoods, projecting just above the upper deck, with range-finders for the battery guns. In addition, a short range-finder is fitted at the upper extremity of the tripod mast, above the upper control top. The principal range-finders are 8-metre (26¼ ft.) instruments. The apertures of the turret range-finders are no longer placed in armoured hoods, projecting above the roof of the gun-house, as in previous ships, but are in the sides of the gun-house, close up to the front armour.

Battle Signal Stations.—Fitted on either side of main deck amidships, there being four 5·9-in. guns before, and four abaft, them. Arrangements probably as in *Kaiser* class (*see* p. 10).

Machinery and Boilers.—

Main Engines.—

Bayern.—Three sets of turbines of Brown-Curtis type made by Turbinia Co., Berlin, and arranged in six engine-rooms. Each set consists of one H.P. in one engine-room separated by watertight bulkhead from the L.P. turbine, with which is incorporated one astern turbine in same casing.

Baden.—As in *Bayern*, but of Schichau design.

Sachsen.—Turbines.

Württemberg.—Turbines.

Boilers.—

Fourteen Schulz in nine boiler rooms (*see* Plate 11).

Nos. 1 to 3 burn oil only, and Nos. 4 to 14 coal and oil.

No superheaters are fitted.

Ash ejectors, similar to See's, are supplied.

Auxiliary Machinery includes :—

Two Diesel-driven dynamos and four turbo-dynamos. Each Diesel engine is a 6-cylinder 4-cycle Benz of 450 B.H.P., and each turbo-dynamo of 500 B.H.P. These are in the auxiliary engine-rooms with the evaporating plant—for position, *see* sketch.

Two electrically-driven CO_2 machines for magazine cooling—one forward and one aft.

One steam capstan and two steam reciprocating steering engines—the latter cross-connected.

Rotary circulating pumps with reciprocating drive and fitted with large bilge suctions.

Feed, air, bilge, and fuel pumps are of piston type. Seven electrically-driven ballast pumps of large capacity, arranged from forward to aft on the platform deck.

Propellers.—

Three, direct driven.

Speed, Horse-Power, and Fuel.

Ship.	Speed. (designed).	Horse-Power. (designed).	Fuel. (*a*) Coal. (*b*) Oil.
	Knots.		
Baden / *Bayern*	21	—	(*a*) 4,724 (*b*) Some
Sachsen / *Württemberg*	—	—	—

Steam Trials.—

Ship.	Nature of Trial.	Date.	Speed.	Horse-Power.	Revolutions.	Remarks.
Bayern	Measured mile	-.5.16	22·0	56,000	—	—

BAYERN.
BADEN, SACHSEN and WÜRTTEMBERG similar.

Ordnance Survey, November, 1918.

KÖNIG.
GROSSER KURFÜRST, MARKGRAF and KRONPRINZ WILHELM similar.

Ordnance Survey, November, 1918.

König, Grosser Kurfürst, Markgraf, *and* Kronprinz Wilhelm.

(*See Plates 2 and 12.*)

Ship.	Designation before Launch.	Programme Year.	Built at	Ordered.	Laid Down.	Launched.	Commissioned for Trials.	Completed Trials.
König - -	S. - - - -	1911–12	Imperial Dockyard, Wilhelmshaven.	*–.4.11	–.10.11	1.3.13	26.8.14	*–.11.14
Grosser Kurfürst	Ersatz *Kurfürst Friedrich Wilhelm*	1911–12	Vulcan Works, Hamburg.	*12.8.11	–.10.11	5.5.13	19.8.14	*–.11.14
Markgraf - -	Ersatz *Weissenburg*	1911–12	Weser Yard, Bremen.	*12.8.11	–.11.11	4.6.13	1.10.14	*–.1.15
Kronprinz Wilhelm.	Ersatz *Brandenburg.*	1912–13	Germania Yard, Kiel.	*23.4.12	*8.5.12	21.2.14	*–.12.14	*–.3.15

* Approximate dates.
The *König* was launched with all her boilers and turbines in place.

Cost.—

	£
Hull, machinery, &c. - - - - -	1,467,711
Gun armament - - - - - -	881,115
Torpedo armament - - - - -	67,025
Total - -	£2,415,851

General Remarks.—These were the first German battleships with all centre-line turrets. In other respects they closely resemble their immediate predecessors, the *Kaiser* class.

In these ships the two foremost turrets are not usually fired at night owing to the effect of the flash and blast on the bridge personnel.

The *Kronprinz Wilhelm* was formerly named *Kronprinz*, but was rechristened in June 1918 in honour of the German Crown Prince.

Complement.—Peace complement, about 1,130; war complement, 1,365. A turret's crew, including magazine and shell room parties, numbers three or four officers and 70 men.

General Appearance.—*See* Plate 2 and Silhouette.

Accommodation.—The officers' accommodation is aft.

General Dimensions, &c.—

Length, extreme - - - - -		575 ft. 9 ins.
,, L.W.L. - - - - -		573 ,, 2 ,,
Breadth, extreme - - - - -		96 ,, 9 ,,
Draught, designed load - - - -		27 ,, 3 ,,
Displacement, designed load - - - -		25,390 tons.
Freeboard (approximate) at designed draught - -	at stem - -	22 ft. 0 ins.
	amidships - -	19 ,, 6 ,,
	aft - - -	13 ,, 0 ,,
Height of axis of heavy guns (approximate) at designed draught - - - -	No. 1 turret - -	26 ,, 0 ,,
	No. 2 turret - -	35 ,, 0 ,,
	Nos. 3 and 4 turrets -	27 ,, 0 ,,
	No. 5 turret - -	19 ,, 0 ,,
Height of sighting slits of fore part of C.T. (approx.)		42 ,, 0 ,,
,, fore funnel to L.W.L. ,, -		78 ,, 0 ,,
,, ,, lower mast head to L.W.L. ,, -		100 ,, 0 ,,
Depth from keel to battery deck - - -		39 ,, 11½ ,,

Masts.—

Two light pole masts were originally fitted. In *König*, *Grosser Kurfürst*, and *Kronprinz Wilhelm*, and possibly in *Markgraf* also, a much heavier cigar-shaped

A 3

König, Grosser Kurfürst, Markgraf, *and* Kronprinz Wilhelm—*cont.*

foremast has now been put in, carrying a splinter-proof spotting and control top, with a large screened platform above it; on the upper extremity of the mast a short-base rangefinder is fitted.

Constructive Details.—All considerations of convenience have been subordinated to making the subdivision as complete as possible. Below the armoured deck all main transverse bulkheads are unpierced. All important bulkheads are stiffened in a very thorough manner and tested at 30 ft. head of water.

There are three complete decks above the armoured deck, viz., middle deck, main or battery deck, and upper deck. This is one deck more than in *Kaiser* class.

Armour and other Protection.

Material.—Krupp cemented.

Main Belt.—About 14″ amidships, tapering to the lower edge. Width about 12 ft., of which 5 ft. 7 ins. is below the designed load waterline.

Upper Belt.—About 10″ thick, width 6 ft. 10 ins.

Transverse Bulkheads.—About 10″ thick, one fitted at each end of the main belt, closing the citadel, and one about 20 ft. from the stern.

Belt before and abaft citadel.—6″ thick.

Battery.—7·9″ thick, width 6 ft. 10 ins. Observation slits are cut in the armour at two positions on either side which are used for secondary torpedo control.

Barbettes.—14″ thick.

Turrets.—14″ in front.

Conning and Control Towers.—Fore, 13¾″; after, 10″. Generally similar to those of *Nassau* (*see* p. 17).

Horizontal.—The principal armoured deck, thickness 2·4″ to 3″, is about 1 ft. below the load water line amidships and aft sloping down to lower edge of armour. Forward it descends to a lower level and is flat.

Under-Water Protection—Generally similar to that in *Kaiser* class (p. 9).

Torpedo Nets.—Beam net defence was fitted, but has been discarded since the Battle of Jutland.

Armament.

Guns and Ammunition Supply.

(For approximate arcs of training, *see* Plate 12.)

10—12-in. (30·5 cm.) Q.F. L/50, on model /08 mounting; in pairs in centre line turrets.

Maximum rate of fire about 2½ rounds per gun per minute.

Ammunition supply electric. Loading at fixed angle of elevation.

The gun runs out automatically after recoil, two hydraulic recoil cylinders and a hydro-pneumatic running-out cylinder being fitted.

Elevating gear is hydraulic, alternative hand. Training gear also electric—with one or with two motors at will—alternative hand.

14—5·9 in. (15 cm.), in battery.

Maximum rate of fire about 16 rounds per minute.

4—22-pr. (8·8 cm.) semi-automatic, anti-aircraft guns, on after superstructure.

2—machine guns.

Magazines and Stowage.—The maximum stowage of 12″ ammunition is about 100 rounds per gun. Magazines are below shell rooms.

Torpedo Tubes.—

5—19·7 in. (50 cm.), submerged.

The four broadside tubes are fixed at an angle of 20° before the beam and are fitted with gyro angling gear which can be set for every 15° from 30° before to 60° abaft the direction of the tube.

Torpedoes.—In war time about 20 torpedoes are carried.

Mines.—There is no permanent stowage for mines.

König, Grosser Kurfürst, Markgraf, *and* Kronprinz Wilhelm—*cont.*

Searchlights.—**8**—43·3-in. (110-cm.), worked by hand from a platform immediately below the light. *Grosser Kurfürst* and *Kronprinz Wilhelm* are believed to have been fitted, in addition, with a small searchlight on platform below fore top.

Rangefinder Positions.—On the fore control tower a long rangefinder, probably an 8-metre (26¼-ft.) instrument, is fitted. On the after control tower and in each turret a 5-metre (16½-ft.) instrument is fitted. There are also two rangefinders, probably 3-metre (10-ft.) instruments, mounted in armoured hoods projecting just above the upper deck, for use with the secondary armament. In addition a short rangefinder is fitted at the upper extremity of the foremast above the control top.

Wireless Telegraphy.—The main wireless room is situated on the middle deck, just abaft No. 2 barbette.

Battle Signal Stations.—Fitted on either side of main deck. The general arrangements are similar to those in *Kaiser* class (*see* p. 10).

Anchors.—Three stockless bower anchors and one stern anchor are carried. Steam capstans forward, electric capstan aft.

Machinery and Boilers.—

Main Engines.—

Type.—*König.*—Parsons turbines, modified, 3-shaft.
Grosser Kurfürst.—A.E.G. turbines.
Markgraf.—Bergmann turbines.
Kronprinz Wilhelm.—Germania turbines.

Boilers.—

Fifteen single-ended Schulz boilers of small tube type in nine boiler rooms. Three fitted to burn oil only, the remainder coal and oil.

Auxiliary Machinery.—The main electrical installation consists of four turbo-dynamos and two Diesel-driven dynamos.

Funnels.—Two.

Propellers.—Three.

Speed, Horse-Power, and Fuel.

Speed (designed) Knots.	Horse-Power (designed).	Fuel: (*a*) Coal. (*b*) Oil.
21·0	31,000 (T)	(*a*) 3,543. (*b*) 690.

Endurance.

——	Speed.	Horse-Power.	Daily Consumption.	Radius of Action.	Remarks.
	Knots.		Tons.	Miles.	
At 4/5ths designed power	20·2	24,800	505	4,100	
At maximum continuous seagoing speed.	19·6	21,700	455	4,430	
At 3/5ths designed power	18·8	18,600	400	4,830	
At 2/5ths designed power	16·6	12,400	332	5,140	
At 1/5th designed power	13·0	6,200	212	6,300	
At 10 knots	10·0	2,800	135	7,600	

A 4

Kaiser, Friedrich der Grosse, Kaiserin, König Albert, *and* Prinzregent Luitpold.

(*See Plates* 3 *and* 13.)

Ship.	Designation before Launch.	Programme Year.	Built at	Ordered.	Laid Down.	Launched.	Commissioned for Trials.	Completed Trials.
Kaiser - -	Ersatz *Hildebrand*	1909–10	Imperial Dockyard, Kiel.	*9.09	*10.09	22.3.11	1.8.12	7.12.12
Friedrich der Grosse.	Ersatz *Heimdall*	1909–10	Vulcan Works, Hamburg.	*9.09	26.1.10	10.6.11	15.10.12	22.1.13
Kaiserin - -	Ersatz *Hagen* -	1910–11	Howaldt's Works, Kiel.	*4.10	11.10	11.11.11	14.5.13	13.12.13
König Albert -	Ersatz *Aegir* -	1910–11	Schichau Works, Danzig.	*4.10	17.7.10	27.4.12	31.7.13	8.11.13
Prinzregent Luitpold.	Ersatz *Odin* -	1910–11	Germania Yard, Kiel.	*4.10	1.11	17.2.12	19.8.13	6.12.13

* Approximate dates.

The *Kaiser* was launched with boilers in place and the *Friedrich der Grosse* with turbines as well as boilers. It was stated these were the first cases in which this practice had been adopted for large war vessels in Germany.

Cost.—

	Kaiser, Friedrich der Grosse.	*Kaiserin, König Albert, Prinzregent Luitpold.*
	£	£
Hull, machinery, &c. - - -	1,443,249	1,467,711
Gun armament - - -	814,825	823,141
Torpedo armament - - -	68,493	64,579
Totals - - -	£2,326,567	£2,355,431

General Remarks.—*Friedrich der Grosse* was fitted as fleet flagship, and *Kaiser* and *Prinzregent Luitpold* as flagships. The *Kaiser* was also provided with special accommodation for the Emperor. Up to then it had been customary for this accommodation to be fitted in the fleet flagship.

As compared with the two preceding types, the designs of these ships exhibit a marked improvement in the disposition of the heavy guns, which enables the whole 10 of them to be fired on either broadside, as against 8 out of 12 in the *Nassaus* and *Helgolands*. Although there is a reduction of one turret and two 12-in. guns there is a further increase in displacement. This is partly accounted for by the very heavy armour protection.

In the vessels of this class, the two foremost turrets are not usually fired at night, owing to the effect of the flash and blast on the bridge and searchlight personnel.

Complement.—Peace complement, 1,125 (as private ship). War complement about 1,400. Three Gunnery Officers are carried.

Accommodation.—The officers' accommodation is aft.

General Appearance.—*See* Plate 3 and Silhouette.

General Dimensions.—

Length, L.W.L. - - - - - -		564 ft. 4 ins.
Breadth, extreme - - - - - -		95 „ 2 „
Draught, designed load - - - - -		27 „ 3 „
„ full load - - - - -	about	28 „ 10 „
Displacement, designed load	(except *Prinzregent Luitpold*) -	24,310 tons.
	Prinzregent Luitpold - -	24,410 „
Freeboard (approximate) at designed draught	at stem -	22 ft. 0 ins
	amidships -	19 „ 6 „
	aft - -	13 „ 0 „
Height of axis of heavy guns (approximate) at designed draught - -	in 4 foremost turrets	27 „ 0 „
	in aftermost turret -	19 „ 0 „
Height of fore funnel to L.W.L. - - - -		72 „ 0 „
„ „ lower masthead to L.W.L. - - -		105 „ 0 „

KAISER.

FRIEDRICH DER GROSSE, KAISERIN, KÖNIG ALBERT and PRINZREGENT LUITPOLD similar.

Ordnance Survey, November, 1918.

HELGOLAND.
OSTFRIESLAND, THÜRINGEN and OLDENBURG similar.

Ordnance Survey, November, 1918.

Kaiser, Friedrich der Grosse, Kaiserin, König Albert, *and* Prinzregent Luitpold—*cont.*

Masts.—Two pole masts were originally fitted, but in some, if not all, of these ships a heavy cigar-shaped foremast has now been put in. This carries a splinter-proof spotting and control top with a large screened platform above it.

Constructive Details.—Very broad bilge-keels, extending almost one-third of the length of the ship, are fitted. The *Prinzregent Luitpold*, and probably the other ships of this type also, are fitted with submarine signalling apparatus.

Two longitudinal bulkheads, which extend to the upper deck, run nearly the whole length of the citadel, at about 30 feet from the ship's side. Above the main deck they are pierced at fairly wide intervals. On the middle (armoured) deck the transverse bulkheads are pierced by watertight doors on the inboard side of these longitudinal bulkheads. Thus it is possible to pass on the middle deck nearly from one end of the ship to the other. The ladders from the main to the middle deck lead each into a small watertight compartment.

Armour and other Protection.—

Material.—Krupp cemented.

Main Belt.—Width 9 ft. 9 in., of which about 5 ft. 7 in. is below the waterline. It extends from the foremost to the aftermost turret, is 13¾″ thick, tapering to the lower edge.

Upper Belt.—Width 7 ft. 3 in. It extends from the foremost to the aftermost turret, is 9″ thick.

Main Transverse Bulkheads.—The citadel, formed by the main and upper belts, is closed at either end by a transverse bulkhead, 7¾″ thick, butting on to the base of the barbette.

Belt before and abaft Citadel.—Forward both the main and the upper belt are continued to the stem by 7¾″ armour; aft, the main belt only is continued to within 18 feet of the stern by 7¾″ armour; the after end is closed by a transverse bulkhead of similar thickness.

Battery.—7¾″ thick. It extends from the foremost turret to No. 4 turret, the ends being closed by diagonal bulkheads 7¾″ thick, which butt on the barbettes.

Observation slits are cut in the armour of the foremost and after casemates to enable these positions to be used for secondary torpedo control.

Splinter Bulkheads.—There are two longitudinal bulkheads, ·8″ thick, in rear of the line of 5·9-in. guns on either side, with splinter screens also ·8″ thick between the guns.

Barbettes.—11¾″ thick.

Turrets.—11¾″ in front and 11″ to 11¾″ in rear. The side turrets are 23 ft. from the centre line, as in the *Helgoland* class. The revolving portion of each turret weighs 526½ tons.

Conning and Control Towers.—Foremost 13¾″, after 7¾″, thick. Similar to those of *Nassau* (page 17).

Armoured Hoods.—*See below*, under "Range Finder Positions."

Under-water Protection.—On the Blochmann-Neudeck system, the characteristic feature of which is the provision of two continuous longitudinal protective bulkheads on each side in the region of boilers, engines, and magazines. The outer of the two bulkheads, which is about 6 ft. 7 ins. from the ship's side and about ·3″ thick, extends from the inner bottom to the slope of the armoured deck. The inner or torpedo protection bulkhead is 6 ft. 7 ins. from the outer, and of high tensile steel 1¼″ to 1¾″ thick, or possibly thicker. It extends from the inner bottom to just above the armoured deck, and from there is continued by a splinter protection bulkhead, about 1¼″ thick, up to the battery deck.

In conjunction with this system there is minute subdivision.

Horizontal.—The principal armoured deck extends the entire length of the ship, and is 1¼″ thick on the flat and 4″ on the slopes. Before and abaft the centre line barbettes it descends to a lower level. The decks outside of and on top of the battery are believed to be armoured with armour 1½″ to 2½″ thick.

Torpedo Nets.—Beam net defence was fitted but has been discarded since the Battle of Jutland.

Kaiser, Friedrich der Grosse, Kaiserin, König Albert, *and* Prinzregent Luitpold—*cont.*

Armament.—

Guns and Ammunition Supply.

(For approximate arcs of training, *see* Plate 13.)

10—12-in. (30·5-cm.) Q.F., L/50, in five turrets.
Maximum rate of fire about 2½ rounds per gun per minute.
Ammunition supply electric. Loading at fixed angle of elevation.
Elevating gear hydraulic; training hydraulic or electric, alternative hand.

14—5·9-in. (15-cm.) Q.F., L/45, in battery. Maximum rate of fire about 16 rounds per minute.

4—22-pr. (8·8-cm.) semi-automatic, anti-aircraft guns, on after superstructure.

2—machine guns.

Magazines and Stowage.—The magazines are below the shell-rooms.

The 12-in. magazines are under the turrets, and the 5·9-in. between the echelon turrets.

About 90 rounds per gun are carried for 12-in. guns.

About 200 rounds per gun are carried for 5·9-in. guns.

Magazine cooling arrangements are fitted.

Torpedo Tubes.—

5—19·7-in. (50-cm.) submerged. The 4 broadside tubes are fixed at an angle of 20° before the beam, and are fitted with gyro angling gear which can be set for every 15° from 30° before to 60° abaft the direction of the tube.

Torpedoes.—In war time 16 torpedoes are carried, viz., two for the bow tube, four for each of the foremost, and three for each of the after broadside tubes.

Mines.—There is no permanent stowage for mines.

Searchlights.—

8—43·3-in. (110-cm.), worked by hand from a platform immediately below the light.

Range-Finder Positions.—A 3-metre (10-ft.) or larger range-finder is mounted on each of the two control towers and in each turret. In addition, an armoured hood projecting just above the upper deck is fitted on the starboard side of the battery just before the foremost side turret, and on the port side of the battery just abaft the after side turret; and each of these hoods is fitted with a 10-ft. range-finder for use with the secondary armament.

A shorter high-angle range-finder can be mounted on No. 4 turret.

Wireless Telegraphy.—The main wireless room is on the armoured deck just abaft the foremost funnel casing.

Battle Signal Stations.—

There are two main battle signal stations and two reserve stations.

Main Battle Signal Stations.—On the main deck in two special casemates, one the port side between the midship turret and No. 6 casemate, the other the starboard side between No. 2 casemate and the midship turret.

In action, flag signals are hoisted on the foremast from the starboard station, and on the mainmast from the port station; they are also hoisted on telescopic masts, about 40 feet long, which pass through the upper deck. These masts are operated by hand, the mast being lowered for each fresh signal to be bent on.

Reserve Battle Signal Stations.—In No. 6 casemate starboard and No. 2 casemate port. Signal flags are kept permanently stowed in these positions.

Kaiser, Friedrich der Grosse, Kaiserin, König Albert, *and* Prinzregent Luitpold—*cont.*

Boat Derricks.—One on port side abreast the foremost funnel, and one on starboard side abreast the after funnel.

Steering Gear.—Two parallel balanced rudders are fitted, each having a surface area of 193½ sq. ft., with the centre propeller between them. Diameter of rudder head, 20 in.

Anchors.—Three stockless bower anchors—two on port and one on starboard side; a light stern anchor right aft on centre line.

Coaling Arrangements.—Electric winches are fitted.

Machinery and Boilers.—

Main Engines.—

Turbines.

Type—*Kaiser* and *Kaiserin*, Parsons reaction.

Friedrich der Grosse, A. E. G.-Vulcan.

König Albert, Schichau.

Arrangement—Three complete sets, each consisting of a H.P., L.P., and astern turbine incorporated in L.P. casing. No cruising turbines. No reduction gear.

Prinzregent Luitpold, Parsons reaction, on two wing shafts only. Space has apparently been left to enable a 12,000 B.H.P. Diesel engine to be fitted on a centre shaft. A 6-cylinder 2-cycle engine of this power is said to have been completed at the Germania Works, Kiel, and to have carried out a satisfactory trial run, but, so far as is known, it has not yet been fitted in the ship.

Boilers.—

Sixteen Schulz boilers (except *Prinzregent Luitpold*—fourteen) in ten boiler rooms (*see* Plate 13). All are now fitted to burn coal and oil, two sprayers being provided for each. No superheaters are fitted. The *Kaiserin's* boilers were retubed in 1916.

Oil fuel tanks are situated in the double bottoms between 124–134 and 67½–73.

Auxiliary Machinery includes:—

Centrifugal pumps in the L.P. turbine rooms fitted with large bilge suctions.

Five turbo-generators as follows:

1 between stations 36–41 starboard.
1 ,, ,, 67½–73 ,,
1 ,, ,, 73–80 centre-line.
2 ,, ,, 80–85½ ,,

Two steam-driven air compressors between 99–111.

A steam reciprocating bilge pump in each L.P. turbine room and in Nos. 2, 6, and 9 boiler rooms—they are utilised also for fire main and ash ejectors, one of latter being fitted in each boiler room.

A feed pump in Nos. 1, 3, 5, 7, 8, and 10 boiler rooms.

Propellers.—

Three. Diameter, 11 ft. 8 in.

Lifts.—No lifts are fitted in engine or boiler rooms.

Watches.—Three watches are kept in E.R. Department, but the dog watches are not split.

Boats include:—

Two Diesel-driven 15 knot boats and one steam boat.

Kaiser, Friedrich der Grosse, Kaiserin, König Albert, *and* Prinzregent Luitpold—*cont.*

Speed, Horse-Power and Fuel.—

	Speed (designed).	Horse-Power (designed).	Fuel : (*a*) Coal. (*b*) Oil.
	Knots. 20·5	28,000 (T)	(*a*) 3,543 (*b*) 197*

* Probably increased.

Steam Trials.—

Ship.	Nature of Trial.	Date.	Speed.	Horse-Power.	Revolutions.	Remarks.
Kaiser - - -	Measured mile -	—.9.12	23·46	55,100	—	
" - - -	6 hours full power	—.9.12	22·28	41,516	268	
Friedrich der Grosse -	Measured mile -	—.12.12	22·44	42,113	—	
" "	6 hours full power	—.12.12	21·4	31,721	251	
Kaiserin - -	Measured mile -	—.11.13	22·3	42,501	—	*Kaiserin* did 23 knots during Jutland action.
" - -	6 hours full power	—.11.13	21·15	30,997	247	
König Albert - -	Measured mile -	—.10.13	22·15	39,813	—	
" " -	6 hours full power	—.10.13	21·4	32,965	251	

Endurance.—

—	Speed.	Horse-Power.	Daily Consumption.	Radius of Action.	Remarks.
	Knots.		Tons.	Miles.	
At 4/5ths designed power - -	19·7	22,400	456	3,750	
At maximum continuous seagoing speed.	19·0	19,600	409	4,030	
At 3/5ths designed power - -	18·2	16,800	360	4,400	
At 2/5ths designed power - -	16·1	11,200	300	4,670	
At 1/5th designed power - -	12·8	5,600	192	5,800	
At 10 knots - - - - -	10·0	2,800	135	6,450	

Helgoland, Ostfriesland, Thüringen, *and* Oldenburg.

(*See Plates* 4 *and* 14.)

Ship.	Designation before Launch.	Programme Year.	Built at	Ordered.	Laid Down.	Launched.	Commissioned for Trials.	Completed Trials.
Helgoland -	Ersatz *Siegfried*	1908–09	Howaldt's Works, Kiel.	*20.6.08	24.12.08	25.9.09	23.8.11	19.12.11
Ostfriesland	Ersatz *Oldenburg*	1908–09	Imperial Dockyard, Wilhelmshaven.	*19.4.08	19.10.08	30.9.09	1.8.11	15.9.11
Thüringen -	Ersatz *Beowulf*	1908–09	Weser Yard, Bremen.	*20.6.08	7.11.08	27.11.09	1.7.11	10.9.11
Oldenburg -	Ersatz *Frithjof*	1909–10	Schichau Works, Danzig.	* 8.4.09	—.1.09 (?)	30.6.10	1.5.12	1.7.12

* Official dates.

Cost.—

	£
Hull, machinery, &c. - - - - - -	1,307,263
Gun armament - - - - - -	929,550
Torpedo armament - - - - - -	68,493
Total - - - -	£2,304,306

Helgoland, Ostfriesland, Thüringen, *and* Oldenburg—*cont.*

General Remarks.—

Ostfriesland is fitted as flagship.

These ships, in many respects, are a distinct improvement on the *Nassau* type. Their displacement is 3,840 tons greater, they have a heavier armament (12-in.) and slightly higher speed, but their heavy guns are still disposed as in *Nassau* (*see* p. 17). The funnels were raised about 6 feet in 1914–15.

The foremost turret is not usually fired at night owing to the effect of the flash and blast on the bridge personnel.

Complement.—Peace complement, 1,106 (as private ship), except *Ostfriesland*, which, when flagship, has 1,097, not including flag officer, staff and retinue. War complement, about 1,400.

General Appearance.—*See* Plate 4 and Silhouette.

General Dimensions, &c.—

Length, L.W.L.		546 ft. 3 ins.
Breadth, extreme		93 „ 6 „
Draught, designed load		26 „ 11 „
Displacement, designed load		22,440 tons.
Freeboard (approximate), at designed draught	forward	22 ft. 0 ins.
	amidships	17 „ 9 „
	aft	19 „ 6 „
Height of axis of heavy guns (approximate), at designed draught	forward	27 „ 0 „
	aft	25 „ 0 „
Height of fore funnel to L.W.L.		67 „ 0 „
„ „ lower masthead		120 „ 0 „

Masts.—There are two steel pole masts. The foremast (and probably the mainmast also) is insulated by a teak covering to well above the level of the searchlights.

Constructive Details.—A fore bridge is not specially fitted, the top of the fore superstructure being arranged as such, with wing extensions on either side, pivoted, so that they can be swung aft when the ship is cleared for action.

Docking and bilge keels are fitted.

Armour and other Protection.—

Material.—Krupp cemented.

The battery armour, turrets, and conning towers were supplied by Krupp, the belt by Dillingen.

Belt.—Complete, except for about 12 ft. at the stern, where a transverse bulkhead 7¾″ thick connects the ends.

Thickness 11·8″ amidships at the waterline, 7¾″ forward and aft. From 4 ft. above the waterline amidships it commences to taper, reaching a thickness of 7¾″ at the battery deck.

Between the foremost and aftermost turrets the armour extends from 5 ft. 7 in. below the waterline up to the battery deck, having a total width of 16 ft. 1 in. Both ends of this midship section are closed by 7¾″ bulkheads, which butt on to the bases of the barbettes, thus forming the citadel. The belt is continued to the bow and stern, the upper edge being carried up to the battery deck forward and to the lower edge of the scuttles aft.

Battery.—The battery armour is 7¾″ thick, and extends from the foremost to the aftermost turrets, being closed by diagonal bulkheads 7¾″ thick which butt on the barbettes.

Observation slits are cut in the armour of the foremost and after casemates to enable these positions to be used for secondary torpedo control.

Splinter Bulkheads.—There are two longitudinal bulkheads, ·8″ thick, in rear of the line of 5·9-in. guns on either side, with splinter screens; also ·8″ thick between the guns, forming casemates.

Barbettes.—11¾″ thick. The trunks of the side turrets are not armoured behind the side armour, and do not pierce the slope of the armoured deck. The manner in which this is avoided, namely, by placing them nearer the centre line, is shown in Plate.

B 3

Helgoland, Ostfriesland, Thüringen, *and* Oldenburg—*cont.*

Armour and other Protection—*cont.*

Turrets.—11¾″ thick. The turrets pivot on the lower platform deck. It is noteworthy that both in *Helgoland* and *Kaiser* types the side turrets are situated at the same distance from the centre line, viz., 23 feet.

Conning and Control Towers.—Fore, 12″ or over; after, 7¾″ thick. Similar in shape and arrangement to those of *Nassau* (*see* p. 17).

Armoured Hoods.—Fitted one on either side of the battery, projecting just above upper deck, and take a range-finder for use with the secondary armament.

Torpedo Protection Bulkheads.—1½″ thick, and distant about 14 ft. from ship's side.

Horizontal.—The principal armoured deck extends the entire length of the ship and is 1½″ thick on the level and 2¾″ on the slopes. Before and abaft the middle line barbettes it descends to a lower level. The deck on top of the battery is believed also to be armoured.

Torpedo Nets.—All ships of this type were fitted with beam net defence, but this has been discarded since the Battle of Jutland. The net shelves are on a level with the battery deck.

Armament.—

Guns and Ammunition Supply.—

(For approximate arcs of training, *see* Plate 14.)

12—12-in. (30·5 cm.) Q.F., L/50, in six turrets.
Loading at fixed angle of elevation.

14—5·9-in. (15 cm.) Q.F., L/45, on C.P. mounting model 02/06 in battery.

2 *or* **4**—22-pr. (8·8 cm.) semi-automatic, anti-aircraft guns, on after superstructure.

2—machine guns.

Magazines and Stowage.—The 12-in. magazines are situated within and at the base of the trunks of the turrets below the armoured deck. The 5·9-in. magazines are on each side of the turret magazines.

Torpedo Tubes—

6—19·7-in. (50 cm.) submerged. The 4 broadside tubes are fixed at an angle of 20° before the beam and fitted with gyro angling gear, which can be set for every 15° from 30° before to 60° abaft direction of tube.

Mines.—There is no permanent stowage for mines.

Searchlights.—

8—43·3-in. (110 cm.). The searchlights are carried at heights varying from 40 to 55 ft. above load waterline.

Range-finder Positions.—The same system as in *Kaiser* (page 10).

Boat Derricks.—One on either side directly abaft the third funnel. The derrick masts are connected transversely by a cross-piece near the top.

Steering Gear.—Two parallel balanced rudders are fitted.

Anchors.—Three stockless bower anchors—two on port and one on starboard side; also a light stern anchor.

Coaling Arrangements.—Electric winches are fitted.

Machinery and Boilers.—

Main Engines.—

Three sets, vertical, 4-cylinder, triple-expansion, abreast, in separate watertight compartments.

Helgoland, Ostfriesland, Thüringen, *and* Oldenburg—*cont.*

Machinery and Boilers—*cont.*

Boilers.—

Fifteen Schulz, Navy type, in nine separate boiler-rooms. All are now fitted to burn coal and oil.

Funnels.—Three. The first two are circular and the after one oval in section.

Auxiliary Machinery includes Zoelly turbines made by Escher, Wyss & Co. for driving dynamos.

Propellers.—Three in number.

Speed, Horse-Power, and Fuel.

	Speed (designed).	Horse-Power (designed).	Fuel : (*a*) Coal. (*b*) Oil.
	Knots. 20·5	25,000	(*a*) 2,950 (*b*) 197*

* Probably increased.

Steam Trials.

Ship.	Nature of Trial.	Date.	Speed.	Horse-Power.	Revolutions.	Remarks.
			Knots.			
Ostfriesland - -	Measured mile -	—.6.11	21·23	35,500	126	Means of best performances on measured mile.
Helgoland - -	„ -	—.8.11	20·8	31,258	125	
Thüringen - -	„ -	—.4.11	21·07	34,944	114	
Oldenburg - -	„ -	—.1.12	21·3	31,394	—	

Endurance.

——	Speed.	Horse-Power.	Daily Consumption.	Radius of Action.	Remarks.
	Knots.		Tons.	Miles.	
At ⅘ths designed power - -	18·9	20,000	471	2,950	
At maximum continuous seagoing speed.	18·3	17,500	418	3,200	
At ⅗ths designed power - -	17·6	15,000	370	3,500	
At ⅖ths designed power - -	15·7	10,000	250	4,600	
At ⅕th designed power - -	12·3	5,000	149	6,050	
At 10 knots - - - - -	10·0	2,700	93	7,900	

Nassau, Westfalen, Rheinland, *and* Posen.

(*See Plates 5 and 15.*)

Ship.	Designation before Launch.	Programme Year.	Built at	Ordered.	Laid Down.	Launched.	Commissioned for Trials.	Completed Trials.
Nassau -	Ersatz *Bayern* -	1906–07	Imperial Dockyard, Wilhelmshaven.	31.5.06	*8.07	7.3.08	1.10.09	3.5.10
Westfalen	Ersatz *Sachsen* -	1906–07	Weser Yard, Bremen	30.10.06	12.8.07	1.7.08	16.11.09	3.5.10
Rheinland	Ersatz *Württemberg*	1907–08	Vulcan Works, Stettin	†2.4.07	*8.07	26.9.08	30.4.10	21.9.10
Posen -	Ersatz *Baden* -	1907–08	Germania Yard, Kiel	†2.4.07	*8.07	12.12.08	31.5.10	21.9.10

* Official date of *commencement*. † Official date of order.

B 4

Nassau, Westfalen, Rheinland, *and* Posen—*cont.*

Cost.—

	£
Hull, machinery, &c. - - - - -	1,089,531
Gun armament - - - - - -	660,469
Torpedo armament - - - - -	48,433
Total - - - -	£1,798,433

The above sum includes 195,695*l.* for ammunition and trials.

General Remarks.—

Westfalen and *Posen* are fitted as flagships.

The *Nassau* and *Westfalen* were the first ships of Dreadnought type to be built by Germany.

With two guns more than the British *Dreadnought*, these ships can only bring the same number of guns to bear on the broadside.

The foremost turret is not usually fired at night, owing to the effect of the flash and blast on the bridge personnel.

The *Rheinland* stranded on the Aland Skerries in January 1918. She was refloated with difficulty in July, and is at present being repaired (August 1918).

Complement.—Peace complement, 966 (as private ship), except *Westfalen*, which has 957 (as private ship). War complement, about 1,300. A turret's crew, including magazine and shell room parties, number 3 or 4 officers and 54 men.

Accommodation.—The accommodation for both officers and men is reported to be very poor. The officers' accommodation is aft. The sick bay is in the forecastle, and contains 12 swinging cots.

General Appearance.—*See* Plate 5 and Silhouette.

General Dimensions, &c.—

Length	between perpendiculars - - - -	451 ft. 9 ins.
	L.W.L. - - - - - -	478 „ 0 „
Breadth, extreme - - - - - -		88 „ 3 „
Draught, designed load - - - - -		26 „ 7 „
Displacement, designed load - - - -		18,600 tons.
Freeboard (approximate), at designed draught	forward -	22 ft. 0 ins.
	amidships -	17 „ 9 „
	aft - -	19 „ 6 „
Height of axis of 11-in. guns (approximate), at designed draught forward and aft - - - - -		26 „ 0 „
Height of fore funnel to L.W.L. - - - -		63 „ 0 „
„ „ lower masthead to L.W.L. - - -		120 „ 0 „

Masts.—There are two steel pole masts, insulated by a teak covering—the foremast to well above the top of the fore funnel, and mainmast to a few feet above the level of the standard compass.

Constructive Details.—The stern consists of a single steel casting.

There is a very elaborate system of subdivision. Six decks are fitted, viz.:—

(1) superstructure decks,
(2) upper deck,
(3) battery or main deck,
(4) armoured and middle decks,
(5) upper platform deck,
(6) lower platform deck.

The exposed decks are covered with teak, the unexposed with corticene.

Steam galleys and a bakery are provided.

An extensive refrigerating room is situated on the lower platform deck forward.

Docking keels are fitted.

Plate 5.
C.B. 1182 P.
October, 1918.

NASSAU.
WESTFALEN, RHEINLAND and POSEN similar.

Ordnance Survey, November, 1918.

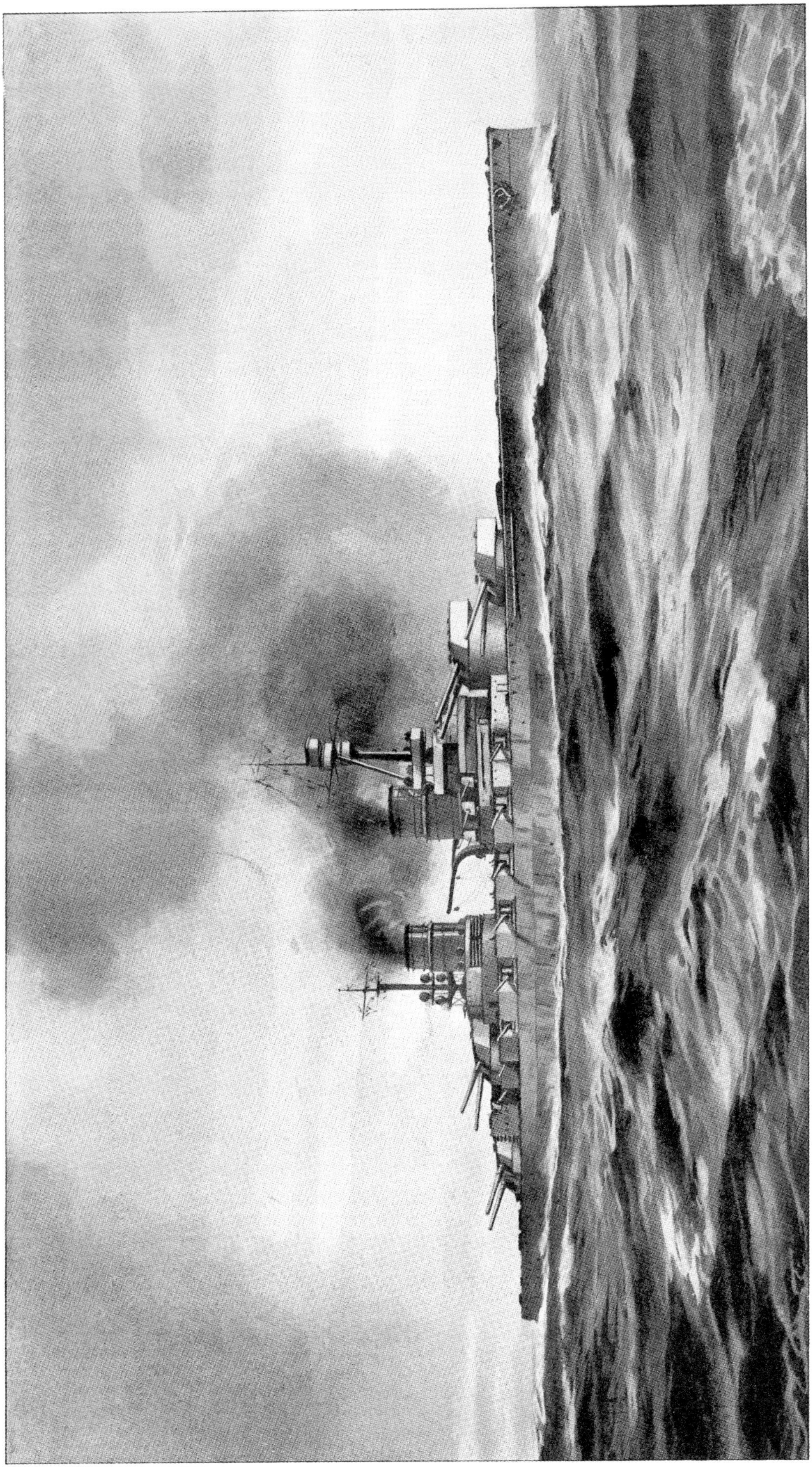

HINDENBURG.
MACKENSEN similar.

Ordnance Survey, November, 1918.

Nassau, Westfalen, Rheinland, *and* Posen—*cont.*

Armour and other Protection.—

Material.—Krupp cemented.

Belt.—*Nassau* and *Westfalen* 11·4″; *Rheinland* and *Posen* 11·8″ thick at the waterline; reduced to 6″ forward and 4″ aft, and tapering amidships to 4¾″ at the lower edge and to 7¾″ at the battery deck.

The belt is complete except for about 5 feet at the stern, where a transverse armoured bulkhead connects the ends. It extends from about 5 feet below the load-line to the battery deck. The midship section is closed by armoured bulkheads 7¾″ thick, which butt on to the bases of the barbettes, thus forming the citadel. At the stem the lower edge of the belt is carried to about 9 feet below the waterline.

Battery.—6½″ thick. Extends from the foremost to the aftermost turrets and is closed by diagonal bulkheads 6½″ thick which butt on to the barbettes. Observation slits are cut in the armour of the foremost and after casemates to enable these positions to be used for secondary torpedo control.

Splinter Bulkheads.—There are two longitudinal splinter-proof bulkheads, ·8″ thick, in rear of the line of 5·9-in. guns on either side, with splinter screens, also ·8″ thick, between the guns, forming casemates.

Barbettes.—11″ thick.

Turrets.—11″ thick.

Conning and Control Towers.—Fore, 12″; after, 7″.

The fore conning and control tower is roughly semicircular in transverse section —internal dimensions 16 ft. (max.) athwartships and 13 ft. fore and aft. In elevation it is roughly an inverted truncated cone. It is divided into two parts by a transverse bulkhead. The after part, which is used as control position, projects above the fore or navigating position. In the control position gratings are fitted well above the level of the superstructure deck, as a platform for the control officers and range-taker; the space below these gratings forms the upper transmitting station. A communication tube leads to the lower conning tower below the armoured deck.

The after conning and control tower is circular in section and nearly 8 ft. in diameter. In this tower there is no dividing bulkhead.

Armoured Hoods.—Are fitted one on either side of battery, projecting just above the upper deck between the broadside turrets, and take a range-finder for use with the secondary armament.

Torpedo Protection Bulkhead.—Reported to be 1″ thick and distant 6 ft. 7 in. from the ship's side. It extends from the double bottom to the armoured deck, and in a fore and aft direction as far as the ends of the citadel, where it is closed by transverse bulkheads.

Horizontal.—The principal armoured deck extends the entire length of the ship and is about 2½″ thick on the slopes. Before and abaft the centre line barbettes it descends to a lower level. The deck on top of the battery is believed also to be armoured.

Torpedo Nets.—Beam net defence was fitted, but has been discarded since the Battle of Jutland.

Armament.—

Guns and Ammunition Supply.—

(For approximate arcs of training, *see* Plate 15.)

12—11-in. (28-cm.) Q.F., L/45, in six turrets, on model /06 mountings. Loading at fixed angle of elevation.

12—5·9-in. (15-cm.) Q.F., L/45, on central pivot mounting model 02/06.

2 or **4**—22-pr. (8·8-cm.), semi-automatic, anti-aircraft guns, on after superstructure.

2—machine guns.

Magazines and Stowage.—

The 11-in. magazines are situated at the bases of the trunks of each turret.
80 rounds per gun are carried for 11-in. guns.
180 „ „ „ 5·9-in. guns.

Nassau, Westfalen, Rheinland *and* Posen—*cont.*

Armament—*cont.*

Torpedo Tubes.—

6—17·7-in. (45 cm.) submerged. The 4 broadside tubes are fixed at an angle of 20° before the beam and are fitted with gyro angling gear which can be set for every 15° from 30° before to 60° abaft the direction of the tube. At ordinary draught the stern tube is about 4 ft. under water.

Mines.—

There is no permanent stowage for mines.

Searchlights.—

8—43·3-in. (110-cm.).

Range-finder Positions.—

The same general system as in *Kaiser* (page 10), but probably shorter-base instruments are fitted.

Wireless Telegraphy.—The wireless room is on the battery deck abaft the second funnel.

Boat Cranes.—There are two cranes for hoisting boats, one on each side between the side turrets.

Steering Gear.—Two parallel balanced rudders are fitted, between which the centre propeller revolves. Each rudder is actuated by a separate steering engine.

Anchors.—Three stockless bower anchors, two on starboard and one on port side; also a light stern anchor.

Coaling Arrangements.—Electric winches are fitted for coaling.

Machinery and Boilers.—

Main Engines.—

Three sets, vertical, triple expansion, abreast, in separate watertight compartments, situated under the mainmast.

Boilers.—

Twelve Schulz, in three separate transverse boiler-rooms, each containing four boilers. All are now fitted to burn coal and oil, there being two Körting sprayers to each boiler. The fore boiler-room is situated before the foremost side turrets, whilst the other two boiler-rooms are immediately abaft these and before the aftermost side turrets.

Escapes.—The escapes leading from the various boiler-rooms to the armoured deck are watertight, and pass through the bunkers. They are provided with self-acting watertight doors.

Funnels—Two. Height above bars, about 80 feet.

Auxiliary Machinery.—

Three auxiliary machinery rooms, situated abaft the three engine rooms, contain the condensers, air pumps, circulating pumps, &c.

Bunkers.—There are 10 wing bunkers on either side of the ship extending up to the armoured deck, and from the foremost to the aftermost turrets. The sliding bunker doors can be closed from the upper decks. If required, coal can be delivered directly in front of the furnaces through special shafts.

Propellers.—Three, bronze.

Nassau, Westfalen, Rheinland *and* Posen—*cont.*

Speed, Horse-Power, and Fuel.—

	Speed (designed).	Horse-Power (designed).	Fuel : (*a*) Coal. (*b*) Oil.	
	Knots. 19	20,000	(*a*) 2,658 (*b*) 197*	

* Probably increased.

Steam Trials.—

Ship.	Nature of Trial.	Date.	Speed.	Horse-Power.	Revolutions.	Remarks.
			Knots.			
Nassau - - -	Measured mile	–.4.10	20·06	25,850	124	In deep water.
Westfalen - -	"	–.4.10	20·3	27,477	125	
Rheinland - -	"	–.8.10	20·03	27,498	124	
Posen - - -	"	–.8.10	20·12	28,117	127	

Endurance.—

——	Speed.	Horse-Power.	Daily Consumption.	Radius of Action.	Remarks.
	Knots.		Tons.	Miles.	
At $\frac{4}{5}$ths designed power - -	18·3	16,000	380	3,200	
At maximum continuous seagoing speed.	17·7	14,000	335	3,540	
At $\frac{3}{5}$ths designed power - -	17·0	12,000	296	3,800	
At $\frac{2}{5}$ths " " - -	15·2	8,000	200	5,050	
At $\frac{1}{5}$th " " - -	12·0	4,000	110	7,250	
At 10 knots - - - - -	10·0	2,500	83	8,000	

BATTLE CRUISERS.

Graf Spee *and* Prinz Eitel Friedrich.

(*See Plate* 16.)

Ship.	Designation before Launch.	Programme Year.	Building at.	Ordered.	Laid Down.	Launched.	Commissioned for Trials.	Completed Trials.
Graf Spee - -	Ersatz *Freya*	1916–17	Schichau, Danzig -	—	1916	15.9.17	—	—
Prinz Eitel Friedrich	Ersatz *Vineta*	1916–17	Imperial Dockyard, Wilhelmshaven.	—	*-.5.16	—	—	—

* Approximate date.

General Remarks.—

These ships may be of the type shown in Plate 16, with a primary armament of 6—15-in. guns mounted in three centre-line turrets.

It is possible, however, that they are merely improved *Hindenburgs*.

It appears that all work on them has been stopped (October 1918).

Hindenburg *and* Mackensen.

(*See Plates* 6 *and* 17.)

Ship.	Designation before Launch.	Programme Year.	Built or Building at	Ordered.	Laid down.	Launched.	Commissioned for Trials.	Completed Trials.
Hindenburg -	Ersatz *Hertha* -	1913–14	Imperial Dockyard, Wilhelmshaven.	*20.4.13	*9.6.13	1.8.15	*-.5.17	—
Mackensen -	Ersatz *Victoria Louise.*	1915–16	Blohm & Voss Works, Hamburg.	—	1915	21.4.17	*-.9.18	—

* Approximate dates.

General Remarks.—*Hindenburg* generally resembles *Derfflinger* (*see* p. 21), but has two more 5·9-in. guns (16 instead of 14). *Mackensen* is believed to be similar. The completion of the *Hindenburg* was much delayed owing to the removal of several armour plates, two of her turrets, and all her 12-in. guns, to make good damage in other ships after the Battle of Jutland.

General Appearance.—*See* Plate 6 and Silhouette.

General Dimensions, &c. (*These figures are doubtful.*)—

Length, L.W.L. - - - - - - -	689 ft. 0 ins.
Breadth, extreme - - - - - - -	95 ,, 2 ,,
Draught, designed - - - - - -	27 ,, 3 ,,
Displacement, designed - - - - -	26,200 tons.

Armour.—*See Derfflinger*, p. 22.

Armament.—

8—12-in. (30·5-cm.) L/50, in pairs, in turrets on centre line.

16—5·9-in. (15-cm.) L/50, in battery.

8—22-pr. (8·8-cm.) semi-automatic, anti-aircraft guns, fitted with curved shields.

There is no reliable information as to the torpedo armament.

Speed and Horse-Power.—The designed S.H.P. is reported to be 85,000, and designed speed, 28 knots.

Derfflinger.

(*See Plates 7 and 18.*)

Ship.	Designation before Launch.	Programme Year.	Where Built.	Ordered.	Laid Down.	Launched.	Commissioned for Trials.	Completed Trials.
Derfflinger	K. - - -	1911–12	Blohm & Voss, Hamburg.	*–.6.11	–.3.12	1.7.13	*–.8.14	*–.11.14

* Approximate dates.

Cost.—

	£
Hull, machinery, &c. - - - -	1,450,586
Gun armament - - - -	735,322
Torpedo armament - - - -	47,456
Total - - - -	2,233,364

General Remarks.—The *Lützow*, which was a sister ship to the *Derfflinger*, was sunk in action 31st May—1st June 1916. These were the first German battle cruisers with all centre-line turrets.

The two foremost turrets are not usually fired at night, owing to the effect of the flash and blast on the bridge personnel.

A noticeable feature in the *Derfflinger* is the continuous deck running with a marked sheer from the stem to the stern and forming amidships the battery deck. The 5·9-in. battery is thus practically one deck higher than in previous battle cruisers.

Complement.—Peace complement, 1,125 ; war complement, about 1,400.

General Appearance.—*See* Plate 7 and Silhouette.

General Dimensions, &c.—

Length, L.W.L. - - - - - -	689 ft. 0 ins.
Breadth, extreme - - - - -	95 „ 2 „
Draught, designed load - - - -	27 „ 3 „
Displacement, designed load - - - -	26,180 tons.
Freeboard (approximate) at designed draught - stem - - -	24 ft. 0 ins.
aft - - -	15 „ 0 „
Heights of axis of guns above waterline—	
1st turret - - - - - -	26 ft. 11 ins.
2nd „ - - - - - -	35 „ 7 „
3rd „ - - - - - -	30 „ 4 „
4th „ - - - - - -	20 „ 9 „
5·9-in. guns (*average*) - - - -	18 „ 0 „
Height of fore funnel to L.W.L. - -	about 69 „ 0 „
„ „ lower masthead to L.W.L. - -	„ 102 „ 0 „

Constructive Details.—

Below the armoured deck the framing is of the longitudinal type except at the ends of the ship. Above the armoured deck it is of the transverse type.

The distance between frame stations is 25·2 ins. For the greater part of the length of the ship bracket frames are worked at every third frame station; at the intermediate stations the vertical keel is stiffened by triangular plates.

The vertical keel is ·7″ thick. Docking keels are fitted, but no bilge keels.

From abreast the fore side of the first turret to just abaft the fourth turret two continuous longitudinal protective bulkheads are fitted. The outer one, at a distance of about 6 ft. 7 ins. from the ship's side, is about ·3″ thick. The inner, which is referred to below under "Armour" as the torpedo protection bulkhead, is about

C 3

Derfflinger—*cont.*

Constructive Details—*cont.*

13 ft. from the ship's side and is from 1·2″ to 1·8″ in thickness. The space between these two bulkheads forms the protective coal bunkers. The supply bunkers are on the inner side of the torpedo protection bulkhead and are about 10 ft. in width. A middle line bulkhead is fitted for the whole length of the ship between the turrets, being interrupted only in way of turret ammunition trunks.

The average thickness of the hull plating is ·6″. The thickness is increased to ·9″ in way of the stern framing, &c. Immediately below the armour the plating is doubled.

Armour and other Protection.—

Material.—Krupp cemented.

Main Belt.—Extends from the fore side of the first to 10 ft. abaft the fourth turret. Length, 400 ft. Width, amidships, 19 ft.; extreme width at foremost end, 23 ft. 10 ins. Throughout its length it extends 5 ft. 7 ins. below the waterline. From the foremost end as far as No. 3 casemate the belt consists of two rows of plates worked horizontally, the junction being 3 ft. 9 ins. above the waterline at No. 3 casemate and 6 ft. 3 ins. above the waterline at the foremost end. Abaft No. 3 casemate the belt consists of plates worked vertically. The maximum thickness of the belt is believed to be about 12″, armour of this thickness extending about 14 ins. below and about 4 ft. above the waterline amidships, thence tapering downwards to the lower edge and upwards to the battery deck.

Transverse Bulkheads.—At either end of the main belt a transverse bulkhead is fitted, closing the citadel. The foremost bulkhead butts on to the base of No. 1 barbette. The thickness in both cases is about 9″.

Belt before and abaft Citadel.—The main belt is continued forward to the stem, and aft to within 15 ft. of the stern, by armour believed to have a maximum thickness of 5″. At the after end a transverse bulkhead is fitted, about 5″ thick.

Battery.—The battery armour extends between No. 2 and No. 3 turrets and is closed at either end by diagonal bulkheads which butt on the barbettes. The thickness is believed to be about 7″. Observation slits are cut in the armour at two positions on either side which are used for secondary torpedo control.

Splinter Bulkheads.—There are two longitudinal bulkheads, ·8″ thick, in rear of the line of 5·9-in. guns, on either side, with splinter screens, also ·8″ thick, between the guns, forming casemates.

Barbettes.—Believed to be about 10″ in thickness.

Turrets.—Front armour believed to be about 11″ in thickness.

5·9-in. Gun Shields.—3·1″ thick.

Conning and Control Towers.—Believed to be generally similar to those in *Seydlitz* (*see* p. 27).

Armoured Hoods.—Fitted, one on either side of battery, and projecting just above the upper deck, and intended for range-finders for use with the secondary armament.

Torpedo Protection Bulkhead.—1·2″ to 1·8″ thick, is fitted on either side of the ship, extending for the same longitudinal distance as the main belt, viz., 400 ft. Vertically it extends from the outer bottom to 28 ins. above the armoured deck, whence it is continued as a wing splinter bulkhead, 1·2″ thick, up to the battery deck.

Horizontal:—

Principal Armoured Deck.—Is generally 3 ft. 9 ins. above the waterline amidships, 4 ft. 4 ins. below the waterline forward, and about 2 ins. above it aft. Forward the deck is flat; amidships and aft it slopes down at an angle of 30° to 4 ft. 4 ins. below the waterline at the sides. Forward and over the magazines it is 2″ thick; amidships it is 1·2″ thick; aft it is about 3·2″ on the flat and 2″ on the slopes.

Battery Deck.—Probably armoured only where shown in Plate 18; thickness ·8″ to 1·2″, with doubling plates ·8″ thick in certain places.

Upper Deck.—Thickness, 1″ to 2″.

Plate 7.
C.B. 1182 P.
October, 1918.

DERFFLINGER.

Ordnance Survey, November, 1918.

SEYDLITZ.

Ordnance Survey, November, 1918.

Derfflinger—*cont.*

Torpedo Nets.—

Beam net defence was originally fitted, but has been discarded since the Battle of Jutland.

Armament.—

Guns and Ammunition Supply.—

(For approximate arcs of training, *see* Plate 18.)

8—12-in. (30·5-cm.) Q.F. L/50, mounted in pairs in turrets on centre line.
Training electric, elevation hydraulic, ammunition hoists electric.
Rate of delivery of projectiles and charges, 3½ rounds a minute.
Loading at fixed angle of elevation.

14—5·9-in. (15-cm.) Q.F. L/45, on C.P. mounting, with 3·1-in. shields, in battery.
Ammunition hoists electric, alternative hand.

8—22-pr. (8·8-cm.), semi-automatic, anti-aircraft guns, fitted with curved shields.

2—machine guns.

Magazines and Stowage.—

The main magazines are situated directly below each turret. The central magazine, between the two boiler rooms, supplies the 5·9-in. battery.

A middle line passage immediately under the armoured deck allows of an interchange of ammunition between the third turret supply and the fourth turret supply.

90 rounds per gun are carried for 12-in. guns.
160 ,, ,, ,, ,, 5·9-in. ,,

Torpedo Tubes.—

4—19·7-in. (50-cm.) submerged tubes. The bow tube, depression nil, is fitted 15 ft. 5 ins. below waterline; the stern tube, depression 1° 45′, 5 ft. 5 ins. below waterline. The two broadside tubes, depression 2°, are fixed at an angle of 20° before the beam and are fitted with gyro angling gear which can be set for every 15° from 30° before to 60° abaft the direction of the tube ; they are about 18 ft. below waterline.

Torpedoes.—

Sixteen torpedoes are carried, three being stowed near the bow and three near the stern tube, and five near each broadside tube.

Warhead magazines are fitted both forward and aft.

Mines.—There is no permanent stowage for mines.

Searchlights.—

8—43·3-in. (110-cm.) searchlights are carried in position, and two spare. In addition, a smaller searchlight is carried on a platform under fore top, Of the eight principal searchlights, two are mounted on platforms on tripod mast, two on fore funnel, two on platform on fore side of after funnel, and two on platform on fore side of mainmast. The searchlights are worked by hand from control positions below the lights. Switching on and off is effected from the control position.

Stowage for all the searchlights in use as well as for the two spare searchlights is provided behind armour. An electrically controlled disappearing platform is reported to be fitted for each spare searchlight.

The searchlights are fed by four groups of special motor generators, 80 volts, 200 ampères.

Communications.—

A transmitting station is fitted below the armoured deck, just abaft after conning and control tower.

For the whole length of the boiler rooms there are two fore and aft passages, one on either side of the middle line bulkhead immediately below the armoured deck. These passages carry cables, voice pipes, and other means of communication between the fore part of the ship and the after part.

Derfflinger—*cont.*

Range-finders.—3 metre (10-ft.) or larger range-finders are mounted, one on each control tower and one in each of the turrets; formerly an additional range-finder, for use with the secondary armament, was fitted in each of the two armoured hoods projecting just above the upper deck, but these instruments are reported to have been removed (September 1918), being found useless owing to spray.

Battle Signal Stations.—

The main battle signal stations are situated on each side of the main deck. Details as in *Kaiser* class (*see* p. 10).

Derricks.—There are two derricks on stump masts, each capable of lifting 12 tons.

Steering Gear.—

Two rudders are fitted, one abaft the other. Area of after rudder, 344 sq. ft.; foremost rudder, 205 sq. ft. Foremost rudder works on ball bearings.

Maximum angle of helm is 40°.

Each rudder has a special electric steering engine and controls, but the two rudders can be worked simultaneously and by either of the steering engines alone. The hand steering arrangements also permit of the two rudders being worked independently or together.

Bridge.—

The navigating bridge is formed by the fore superstructure deck with a swinging bridge extension on either side. A portable shelter and a compass and steering wheel are fitted before the conning tower. Portable shelters for signalmen are fitted on either side of the bridge. The charthouse is abaft the conning tower.

A signalling bridge is fitted above the charthouse, and is so arranged as not to interfere with the view from the conning tower.

Compasses.—

The following are fitted:—

(*a*) 4 magnetic compasses, including a master magnetic compass fitted just above the inner bottom forward.

(*b*) 3 magnetic repeaters, viz., one on the fore bridge, one in the fore conning tower, and one in the fore lower conning tower.

(*c*) A master gyro compass in the fore lower conning tower.

(*d*) 4 gyro repeaters, viz., one on the fore bridge, one in the fore conning tower, one in the upper steering engine flat, and one by the hand wheel aft.

The standard compass ships on a portable platform on the quarterdeck.

Anchors and Cables.—

Two bower anchors, each of 7·5 tons, and one stern anchor of 3·75 tons, are carried.

For the bower anchors, 21 shackles of 2·8-in. cable are carried; for the stern anchor, a 3½-in. wire hawser.

The two capstans for the bower anchors are worked independently of each other by a steam engine. The capstan for the stern anchor is worked by an electric motor.

Coaling Arrangements.—

Coaling stays are fitted on either side of the ship with 18 whips, worked by nine portable electric winches with double drums. The foremost and after capstans and the boat hoists are also utilised.

There are bunkers on either side of the wing splinter bulkhead *above* the armoured deck; protective bunkers on the one side and supply bunkers on the other side of the torpedo protection bulkhead *below* the armoured deck. Openings in the wing splinter bulkhead enable coal to be passed from the upper outer into the upper inner bunkers. The protective bunkers are filled from the scuttles of the upper outer bunkers.

Coal is passed from the upper deck to the scuttles of the battery deck through shoots which can be unshipped where they obstruct the fire of the casemate guns. A separate scuttle is fitted for every 40–50 tons of coal.

Derfflinger—*cont.*

Coaling Arrangements—*cont.*

The scuttles of the after bunkers, which extend abreast the engine rooms, allow of the passage of a full bag, as the coal in these bunkers is kept stowed in bags to enable it to be transported forward when necessary. These after bunkers constitute reserve and protective bunkers only.

There are no coaling scuttles in the casemates.

Machinery and Boilers.—*Main Engines*—Turbines, Navy type, on four shafts.

Boilers.—18 Schulz, Navy type, in ten boiler rooms. Several boilers are fitted for oil-firing only, the remainder for coal and oil.

Propellers.—Four.

Speed, Horse-Power, and Fuel.—

Ship.	Speed (designed).	Horse-Power (designed).	Fuel : (*a*) Coal. (*b*) Oil.
Derfflinger	Knots. 26·5	63,000	(*a*) 4,625 (*b*) 984

Endurance.—

——	Speed.	Horse-Power.	Daily Consumption.	Radius of Action.	Remarks.
	Knots.		Tons.	Miles.	
At $\frac{4}{5}$ths designed power	25·1	50,400	1,026	3,350	
At maximum continuous seagoing speed.	24·3	43,100	905	3,670	
At $\frac{3}{5}$ths designed power	23·3	37,800	810	3,930	
At $\frac{2}{5}$ths designed power	20·7	25,200	567	5,000	
At $\frac{1}{5}$th designed power	16·4	12,600	351	6,400	
At 10 knots	10·0	3,200	154	8,830	

Seydlitz.

(*See Plates 8 and 19.*)

Ship.	Designation before Launch.	Programme Year.	Built at	Ordered.	Laid Down.	Launched.	Commissioned for Trials.	Completed Trials.
Seydlitz	J.	1910–11	Blohm and Voss, Hamburg.	*–.4.10	4.2.11	30.3.12	22.5.13	17.8.13

* Approximate date.

Cost.—

	£
Hull, machinery, &c.	1,450,586
Gun armament	690,557
Torpedo armament	45,011
Total	£2,186,154

General Remarks.—The *Seydlitz*, an improved *Moltke*, differs in having her main armament on three instead of two decks, the forecastle and fore turret being a deck above the main portion of the upper deck, whilst the quarterdeck and aftermost turret are a deck below it.

Seydlitz—*cont.*

General Remarks—*cont.*

The foremost turret is not usually fired at night, owing to the effect of the flash and blast on the bridge personnel. The searchlights' crews are therefore supplied by these turrets.

Complement.—Peace complement, 1,108 (as private ship); war complement, about 1,400. A turret's crew, including magazine and shell-room parties, numbers one officer, one warrant officer, and about 60 men.

General Appearance.—*See* Plate 8 and Silhouette.

General Dimensions, &c.—

Length, L.W.L. - - - - -	656 ft. 2 ins.
Breadth { L.W.L. - - -	93 ,, 2 ,,
Breadth { extreme, below waterline	93 ,, 6 ,,
Draught, designed load - - - -	26 ,, 11 ,,
Displacement, designed load - - -	24,610 tons.
Height from base to waterline - - -	26 ft. 10 ins.
,, of upper deck above base line -	45 ,, 6 ,,
,, of battery (main) deck above base line -	38 ,, 10½ ,,
Freeboard (approximate) at designed draught { stem - -	26 ,, 0 ,,
Freeboard (approximate) at designed draught { amidships -	17 ,, 6 ,,
Freeboard (approximate) at designed draught { aft - -	10 ,, 6 ,,
Height of axis of guns above waterline:—	
No. 1 Turret - - - - -	34 ft. 0 ins.
Nos. 2 and 3 Turrets - - - -	26 ,, 9 ,,
No. 4 Turret - - - - -	27 ,, 8 ,,
No. 5 Turret - - - - -	19 ,, 8 ,,
5·9-in. guns - - - - -	15 ,, 9 ,,
Height of fore funnel to L.W.L. - - -	69 ,, 0 ,,
,, ,, lower masthead to L.W.L. - -	102 ,, 0 ,,

Constructive Details.—Frahm's anti-rolling tanks were fitted, but are probably now used for stowing additional coal, as in the case of *von der Tann* (*see* page 34).

Armour and other Protection.—

Material.—Krupp cemented.

Main Belt.—Extends from the foremost to the aftermost turret. Length 376 ft. 4 in., width 9 ft. 10 in., of which 5 ft. 3 in. is below the waterline. It consists of 22 plates, each 9 ft. 10 in. wide, with a mean length of 17 ft. 5 in. Thickness 11·8″ from the upper edge to 1 ft. below the waterline, whence it tapers to 5·9″ at the lower edge.

Upper Belt.—Extends from the foremost to the aftermost turret. Length 376 ft. 4 in. It varies in width from 9 ft. 2 in. in way of the battery (where it overlaps the battery deck and is carried up to the battery port sills) to about 7 ft. before and abaft this section (where it only extends up to the battery deck). In way of the battery it consists of 8 plates having a mean width of 9 ft. 2 in., and from 17 ft. 5 in. to 19 ft. 8 in. long. Maximum thickness 11·8″ at the lower edge, tapering to 9″ at the battery deck, and, in the case of the plates which overlap, further tapering to 7·9″ at the battery port sills.

Main Transverse Bulkheads.—The citadel, formed by the main and upper belts, is closed at either end by transverse bulkheads. The foremost one rests on the lower portion of the armoured deck, which is 3 ft. 11 in. below waterline, and is uniformly 8·7″ thick. It butts on to the base of No. 1 barbette. The after transverse bulkhead is similar, but the middle portion, which is curved, is 9″ thick instead of 8·7″, and a door is cut on the port side. It likewise rests on the lower part of the armoured deck, which is there, however, only 6 in. below the waterline. (*See* Plate.)

Belt before and abaft Citadel.—The main and upper belts are continued forward (for about 59 ft.) by 4·7″, and to the stem (about 88½ ft.) by 3·9″ armour, and aft to

Seydlitz—*cont.*

Armour and other Protection—*cont.*

within 10 ft. of the stern by 3·9″ armour. A transverse bulkhead 3·9″ thick closes the after end. The belt forward consists of 6 plates of 4·7″ and 9 plates of 3·9″ armour, each 17 ft. 5 in. wide and 9 ft. 10 in. long. The belt aft consists of 13 plates of 3·9″ armour, each 13 ft. 6 in. wide, with a mean length of 9 ft. 10 in.

Battery.—The battery above the upper belt is protected as follows:—

In way of the battery the armour plates forming the upper belt (*see* page 26) do not terminate at the battery deck, but continue above it up to the battery port sills. The overlap tapers from 9″ to 7·9″ at the upper edge of the plates. On these plates rests the battery armour proper, which is uniformly 5·9″ thick.

The battery is 236 ft. long, and is closed at the fore end by diagonal bulkheads 5·9″ thick, which butt on the base of the conning tower (3·1″ thick), and at the after end by an athwartship bulkhead 5·9″ thick.

Observation slits are cut in the armour of the second and aftermost casemates on either side to enable these positions to be used for secondary torpedo control.

Splinter Bulkheads.—There are two longitudinal splinter-proof bulkheads, ·8″ thick, in rear of the line of 5·9-in. guns, on either side, with splinter screens, also ·8″ thick, between the guns, forming casemates.

Barbettes.—

No. 1.—28 ft. in diameter, consists of 9 plates, each 18 ft. 4 in. high and with a mean width of 9 ft. 2 in., extending to battery deck. These plates are 9″ thick, except the rear plate, which is 7·9″ thick. Below the battery deck the barbette is 1·2″ thick except on the fore side, where it is 9″ thick and forms part of the transverse bulkhead. This portion consists of two plates, each 15 ft. 5 in. high and 9 ft. 10 in. wide.

Nos. 2 and 3.—Of same diameter as No. 1, but cylindrical only above upper deck and like an inverted irregular truncated cone between that deck and the battery deck. It consists of 5 plates each 3 ft. 11 in. high by 17 ft. 0 in. wide and 9″ thick above upper deck, and of 4 plates, each 7 ft. 3 in. high and with a mean width of 7 ft. 3 in., and 3·9″ thick between that and the battery deck; and is 1·2″ thick between the battery and armoured decks.

No. 4.—Above the battery deck it consists of 9 plates, each 12 ft. 2 in. high with a mean width of 9 ft. 2 ins., 8 of which are 9″ thick, whilst the ninth on the after side, where it is partially protected by No. 5 barbette, is only 7·9″ thick. Between the battery and armoured decks it is only 1·2″ thick.

No. 5.—Consists of 4 plates, each 4 ft. 3 in. high and 17 ft. 5 in. wide, extending to battery deck. These plates are 9″ thick. Below the battery deck the barbette is 1·2″ thick, except on the after side, where it is 9″ thick and forms part of the transverse bulkhead. This portion consists of two plates, each 16 ft. 5 in. high by 8 ft. 6 in. wide.

Turrets.—Front 9·8″, sides and rear 7·9″ thick. Roof 3·9″ on slope, 2·75″ on flat. Floor 3·9″ and 2″ thick.

Two ports are cut in the rear wall of each turret to permit of the passage of a 22-pr. (8·8-cm.) sub-calibre gun; they are closed by hinged flaps.

5·9-*in. Gun Shields.*—3·1″ thick.

Conning and Control Towers.—

Fore.—In elevation the foremost conning and control tower is roughly an inverted truncated cone. In transverse section it is roughly semi-circular—internal dimensions 16 ft. (max.) athwartships by 13 ft. fore and aft. Front and sides 13·8″ thick, rear 9·8″. Roof 3·1″ thick, and floor 2·75″. Entrance is by door in rear on the centre line. The tower is divided into two parts by a bulkhead 2·4″ thick, in which there are two doors. The after part or control position projects

Seydlitz—*cont.*

Armour and other Protection—*cont.*

Conning and Control Towers—cont.

2 ft. 9 in. above the foremost or navigating position and is 13·8″ thick in front and 9·8″ in rear. Both positions are provided with observation slits, and from either position it is possible to observe, practically without interference, from 0° to 135° each side. In the control position gratings are fitted well above the level of the superstructure deck, as a platform for the control officers and range-taker; the space below these gratings forms the upper transmitting station. A communication tube, square in section, 3¼ ft. × 3¼ ft., and 28 ft. 9 in. long, extends to the armoured deck. It is 7·9″ thick in front and sides and 5·9″ in rear. Conning tower armour, 3·1″ thick, extends down to the battery deck, where there is a door.

After Conning Tower. — Circular: inner diameter 7 ft. 10 in., height 7 ft. 10 in., sides 7·9″ thick, roof and floor 2″. The communication tube is cylindrical—internal diameter 2 ft. 6 in., thickness 5·9″ above, and 3·1″ below, the battery deck. Door in rear on centre line.

Armoured Hoods.—Are fitted, one on either side of the battery, projecting just above the upper deck, immediately before the broadside turret in each case. They originally took 10-ft. rangefinders for use with the secondary armament.*

Torpedo Protection Bulkhead.—Is fitted on either side of the ship and 13 ft. 2 in. distant therefrom, and extending the same longitudinal distance as the midship section of the belt, namely, 376 ft. 4 in. This bulkhead extends from the outer bottom to the battery deck. It is 2″ thick in the wake of the magazines and 1·2″ elsewhere. The ends are closed by ·8″ transverse bulkheads, which are directly below the armoured bulkheads which close the main and upper belts.

Horizontal.—

Armoured Deck.—This is 4 ft. 7 in. above the waterline, for the extent of the midship section of the belt, and drops to 3 ft. 11 in. below the waterline forward and 6 in. below it aft. Amidships it slopes to 3 ft. 11¼ in. below the waterline (or 1 ft. 3¾ in. above the lower edge of the main belt). Forward it is nearly flat; aft it slopes to 4 ft. 3 in. below the waterline, or 11¾ in. above the lower edge of the armour. Amidships it has a uniform thickness of 1·2″; forward 2″; aft 3·1″ on the flat and 2″ on the slopes.

Battery (or Main) Deck.—This is only armoured between the ends of the battery and the armoured bulkheads which close the main and upper belts. It is 1″ thick.

Upper Deck.—2·2″ to 1·4″ thick over the space enclosed by the longitudinal splinter bulkheads (in rear of the 5·9-in. guns) and the ship's side, and 1″ elsewhere over the battery except in the neighbourhood of the echelon turrets, where it is thicker. (*See* Plate.)

Torpedo Nets.—Beam net defence was fitted, but has been discarded since the Battle of Jutland.

Armament.—

Guns and Ammunition Supply.—

(For approximate arcs of training, *see* Plate 19.)

10—11-in. (28-cm.) Q.F., L/50, in pairs in turrets.

Training electric; elevating gear hydraulic. Ammunition supply electric.

Loading at fixed angle of elevation.

12—5·9-in. (15-cm.) Q.F., L/45.

2—22-pr. (8·8-cm.), semi-automatic, anti-aircraft guns.

4—Machine guns.

---

* These rangefinders have probably now been removed, being found useless owing to spray.

Seydlitz—*cont.*

Armament—*cont.*

Magazines and Stowage.—

The magazines are above the shell rooms.

96 rounds per gun for 11-in. guns, centre line turrets.
81 ,, ,, ,, side turrets.
160 ,, ,, 5·9-in. guns.

Torpedo Tubes.—

4—19·7-in. (50-cm.) submerged, situated as follows:—One under the fore foot, one each side immediately before the foremost turret and one aft on the starboard side. The two broadside tubes are fixed at an angle of 20° before the beam and are fitted with gyro angling gear which can be set for every 15° from 30° before to 60° abaft the direction of the tube.

Mines.—There is believed to be no permanent stowage for mines.

Searchlights.—

8—43·3-in. (110-cm.). The forward searchlights are mechanically controlled from the fore bridge, the after ones from the after bridge.

Rangefinders.—As in *Derfflinger* (page 24).

Battle Signal Stations.—

The main battle signal stations are situated one on either side of the main deck. Details as in *Kaiser* class (*see* page 10).

Boat Derricks.—Two, one on either side of after funnel.

Steering Gear.—One balanced rudder is fitted on centre line aft, and two parallel rudders before it (*see* Plate).

Machinery and Boilers.—

Main Engines.—

Parsons turbines on four shafts.

Boilers.—

27 Schulz, Navy type, in five watertight compartments, which are subdivided by means of longitudinal bulkheads into 15 smaller ones (containing one or two boilers each).

Total grate surface—2,806 sq. ft.

Propellers.—Four, diameter about 11 ft. 6 ins.

Speed, Horse-Power, and Fuel.—

Ship.	Speed (designed).	Horse-Power (designed).	Fuel: (a) Coal. (b) Oil.
Seydlitz	Knots. 26·5	63,000	(a) 3,543 (b) 197

Steam Trials.—

Ship.	Nature of Trial.	Date.	Speed.	Horse-Power.	Revolutions.	Remarks.
Seydlitz	Measured mile	–.7.13	28·13	89,738		
	6 hours full power	–.7.13	26·75	73,923	293	

Seydlitz—*cont.*

Endurance.—

——	Speed.	Horse-Power.	Daily Consumption.	Radius of Action.	Remarks.
	Knots.		Tons.	Miles.	
At $\frac{4}{5}$ths designed power - -	24·5	50,400	1,026	2,080	
At maximum continuous seagoing speed.	23·7	43,100	905	2,280	
At $\frac{3}{5}$ths designed power - -	22·7	37,800	810	2,440	
At $\frac{2}{5}$ths designed power - -	20·2	25,200	567	3,000	
At $\frac{1}{5}$th designed power - -	16·2	12,600	351	4,020	
At 10 knots - - - -	10·0	3,000	141	6,225	

Moltke.

(*See Plates* 9 *and* 20.)

Ship.	Designation before Launch.	Programme Year.	Built at	Ordered.	Laid Down.	Launched.	Commissioned for Trials.	Completed Trials.
Moltke - -	G.	1908–09	Blohm and Voss, Hamburg.	*17.9.08	†28.12.08	‡7.4.10	30.9.11	31.3.12

* Official dates. † Approximate date. ‡ The *Moltke* was launched with all boilers in place.

Cost.—

(*Moltke.*)	£
Hull, machinery, &c. - - - - - -	1,426,125
Gun armament - - - - - -	684,931
Torpedo armament - - - - -	45,499
Total - - - - -	£2,156,555

General Remarks.—

The *Goeben*, the *Moltke's* sister ship, was ostensibly handed over to Turkey on 11th August 1914, and re-named *Sultan Selim.* She remained, however, under German control throughout the war until 2nd November 1918, when, following the signing of the armistice granted to Turkey by the Allies, the Turks at length took over the vessel entirely.

The *Moltke* is fitted as flagship.

The upper deck is flush from bow to abreast No. 4 Turret, the quarter-deck being one deck lower.

The foremost funnel has been raised to protect the searchlight crews.

The foremost turret is not usually fired at night, owing to the effect of the flash and blast on the bridge personnel.

Complement.—Peace complement, 1,013 (as private ship); war complement, about 1,400.

A turret's crew, including magazine and shell-room parties, consists of 70 officers and men (1 executive officer, 1 warrant officer, 9 petty officers or leading seamen, and 59 seamen).

A 5·9-in. gun's crew consists of 6 men.

General Appearance.—*See* Plate 9 and Silhouette.

MOLTKE.
GOEBEN similar.

Ordnance Survey, November, 1918.

VON DER TANN.

Ordnance Survey, November, 1918.

Moltke—*cont.*

General Dimensions, &c.—

Length, L.W.L.		610 ft. 3 ins.
Breadth, extreme		96 " 10 "
Draught, designed load		26 " 11 "
Displacement, designed load		22,640 tons.
Freeboard (approximate) at designed draught	at stem	24 ft. 0 ins.
	amidships	19 " 0 "
	aft	12 " 0 "
Height (approximate) of axis of heavy guns at designed draught	foremost turret	28 " 0 "
	echelon turrets	26 " 0 "
	aftermost turret	19 " 0 "
Height of fore funnel to L.W.L.		69 " 0 "
" " lower masthead to L.W.L.		127 " 0 "

Armour and other Protection.—

Material.—Krupp cemented.

Belt.—Complete except for about 10 ft. at stern, where it is closed by a transverse bulkhead 5″ thick. The portion between the foremost and aftermost turrets is closed by transverse bulkheads, 8″ thick, butting on bases of barbettes. Maximum thickness 11″, reduced to 6″ and 4″ forward and 4″ aft, and tapering to the battery deck and lower edge. It extends about 5 ft. below the waterline.

Battery.—Extends approximately from foremast to 20 ft. before mainmast. The battery armour and bulkheads are 5″ thick. Observation slits are cut in the armour of the second and after casemates on either side to enable these positions to be used for secondary torpedo control.

Splinter Bulkheads.—There are two longitudinal splinter-proof bulkheads, ·8″ thick, in rear of the line of 5·9-in. guns on either side, with splinter screens, also ·8″ thick, between the guns, forming casemates.

Barbettes.—About 10″ thick.

Turrets.—10″ thick.

5·9-*in. Gun Shields.*—3·1″ thick.

Conning and Control Towers.—Fore 10″, after 7″ thick. Shape and general arrangements as in *Seydlitz* (*see* page 27). The observation slits are about 1 in. wide and 1 ft. long.

Armoured Hoods.—Are fitted one on each side of the battery, projecting just above the upper deck, and originally took rangefinders for use with the secondary armament.*

Torpedo Protection Bulkheads.—Probably similar to *Seydlitz*, page 26.

Horizontal.—Probably similar to *Seydlitz*, page 28.

Torpedo Nets.—Beam net defence was fitted, but has been discarded since the Battle of Jutland.

Armament.—

Guns and Ammunition Supply.—

(For arcs of training, *see* Plate 20.)

10—11-in. (28-cm.) Q.F., L/50; mounted in pairs in turrets. Loading at fixed angle of elevation.

12—5·9-in. (15-cm.) Q.F., L/45, on C.P. mounting, with 3·1″ shield. Ammunition hoists electric, alternative hand.

4—22-pr. (8·8-cm.), semi-automatic, anti-aircraft guns, on after superstructure.

2—Machine guns.

* These rangefinders have probably now been removed, being found useless owing to spray.

D 4

Moltke—*cont.*

Armament—*cont.*

Magazines and Stowage.—

For 11-in. guns, 96 rounds per gun (centre-line turrets).
" " " 81 " " (side ").
" 5·9-in. " 160 " "
" 22-pr. " 287 " " .

Torpedo Tubes.—**4**—19·7-in. (50-cm.) submerged. The two broadside tubes are fixed at an angle of 20° before the beam and are fitted with gyro angling gear, which can be set for every 15° from 30° before to 60° abaft the direction of the tube.

Mines.—There is believed to be no permanent stowage for mines.

Searchlights.—

8—43·3-in. (110-cm.); four on the foremost funnel and four on the mainmast.

The four foremost lights are situated on platforms on the funnel, one group of two lights each side. The lights in each group are situated one vertically over the other. The port group is situated on the port side forward of the funnel, and the starboard group on the starboard side aft of the funnel, *i.e.*, diagonally disposed.

Fore Searchlight Control Platform.—Situated before the lights and at a lower level, completely roofed in both from the weather and from the rays of the lights.

Communications.—The voice-pipe system for turrets consists of one large 8-in. to 10-in. pipe running fore and aft the ship, from which are branched smaller pipes to each turret.

Rangefinders.—As in *Derfflinger* (page 24).

Battle Signal Stations.—

The main battle signal stations are situated one on either side of the main deck. Details as in *Kaiser* class (*see* page 10).

Boat Derricks.—There are two derricks on stump masts, one on each side of the after funnel.

Steering Gear.—Electric.

Wheel-house.—In the form of a segment of a circle round the fore side of the conning tower. Can be unshipped and stowed below in about 10 minutes. The roof and a small portion of the fore screen are made in segments of light iron, secured to the roof of the conning tower, and to the lower or standing part of the fore screen, by steel pins. These pins are pulled out and the segments lifted off, leaving only the lower portion of the screen standing, and thus clearing the field of view of the conning tower.

Look-out Houses.—One each side of the bridge or shelter deck.

Anchors.—Three bower and one stern—all stockless. The stern anchor is carried in a hawsepipe in the centre line, about 4 ft. above the water.

Coaling Arrangements.—A number of electric bollards, fitted on each side of the upper deck, are used for coaling.

Machinery and Boilers.—

Main Engines.—Parsons turbines, on four shafts.

Dimensions.—H.P. rotor 78-in.; blade length, 1st row, 1¾-in.
L.P. rotor 120-in.; blade length, last row, 16-in.

Revolutions (designed)—260.

Boilers.—24 Schulz, Navy type. In four transverse boiler-rooms, six in each.

Propellers.—Four in number.

Moltke—*cont.*

Speed, Horse-Power, and Fuel.—

	Speed (designed).	Horse-Power (designed).	Fuel : (a) Coal. (b) Oil.
	Knots. 25	52,000	(a) 3,050 (b) 197

Steam Trials.—

Ship.	Nature of Trial.	Date.	Speed.	Horse-Power.	Revolutions.	Remarks.
			Knots.			
Moltke - - -	Measured mile -	–.11.11	28·4	85,782	—	
	6 hours forced draught.	–.11.11	27·25	76,795	306	

Endurance.—

——	Speed.	Horse-Power.	Daily Consumption.	Radius of Action.	Remarks.
	Knots.		Tons.	Miles.	
At 4/5ths designed power - -	24·0	41,600	845	2,155	
At maximum continuous seagoing speed.	23·1	36,400	742	2,370	
At 3/5ths designed power - -	22·1	31,200	668	2,510	
At 2/5ths designed power - -	19·6	20,800	490	3,030	
At 1/5th designed power - -	15·6	10,400	322	3,700	
At 10 knots - - - - -	10·0	2,750	141	5,350	

von der Tann.

(*See Plates* 10 *and* 21.)

Ship.	Designation before Launch.	Programme Year.	Built at	Ordered.	Laid Down.	Launched.	Commissioned for Trials.	Completed Trials.
von der Tann -	F.	1907–08	Blohm and Voss, Hamburg.	*30.9.07	*25.3.08	20.3.09	1.9.10	20.2.11

* Approximate dates.

Cost.—

	£
Hull, machinery, &c. - - - - - -	1,272,016
Gun armament - - - - - - -	489,236
Torpedo armament - - - - - -	32,289
Total - - - -	£1,793,541

General Remarks.—

The *von der Tann* was the first battle cruiser and first large turbine ship built for the German Navy. She is fitted as flagship.

The foremost turret is not usually fired at night, owing to the effect of the flash and blast on the bridge personnel.

von der Tann—*cont.*

Complement.—Peace complement, 911 (as private ship); war complement, about 1,300.

Accommodation.—The Admiral's quarters are forward on the starboard side of the upper deck, and the wardroom on the port side forward on the main deck. There is no gunroom. Some four-berth cabins are fitted for midshipmen. The officers' cabins are forward on the upper deck and main deck. The ventilation is very efficient. The men's quarters are on the main and lower decks aft, petty officers' quarters on the main deck before the officers' quarters. There are no messes in the battery.

General Appearance.—*See* Plate 10 and Silhouette.

General Dimensions, &c.—

Length, L.W.L. - - - - -		562 ft. 8 ins.
Breadth, extreme - - - - -		87 „ 3 „
*Draught, designed load - - - -		26 „ 7 „
Displacement, designed load - - -		19,100 tons.
Freeboard (approximate) at designed draught - - - -	at stem -	26 ft. 6 ins.
	amidships -	19 „ 0 „
	aft - -	19 „ 0 „
Height of fore funnel to L.W.L. - - -		65 „ 0 „
„ „ lower masthead to L.W.L. - -		100 „ 0 „

Constructive Details.—Frahm's anti-rolling tanks were proposed and fitted after construction was well advanced, and had therefore to be adapted to the existing structure. They do not appear to have been very effective, as they are now used for storing additional coal. About 200 tons can be stowed in them for use in an emergency.

Armour and other Protection.—

Material.—Krupp cemented.

Belt.—Complete, except for about 10 ft. at stern, where it is closed by a transverse bulkhead 4″ thick. The portion between the foremost and aftermost turrets is closed by transverse bulkheads 7″ thick. Maximum thickness of belt 9·8″, reduced to 6″ and 4″ forward and 4″ aft, and tapering to the battery deck and lower edge. It extends about 5 ft. below the waterline.

Battery.—Battery armour and bulkheads, 5″ thick. The former extends from just abaft the fore funnel to abreast the mainmast. Observation slits are cut in the foremost and after casemates to enable these positions to be used for secondary torpedo control.

Splinter Bulkheads.—There are two longitudinal splinter-proof bulkheads, ·8″ thick, in rear of the line of 5·9″ guns on either side, with splinter screens, also ·8″ thick, between the guns, forming casemates.

Barbettes.—9″ thick.

Turrets.—Fronts, 9″; hoods, 3″ thick.

Conning and Control Towers.—Fore 10″, after 7″ thick.—Shape and general arrangements as in *Seydlitz* (*see* page 27). The observation slits are about 2 ft. by 3 ins.

Armoured Hoods are fitted one on either side of the battery, projecting just above the upper deck, between the broadside turrets and the ship's side, and originally took rangefinders for use with the secondary armament.†

Torpedo Protection Bulkheads.—Probably as in *Nassau* (page 17).

Horizontal.—Probably similar to *Seydlitz* (page 28).

* After commissioning, the vessel was reported to have a draught of 29 ft. 6 in. forward and aft.

† These rangefinders have now been removed, being found useless owing to the spray.

von der Tann—*cont.*

Torpedo Nets.—Beam net defence was fitted, but has been discarded since the Battle of Jutland.

Armament.—

Guns and Ammunition Supply.—

(For arcs of training, *see* Plate 21.)

8—11-in. (28-cm.) Q.F., L/45, on model /06 mounting; in pairs in turrets. Loading at fixed angle of elevation.

10—5·9-in. (15-cm.) Q.F., L/45, on C.P. mounting 02/06, in battery.

4—22-pr. (8·8-cm.) semi-automatic anti-aircraft guns, on after superstructure.

2—Machine guns.

Torpedo Tubes.—

4—17·7-in. (45-cm.) submerged. The two broadside tubes are fixed at an angle of 20° before the beam, and are fitted with gyro angling gear which can be set for every 15° from 30° before to 60° abaft the direction of the tube.

Mines.—A small number of mines, with moorings, may be carried.

Rangefinders.—As in *Derfflinger* (page 24).

Battle Signal Station.—One on either side of the main deck, just forward of No. 1 casemate.

Submarine Bell Receiver.—A submarine bell telephonic receiver is fitted in the chart house.

Searchlights.—

8—43·3-in. (110-cm.).

Derricks.—Two stump derricks, with standing topping lifts, are fitted, one on either side abreast the after funnel. The purchase—a single whip—is worked electrically, and the guys by hand.

Fore Bridge.—Roomy, triangular in shape, with conning tower at the fore end, around the front of which is a small shelter, with windows, and chart house on the after side amidships. The lateral bridge extensions are pivoted and can be swung aft out of the way.

Steering Gear, &c.—

The ship is fitted with twin rudders.

There are two steam steering engines, with rod control from the steering positions.

Compasses.—There are four gyro receivers forward, one on each side of the fore bridge, one on the upper platform of the fore upper bridge, and one in the conning tower.

Anchors.—Three stockless bower anchors—two on port and one on starboard side; also a light stern anchor right aft on starboard quarter.

Coaling Arrangements.—About eight small electric bollards, fitted on each side of the upper deck, 6 to 10 ft. from the ship's side, are used for coaling.

Machinery and Boilers.—

Main Engines.—

Four sets, Parsons turbines, in four separate compartments.

The H.P. ahead turbines drive the wing shafts, and the L.P. turbines the inner shafts. The cruising turbines are in series, the H.P. cruiser on the starboard inner shaft ahead of the main L.P., and the M.P. cruiser similarly situated on the port side. H.P. astern turbines are fitted on wing shafts and the L.P. astern turbines are embodied in the L.P. casings. The engine-rooms are divided into four by a longitudinal and a transverse bulkhead, the latter situated between the two turbines on each shaft, so that the exhaust pipes from the H.P. to the L.P. turbines pass through it.

H.P. rotor drum 82½″ in diameter; blade length, first row 1⅜″.

von der Tann—*cont.*

Machinery and Boilers—*cont.*

Main Engines—cont.

L.P. rotor drum 115″ to 111″ in diameter, stepped; blade length, last row 18½″.

Revolutions—280 (designed).

Boilers.—

18 Schulz, Navy type. Diameter of tubes, 1·4″. No superheater.

Date when first used at sea—May 1910. All retubed after the Battle of Jutland.

Grate surface—1,935 square feet.

Steam guaranteed—625,000 lbs. per hour.

Ventilation—Through louvres in the superstructure at base of funnel which form the downtakes to the stokeholds. There is also a big ventilating trunk under the fore bridge to the foremost side of the fore funnel. The internal ventilation is by air trunks and electric fans.

Auxiliary Machinery.—Includes eight turbo-dynamos for electric lighting and turret working, voltage 220.

Two steam steering engines.

Propellers.—Four in number.

Speed, Horse-Power, and Fuel.—

Ship.	Speed (designed).	Horse-Power (designed).	Fuel: (*a*) Coal. (*b*) Oil.
von der Tann - - - - - - -	Knots. 24	43,600	(*a*) 2,760* (*b*) 197

* In addition, about 200 tons can be stowed in the Frahm's anti-rolling tanks (*see* page 34).

Steam Trials.—

Ship.	Nature of Trial.	Date.	Speed.	Horse-Power.	Revolutions.	Remarks.
von der Tann - -	Measured mile	-.10.10	27·75	79,802	339	Mean of best runs.
" " - -	—	—	26·2	62,000	—	After bilge keels fitted.

In 1911, when returning from a cruise to South America, this ship maintained a speed of 24 knots from Teneriffe to Heligoland, a distance of 1,913 miles.

During the Battle of Jutland she touched 27·6 knots.

Endurance.—

——	Speed.	Horse-Power.	Daily Consumption.	Radius of Action.	Remarks.
	Knots.		Tons.	Miles.	
At ⅘ths designed power - -	23·3	34,800	690	2,330	
At maximum continuous seagoing speed.	22·5	30,000	626	2,500	
At ⅗ths designed power - -	21·4	26,160	532	2,795	
At ⅖ths designed power - -	18·8	17,140	386	3,110	
At ⅕th designed power - -	15·0	8,720	260	4,000	
At 10 knots - - - - -	10·0	2,600	134	5,130	

OBSOLETE BATTLESHIPS.

It is not likely that any of these vessels will again be employed in the line of battle, but details of them are given on the chance of their being used for some special purpose, such as blocking.

Deutschland, Hannover, Schleswig-Holstein, *and* Schlesien.

(*See Plate* 22.)

Name.	Designation before Launch.	Programme Year.	Where Built.	Laid Down.	Launched.	*First Commis-sioned.
Deutschland - -	N	1903–04	Germania Yard, Kiel -	20.7.03	19.11.04	3.8.06
Hannover - - -	P	1904–05	Wilhelmshaven Dockyard	–.4.04	29.9.05	1.10.07
Schleswig-Holstein -	Q	1905–06	Germania Yard, Kiel -	–.5.05	17.12.06	6.7.08
Schlesien - - -	R	1905–06	Schichau Works, Danzig	–.5.05	28.5.06	5.5.08

General Remarks.

All these ships have been disarmed with the exception of *Hannover*, which is employed as guardship in the Sound. The *Deutschland, Schlesien,* and *Schleswig-Holstein* are stationary ships employed in connection with the Submarine and Training Services.

The fifth vessel of this type—the *Pommern*—was sunk in action on the night of 31st May–1st June 1916.

Complement.—Peace complement, 743 (as private ship).

General Appearance.—*See* Silhouette.

General Dimensions, &c.—

Length between perpendiculars - - - -	398 ft. 7 ins.
,, L.W.L. - - - - -	413 ,, 1 ,,
,, extreme - - - -	419 ,, 0 ,,
Breadth, extreme - - - - -	72 ,, 10 ,,
Draught, designed load - - - - -	25 ,, 3 ,,
Displacement { Designed load - - - -	13,040 tons.
Displacement { Full load - - - -	14,224 ,,
Height of fore funnel to L.W.L. - - -	72 ft. 0 ins.
,, ,, lower masthead to L.W.L. - -	94 ,, 0 ,,

Constructive Details.—There are very few watertight doors, all compartment bulkheads being practically intact. There is a central watertight passage under the armoured deck, passing between the funnel uptakes, and containing leads, voice tubes, steering shafts, engine-room telegraph shafting, &c.

Armour.—

Material.—Krupp cemented.

Main Belt.—7 ft. 6 in. wide, 9·4″ thick (*Deutschland* 9″), tapering to 6·7″ at lower edge.

Upper Belt.—7 ft. 6 in. wide, 8″ thick (*Deutschland* 7½″).

Transverse Bulkheads.—The citadel is closed at the ends by diagonal bulkheads 6·7″ to 8″ thick, extending from hull plating to barbettes.

Belt before and abaft Citadel.—5·9″ and 3·9″ thick.

Battery.—7 ft. 6 in. wide, 6·7″ thick (*Deutschland* 6·3″), and closed by diagonal bulkheads 6·7″ thick (*Deutschland* 6·3″), which connect to barbettes.

F 2

Deutschland, Hannover, Schleswig-Holstein, *and* Schlesien—*cont.*

Armour—*cont.*

Upper Deck Casemates.—6·7″ thick in front and 5·5″ to 4·7″ in rear.

Barbettes and Ammunition Tubes.—

Forward barbette—11″ thick before diagonal bulkheads, and 4·9″ thick abaft them.

After barbette—9·8″ thick abaft diagonal bulkheads; 4·9″ before them.

Turrets.—Fronts and sides 11″, rear 9·8″ thick.

Conning Towers.—

Fore—11·8″ to 9·8″. Roof, 2″.

After—5·5″. Roof, 1″.

Communication Tubes.—Fore, 9·8″, rectangular in section.
After, 4·7″.

Horizontal.—

Principal armoured deck: Inside citadel, 1·6″ on flat, 2·6″ on slopes; Outside citadel, 2·4″ to 1·6″; on glacis forward and aft, 3·8″.
Lower deck, outside citadel, on top of belt, ·8″.
Roof of battery—1·4″.
Roof of upper deck casemates—1·2″.

Armament.—

(All these ships, with the exception of *Hannover*, have been disarmed.)

Guns and Ammunition Supply.

(For arcs of training, *see* Plate.)

4—11-in. (28-cm.) Q.F., L/40, in pairs in turrets.

14—6·7-in. (17-cm.) Q.F., L/40.

2—22-pr. (8·8 cm.), semi-automatic, anti-aircraft guns.

2—Machine guns.

Torpedo Tubes.—

6—17·7-in. (45-cm.) submerged.

Searchlights.—

8—35·4-in. (90-cm.).

Steering Gear.—Steam and hand, two engines aft in tiller compartment.

Machinery and Boilers.—

Main Engines.—

Three sets, vertical, triple expansion, in separate watertight compartments, the central engine being abaft the other two, which are abreast.
Revolutions at full power, 120 (designed).

Boilers.—

Deutschland has same number and arrangement of boilers as *Braunschweig* type (*see* page 40).
The remaining ships have—

12 boilers in three separate boiler rooms, four boilers in each.

Type—Schulz, Navy type (eight large and four small), fitted to burn coal and oil.

Tube heating surface of the eight large boilers	- -	34,560 sq. ft.
,, ,, ,, four small boilers	- -	15,560 ,,
Total grate surface	- - - - -	1,000 ,,

Deutschland, Hannover, Schleswig-Holstein, *and* Schlesien—*cont.*

Machinery and Boilers—*cont.*

Propellers.—Three, bronze. Outer—three-bladed. Centre—four-bladed.

Speed, Horse-Power, and Fuel.—

Speed (designed).	Horse-Power (designed).	Fuel : (*a*) Coal. (*b*) Oil.
Knots. 18	16,000	(*a*) 1,771 (*b*) 197

Braunschweig, Elsass, Preussen, Hessen, *and* Lothringen.

(*See Plate* 23.)

Name.	Designation before Launch.	Programme Year.	Where Built.	Laid Down.	Launched.	First Commissioned.
Braunschweig	H	1901–02	Germania Yard, Kiel	24.10.01	20.12.02	15.10.04
Elsass	J	1901–02	Schichau Works, Danzig.	5.10.01	26.5.03	29.11.04
Preussen	K	1902–03	Vulcan Works, Stettin	14.6.02	31.10.03	12.7.05
Hessen	L	1902–03	Germania Yard, Kiel	15.4.02	18.9.03	19.9.05
Lothringen	M	1903–04	Schichau Works, Danzig.	1.4.03	27.5.04	18.5.06

General Remarks.—All these vessels, with the exception of *Lothringen*, which is believed to have only had four 6·7-in. guns removed, have been disarmed. The present employment of *Lothringen* is uncertain. Of the remainder, *Braunschweig* and *Elsass* are harbour training ships, *Preussen* and *Hessen* are submarine accommodation ships.

Complement.—Peace complement, 743 (as private ship).

General Appearance.—*See* Silhouette.

General Dimensions, &c.—

Length between perpendiculars	398 ft. 7 ins.
,, L.W.L.	413 ,, 5 ,,
,, extreme	419 ,, 0 ,,
Breadth, extreme	72 ,, 10 ,,
Draught, designed load	25 ,, 3 ,,
Displacement { Designed load	12,988 tons.
Displacement { Full load	14,140 ,,
Height of fore funnel to L.W.L.	74 ft. 0 ins.
,, ,, lower masthead to L.W.L.	94 ,, 0 ,,

Armour.—

Material.—Krupp cemented.

Main Belt.—7 ft. 6 in. wide, 8·8″ thick, tapering to 5·9″ at lower edge.

Upper Belt.—7 ft. 6 in. wide, 5·5″ thick.

Transverse Bulkheads.—The citadel is closed by diagonal bulkheads, 5·5″ thick, which connect to barbettes.

Belt before and abaft Citadel.—3·9″.

Battery.—7 ft. 6 in. wide, 5·9″ thick, closed by diagonal bulkheads, 5·9″ thick, which connect to barbettes.

F 3

Braunschweig, Elsass, Preussen, Hessen, *and* Lothringen—*cont.*

Armour—*cont.*

Barbettes.—

For 11-in. guns—forward barbette, 11″ to 5·5″ thick.
after barbette, 9·8″ to 4·9″ thick.

For 6·7-in. guns—6·7″ thick.

	Braunschweig and *Elsass.*	*Preussen, Hessen,* and *Lothringen.*
Turrets.—		
For 11-in. guns - -	9·8″ thick	11″ to 9·8″ thick.
For 6·7-in. guns - -	5·9″ to 4·5″ thick	6·7″ to 4·5″ thick.

Conning Towers.—

Fore, 11·8″ to 9·8″ ; after, 5·5″.

Communication Tubes.—Fore —9·8″ to 4·9″ ; after—4·7″.

Horizontal.—1·5″ on flat, 2·9″ on the slopes. Glacis plates forward and aft, 5·5″. Deck before and abaft the glacis plates, 1·6″. Deck, before and abaft citadel, ·8″. Battery roof, 1·5″.

Former Armament.—

Guns and Ammunition Supply.—

(For arcs of training, *see* Plate.)

4—11-in. (28-cm.) Q.F., L/40, in pairs in turrets.

14—6·7-in. (17-cm.) Q.F., L/40.

2—22-pr. (8·8-cm.) semi-automatic, anti-aircraft guns, on after superstructure.

2—machine guns.

Torpedo Tubes.—

6—17·7-in. (45-cm.) submerged.

Searchlights.—

6—35·4-in. (90-cm.).

Machinery and Boilers.—

Main Engines.—

Three sets, vertical, triple expansion in separate watertight compartments, the central engine being abaft the other two, which are abreast.

Revolutions at full power, 110 (designed).

Boilers.—

Fourteen in number, in three watertight compartments.

Six cylindrical. These are single-ended and are placed fore and aft in the after boiler room. Four of them have four furnaces, the remainder three.

Heating surface - - - 15,070 sq. ft.
Grate surface - - - - 447 ,,

Eight Schulz, Navy type, four placed fore-and-aft in the centre boiler room and four in the foremost one. Fitted to burn coal and oil.

Heating surface - - - 34,176 sq. ft.
Grate surface - - - - 657 ,,

Propellers.—Three.

Speed, Horse-Power, and Fuel.—

Speed (designed).	Horse-Power (designed).	Fuel : (*a*) Coal. (*b*) Oil.
Knots. 18	16,000	(*a*) 1,574 (*b*) 197

Braunschweig, Elsass, Preussen, Hessen, *and* Lothringen—*cont.*

Endurance.—

—	Speed.	Horse-Power.	Daily Consumption.	Radius of Action.	Remarks.
	Knots.		Tons.	Miles.	
At 4/5ths designed power - -	17·3	12,800	303	2,380	
At maximum continuous seagoing speed.	16·8	11,500	273	2,570	
At 3/5ths designed power - -	16·1	9,600	220	3,060	
At 2/5ths designed power - -	14·6	6,400	159	3,840	
At 1/5th designed power - -	11·8	3,200	94	5,250	
At 10 knots - - - - -	10·0	2,000	66	6,340	

Wittelsbach, Wettin, Zähringen, Mecklenburg, *and* Schwaben.

(*See Plate* 24.)

Name.	Designation before Launch.	Programme Year.	Where Built.	Laid Down.	Launched.	First Commissioned.
Wittelsbach - -	C	1899–1900	Wilhelmshaven Dockyard	30.9.99	3.7.00	15.10.02
Wettin - - -	D	1899–1900	Schichau Works, Danzig	10.10.99	6.6.01	1.10.02
Zähringen - - -	E	1899–1900	Germania Yard, Kiel -	21.11.99	12.6.01	25.10.02
Mecklenburg - -	F	1900–01	Vulcan Works, Stettin -	15.5.00	9.11.01	25.6.03
Schwaben - - -	G	1900–01	Wilhelmshaven Dockyard	14.11.00	19.8.01	13.4.04

General Remarks.—With the exception of *Zähringen*, which still retains some or all of her guns for practice purposes, all these vessels have been disarmed. *Zähringen* is now a seagoing training ship; *Wittelsbach*, *Wettin*, and *Schwaben* are harbour training ships, and *Mecklenburg* is a submarine accommodation ship.

Complement.—Peace complement, 683 (as private ship).

General Appearance.—*See* Silhouette.

General Dimensions, &c.—

Length between perpendiculars - - - - -	393 ft. 8 ins.
„ L.W.L. - - - - - - -	410 „ 9 „
„ extreme - - - - - - -	413 „ 5 „
Breadth, extreme - - - - - - -	68 „ 3 „
Draught, designed load - - - - - -	25 „ 3 „
Displacement, designed load - - - -	11,611 tons.
Height of fore funnel to L.W.L. - - - -	74 ft. 0 ins.
„ „ lower masthead to L.W.L. - -	94 „ 0 „

Constructive Details.—

There is a passage under the armoured deck from the fore to the after lower conning tower.

Armour.—

Material.—Krupp cemented.

Main Belt.—7 ft. 8 in. wide, 8·8″ thick amidships, tapering to 5·9″ at lower edge.

Upper Belt.—5·5″ thick.

Transverse Bulkheads.—The citadel is closed by diagonal bulkheads, 5·5″ thick.

Belt before and abaft Citadel.—3·9″.

F 4

Wittelsbach, Wettin, Zähringen, Mecklenburg, *and* Schwaben—*cont.*

Armour—*cont.*

Barbettes and Turrets.—For 9·4-in. guns, turrets, 9·8″ thick; barbettes, 9·8″ to 5·5″ thick. 5·9-in. gun turrets, 5·9″ to 4·5″ thick.

Battery and Casemates.—5·5″ thick.

Conning Towers.—Fore, 9·8″; after, 5·5″ thick.

Control Tower.—5·5″ thick, placed directly above the fore conning tower.

Horizontal.—Two decks. The lower extends from stem to stern, sloping to lower edge of belt, and is 1·6″ thick on the flat and 2·9″ on the slopes. Outside the citadel on top of belt it is ·8″ thick; roof of citadel, outside battery, 1·2″ thick; roof of battery, 1·6″ thick.

Former Armament.—

Guns and Ammunition Supply.—

(For approximate arcs of training, *see* Plate.)

4—9·4-in. (24-cm.) Q.F., L/40, in pairs, in turrets.
18—5·9-in (15-cm.) Q.F., L/40.
12—15-pr. (8·8-cm.) Q.F., L/30.
2—Machine guns.

Torpedo Tubes.—

6—17·7-in. (45-cm.) submerged; one under the ram, two on each broadside, and one in stern on port side.

Searchlights.—

4—35·4-in. (90 cm.).

Machinery and Boilers.—

Main Engines.—

Three sets, vertical, triple expansion, in separate watertight compartments, the central engine being abaft the other two, which are abreast.

Revolutions at full power, 110 (designed).

Boilers.—

Twelve in number. In three compartments. Water-tube boilers are in foremost compartment.

Type—Six cylindrical. Six Schulz, Navy type, fitted to burn coal and oil.

Propellers.—Three.

Speed, Horse-Power, and Fuel.—

	Speed (designed).	Horse-Power (designed).	Fuel: (*a*) Coal. (*b*) Oil.	
	Knots. 18	15,000	(*a*) 1,790 (*b*) 197	

Endurance.—

——	Speed.	Horse-Power.	Daily Consumption.	Radius of Action.	Remarks.
	Knots.		Tons.	Miles.	
At $\frac{4}{5}$ths designed power	17·2	12,000	282	2,825	
At maximum continuous seagoing speed.	16·7	10,500	248	3,170	
At $\frac{3}{5}$ths designed power	16·1	9,000	205	3,650	
At $\frac{2}{5}$ths designed power	14·7	6,000	147	4,660	
At $\frac{1}{5}$th designed power	11·9	3,000	86	6,460	
At 10 knots	10·0	1,790	60	7,850	

Plate 11.
C.B. 1182. P.
October, 1918.

BAYERN

ARMOUR — *Krupp Cemented*

ELEVATION.

Baden only
R.F.
2.S.L.
2.S.L.
R.F.
C.T. 17 3/4"
2-15"
2-15"
2.S.L.
2.S.L.
R.F.
C.T.
2-15"
2-15"
R.F.
15"?
L.W.L.
S.T.
S.T.
S.T.
Torpedo Protection Bulkhead
40 Feet
30 "
20 "
10 "

Scale of Feet

650 Feet 620 600 580 560 540 520 500 480 460 440 420 400 380 360 340 320 300 280 260 240 220 200 180 160 140 120 100 80 60 40 20 0

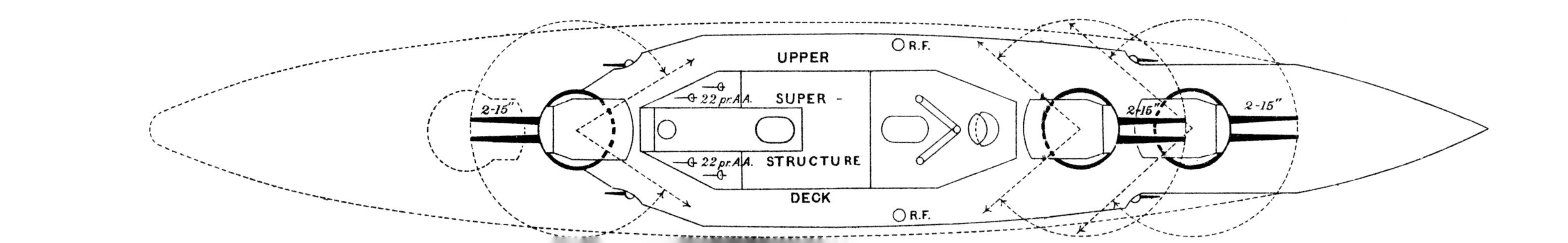

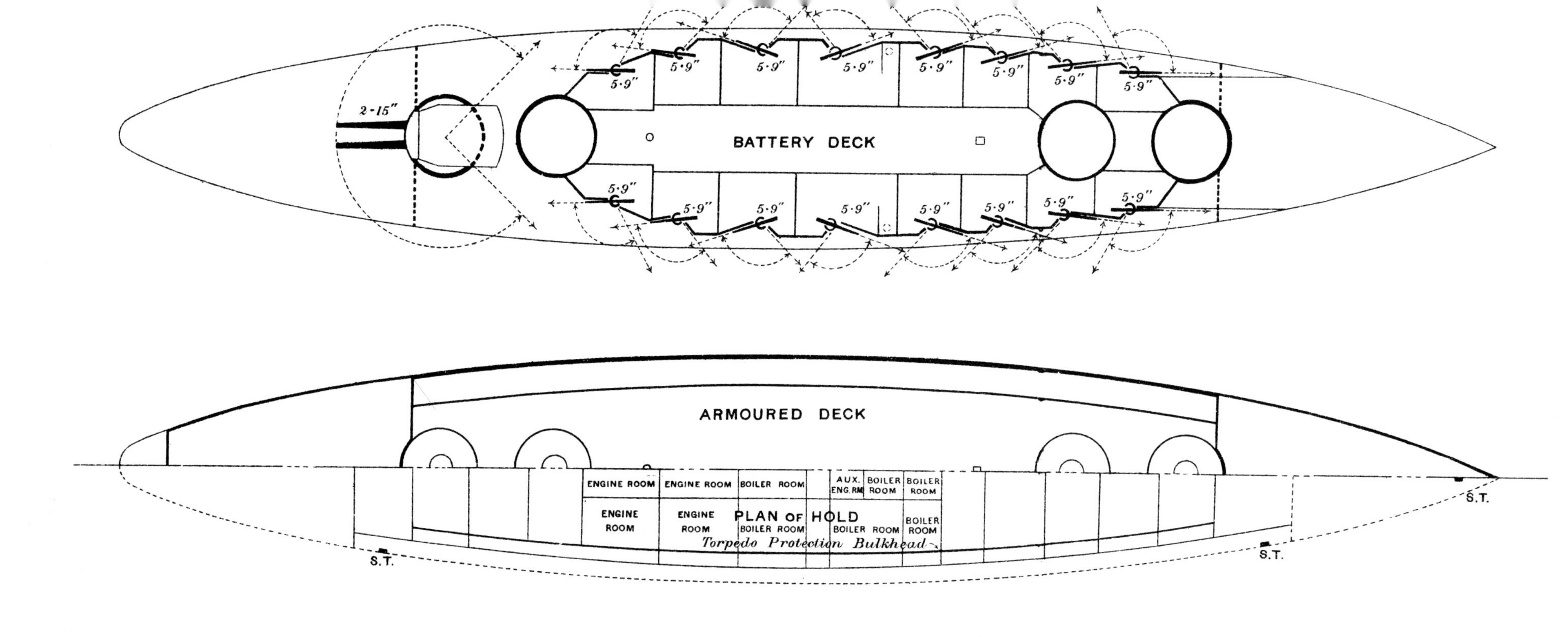

Vessels of Similar Type,

BADEN, SACHSEN, WÜRTTEMBERG.

Plate 12.
C.B. 1182. P.
October, 1918.

KÖNIG, GROSSER KURFÜRST, MARKGRAF, AND KRONPRINZ WILHELM.

ARMOUR - *Krupp Cemented.*

ELEVATION.

Torpedo Protection Bulkhead

Scale of Feet

UPPER DECK

SUPERSTRUCTURE

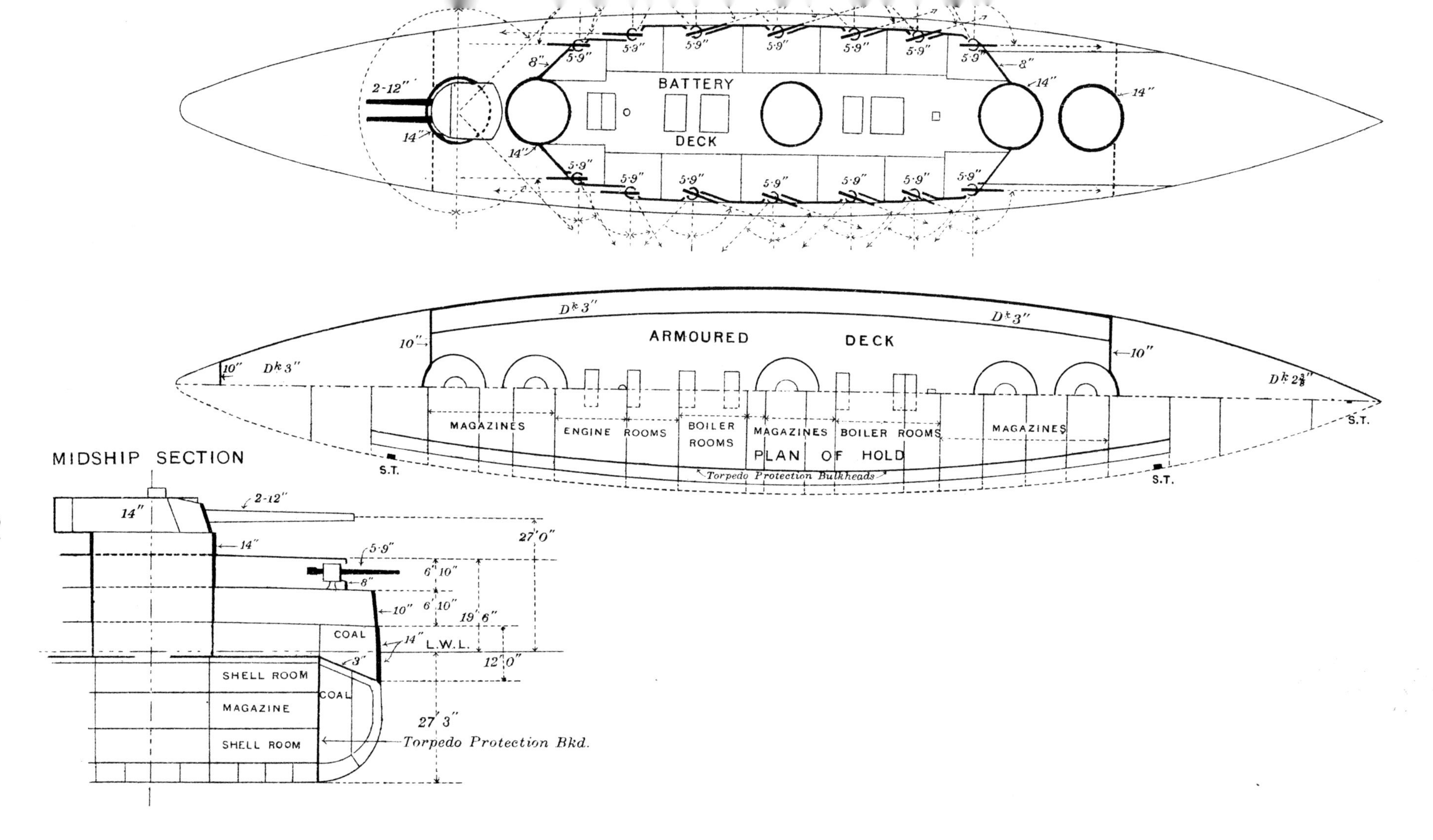

Ordnance Survey, November 1918.

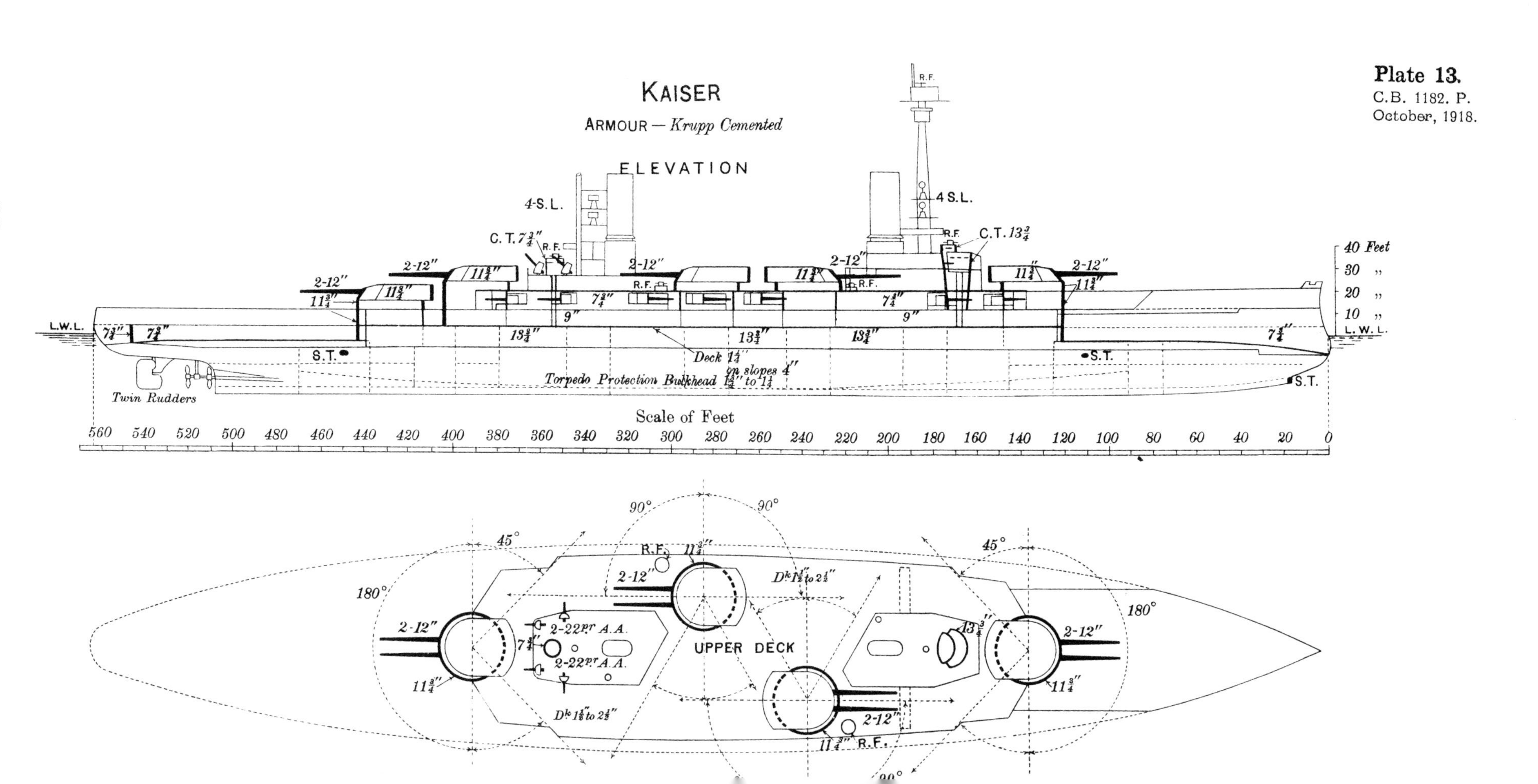

Plate 13.
C.B. 1182. P.
October, 1918.
KAISER
ARMOUR — Krupp Cemented
ELEVATION
4-S.L.
4 S.L.
C.T. 7¾″
C.T. 13¾
R.F.
2-12″
11¾″
7¾″
9″
13¾″
Deck 1¼″
on slopes 4″
Torpedo Protection Bulkhead 1½″ to 1¼
S.T.
L.W.L.
Twin Rudders
40 Feet
30 ″
20 ″
10 ″
Scale of Feet
560 540 520 500 480 460 440 420 400 380 360 340 320 300 280 260 240 220 200 180 160 140 120 100 80 60 40 20 0
UPPER DECK
2-22Pr A.A.
Dk 1¼″ to 2¼″
Dk 1⅛″ to 2⅜″
45°
90°
180°

MAIN DECK

ARMOURED DECK

PLAN OF HOLD

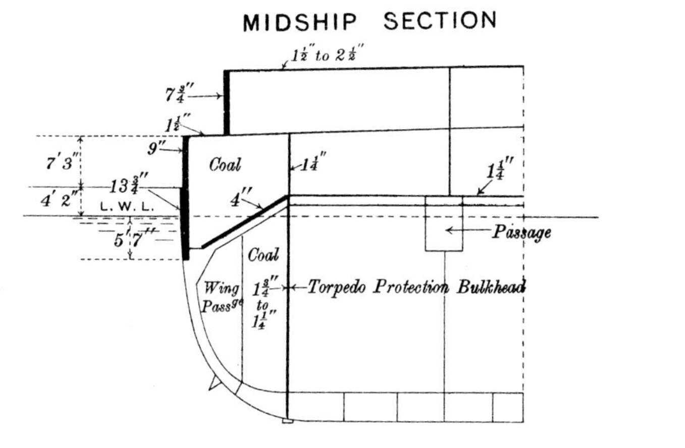

Vessels of Similar Type.

FRIEDRICH DER GROSSE,

KAISERIN.

PRINZREGENT LUITPOLD,

KÖNIG ALBERT.

Ordnance Survey, November, 1918.

Plate 14.
C.B. 1182. P.
October, 1918.

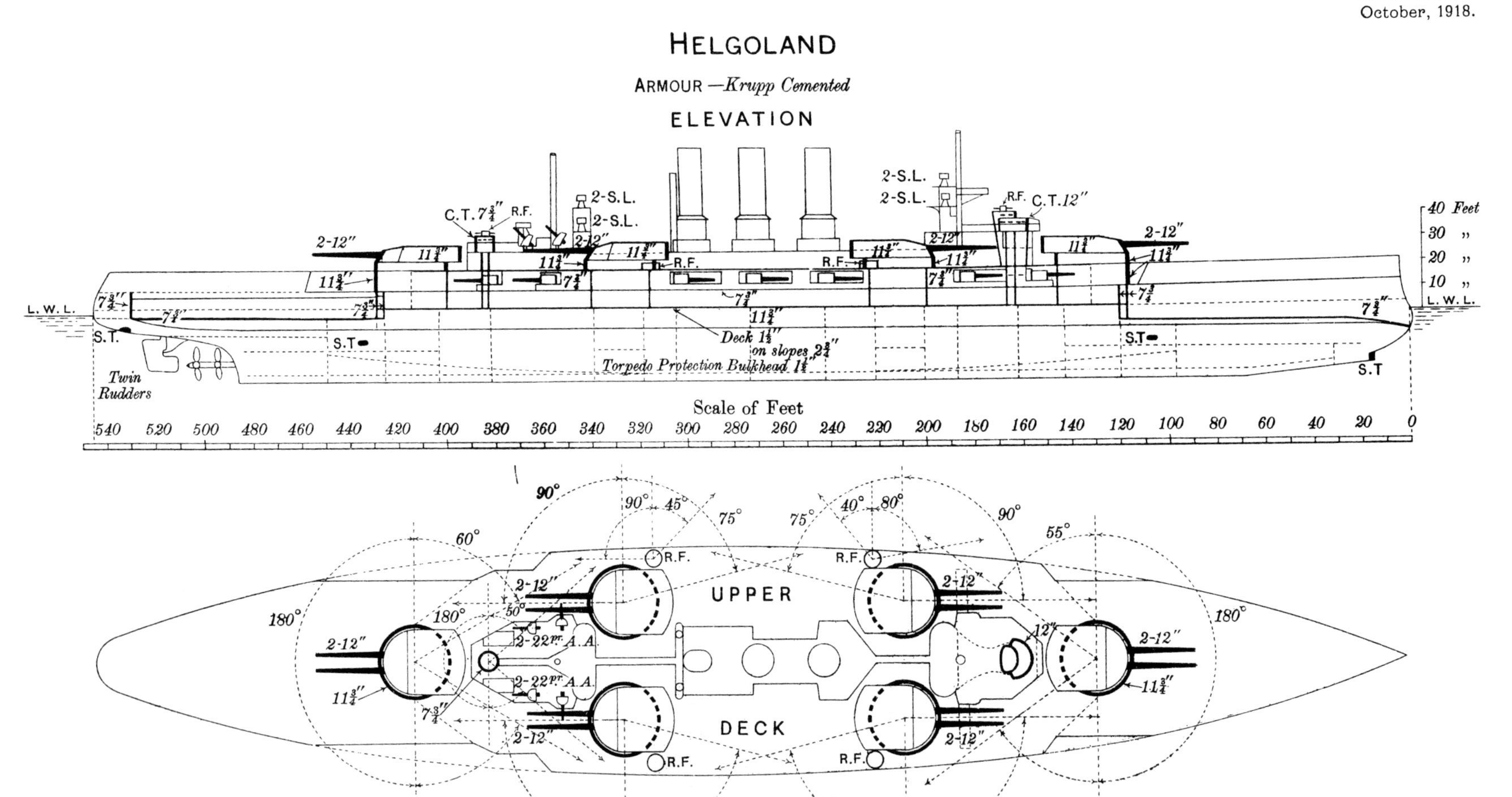

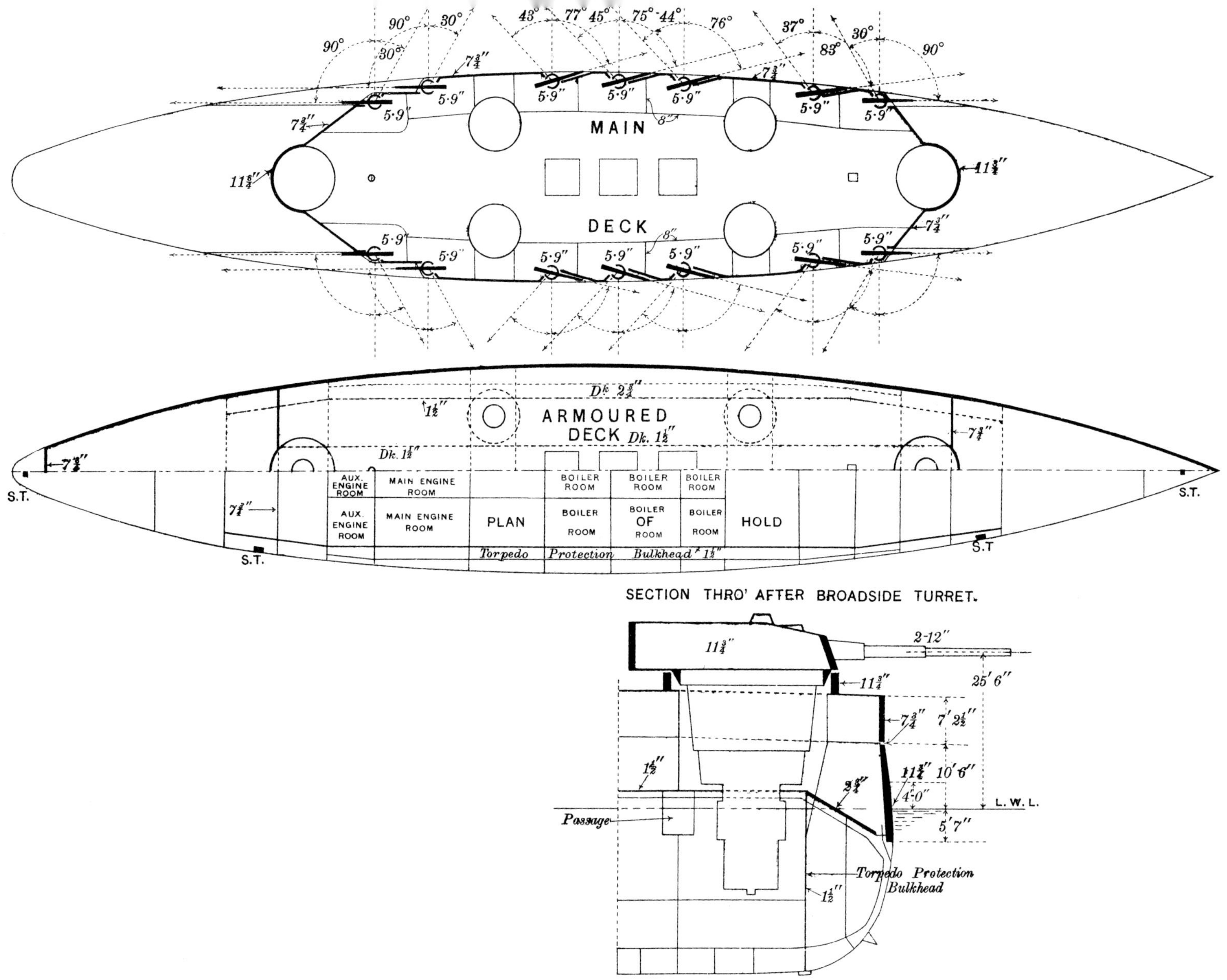

MAIN DECK

ARMOURED DECK PLAN

SECTION THRO' AFTER BROADSIDE TURRET.

Vessels of Similar Type

THÜRINGEN, OSTFRIESLAND, OLDENBURG.

Ordnance Survey November, 1918.

Plate 15.
C.B. 1182. P.
October, 1918.

NASSAU

ARMOUR - *Krupp Cemented*

ELEVATION

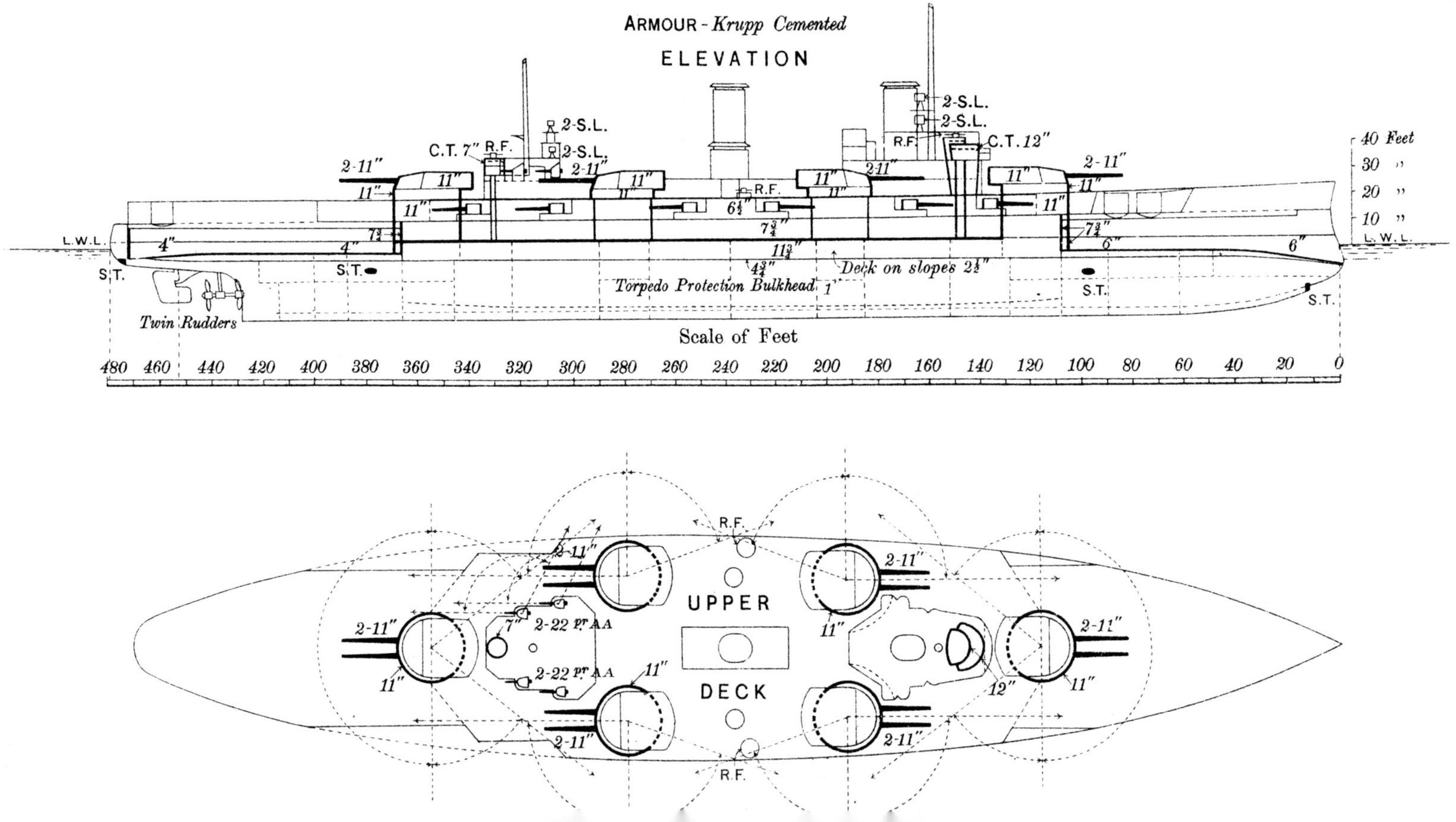

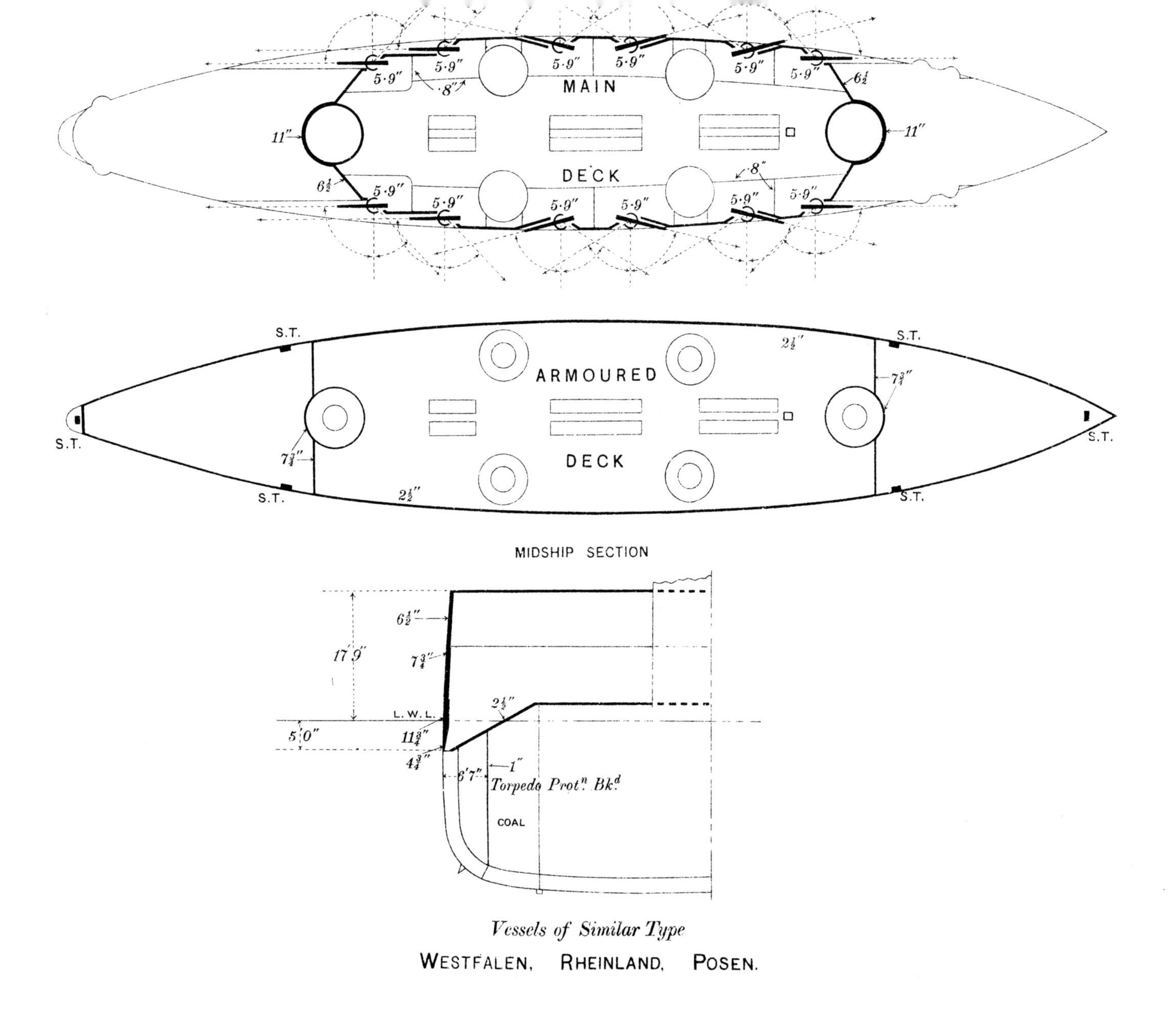

Vessels of Similar Type

WESTFALEN, RHEINLAND, POSEN.

NEW BA

GR

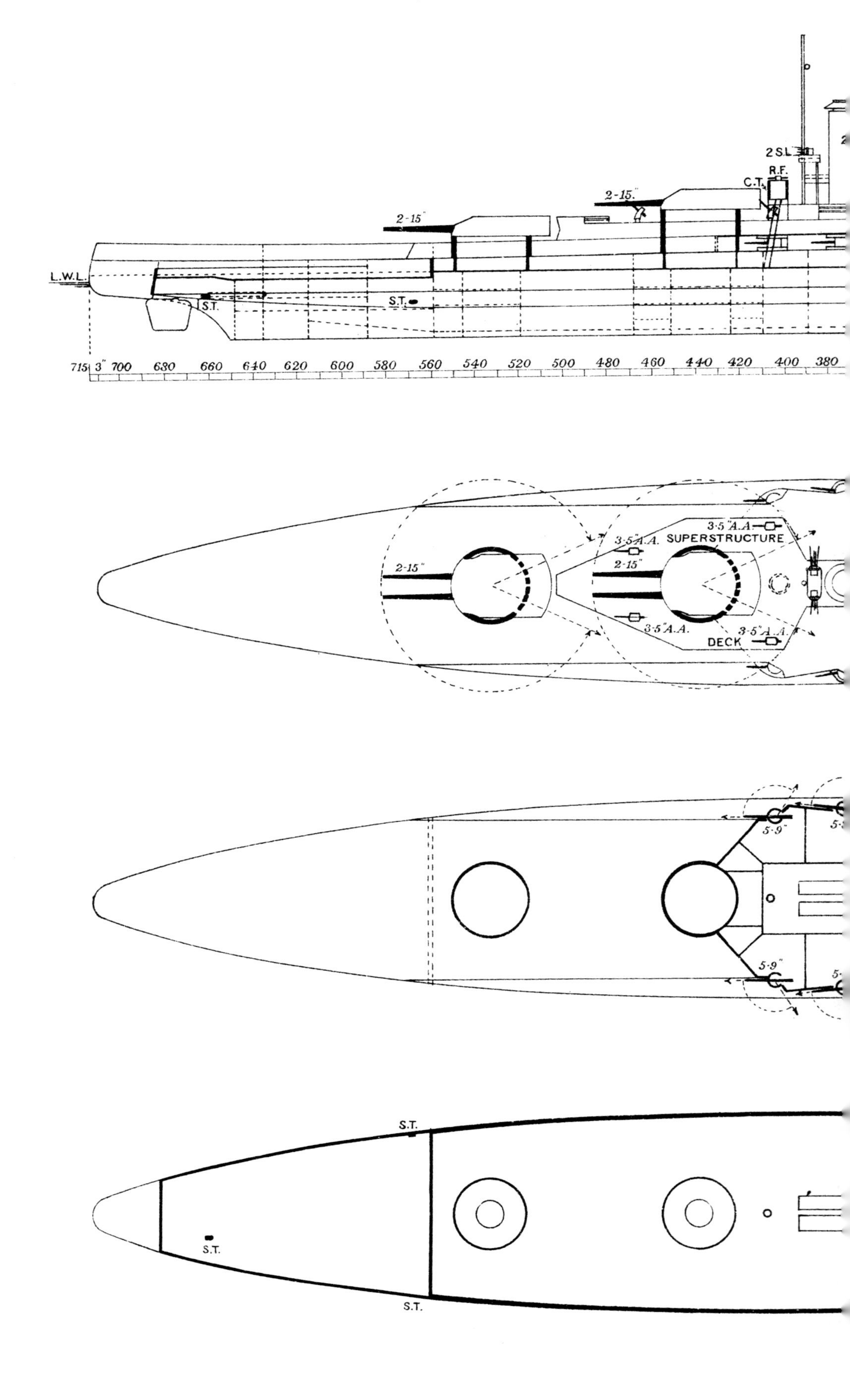
2 SL
R.F.
C.T.
2-15"
2-15"
L.W.L.
S.T.
S.T.
715 3" 700 630 660 640 620 600 580 560 540 520 500 480 460 440 420 400 380
2-15"
3·5"A.A.
3·5"A.A.
SUPERSTRUCTURE
2-15"
3·5"A.A.
DECK
3·5"A.A.
5·9"
5·9"
S.T.
S.T.
S.T.

Plate 16.
C.B. 1182. P.
October, 1918.

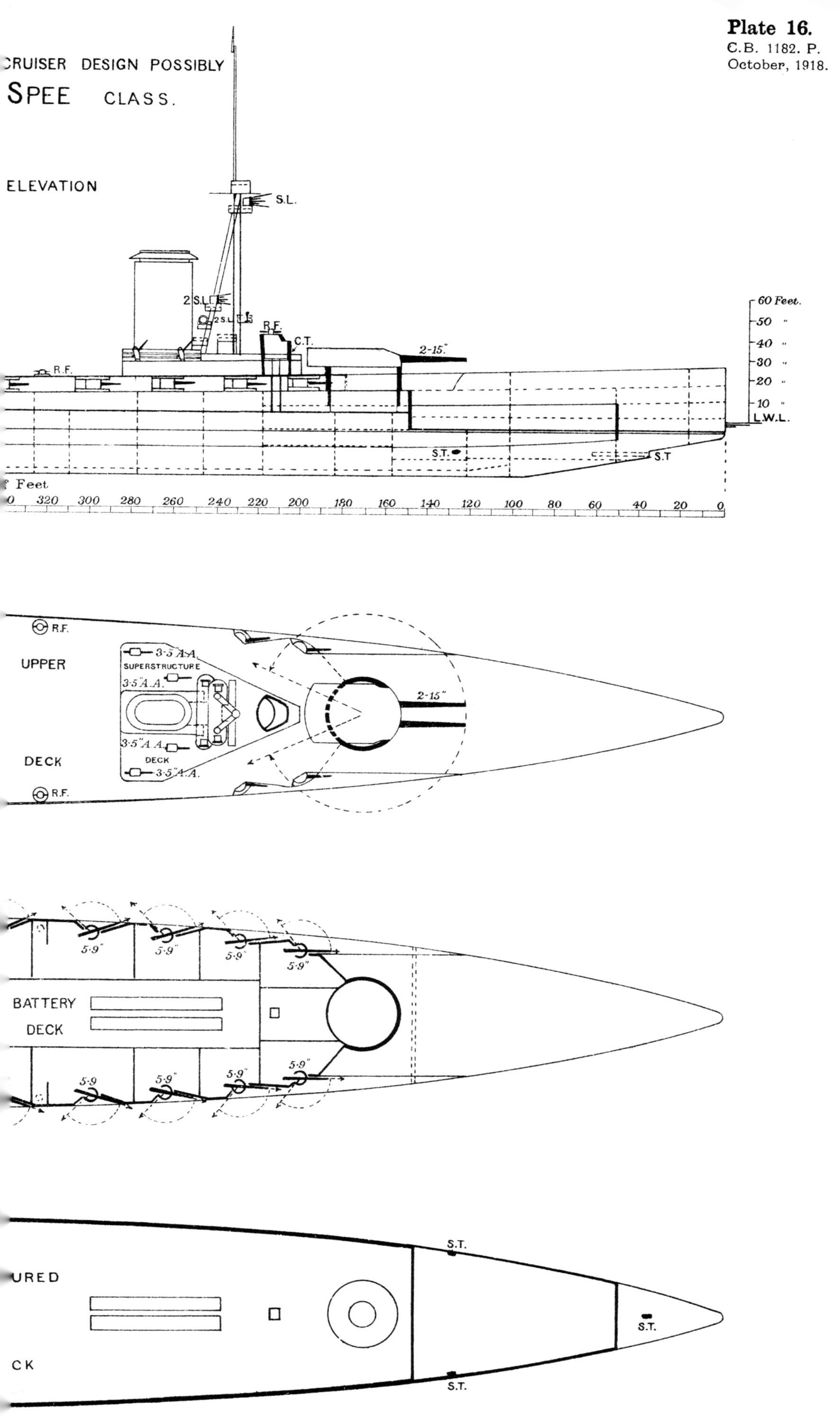

Ordnance Survey, November, 1918.

HINDENBURG

(Draw.

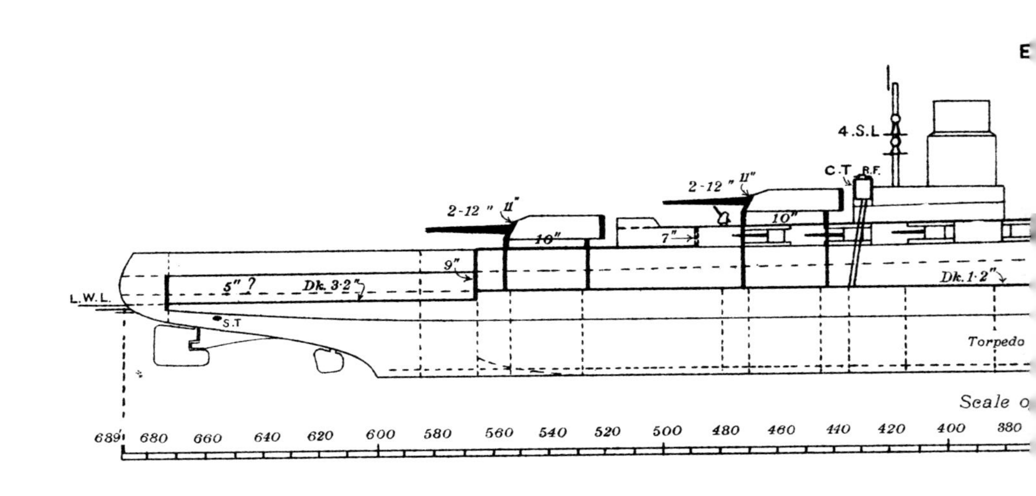

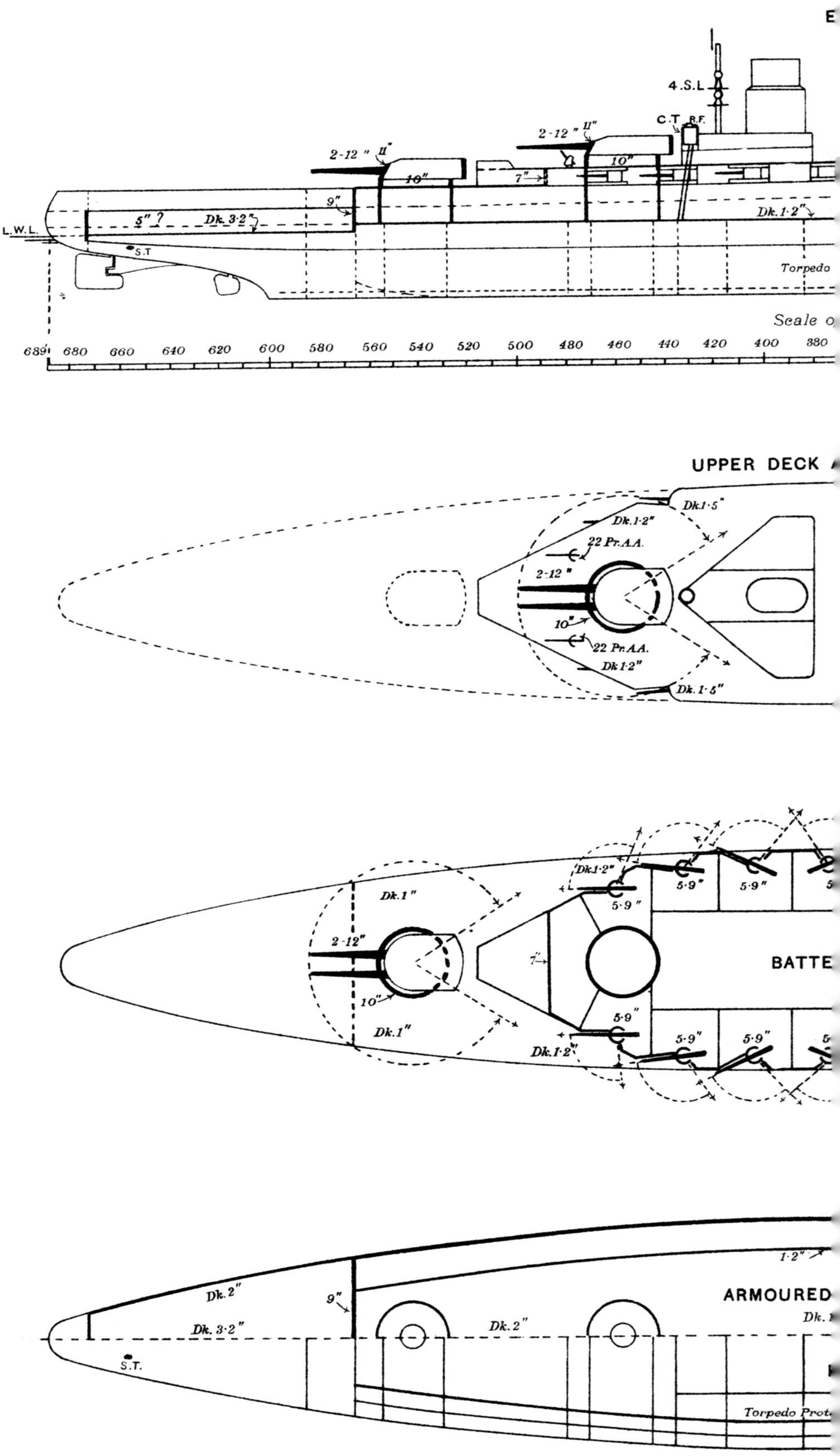

UPPER DECK

Dk.1·5"
Dk.1·2"
22 Pr.A.A.
2-12"
10"
22 Pr.A.A.
Dk.1·2"
Dk.1·5"

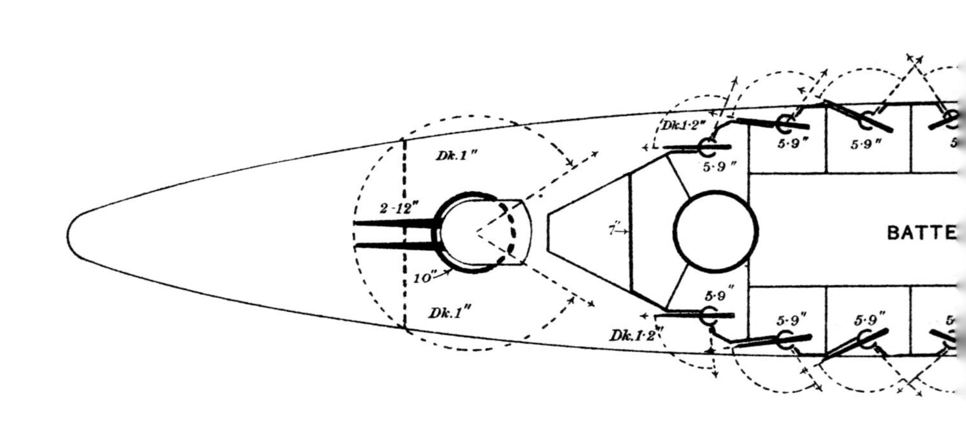

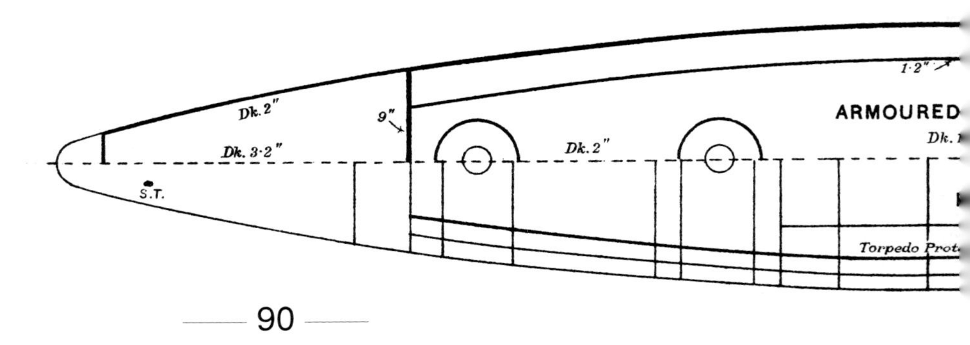

d MACKENSEN.

roximate)

ON.

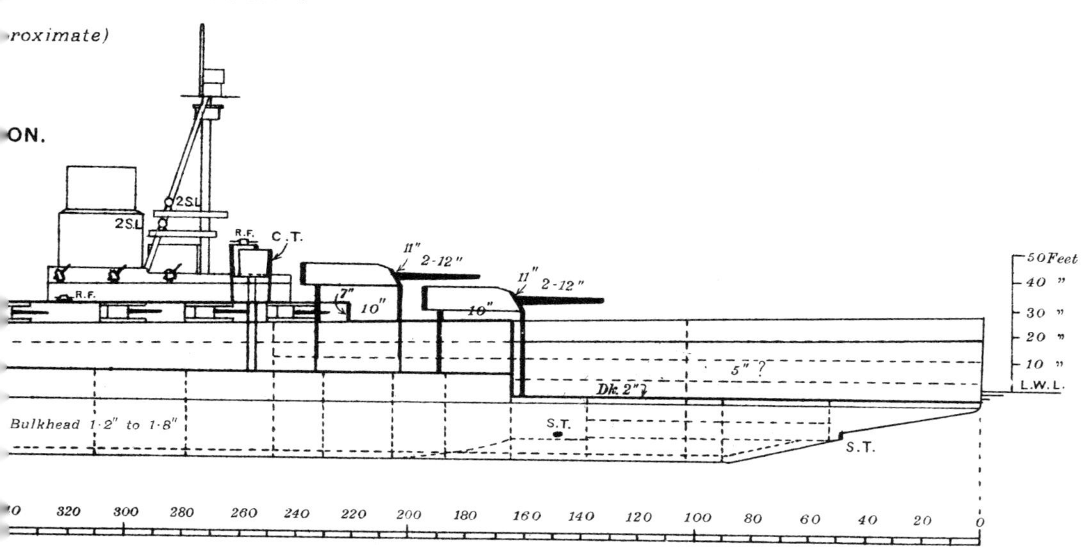

PERSTRUCTURE.

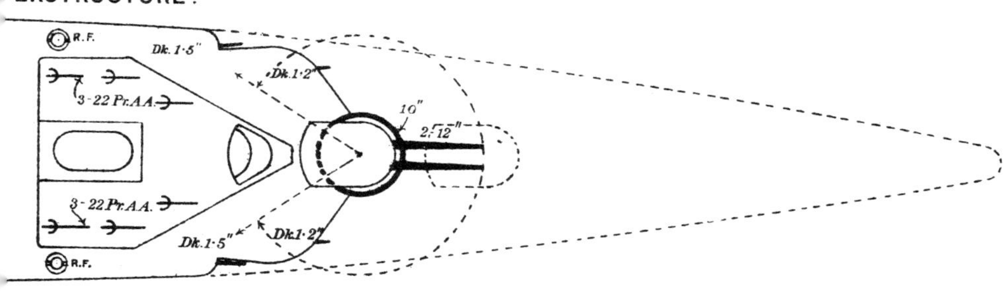

DECK.

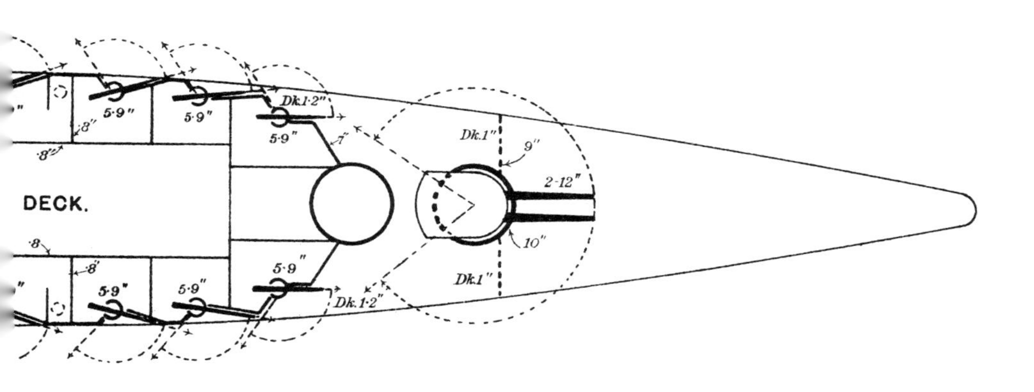

HOLD.

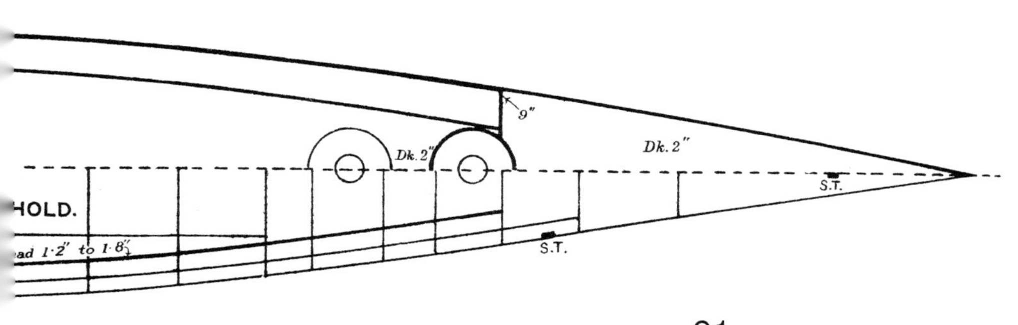

Ordnance Survey, November 1918.

DERFFLINGER.

ARMOUR— *Krupp Cemented*

ELEVATION.

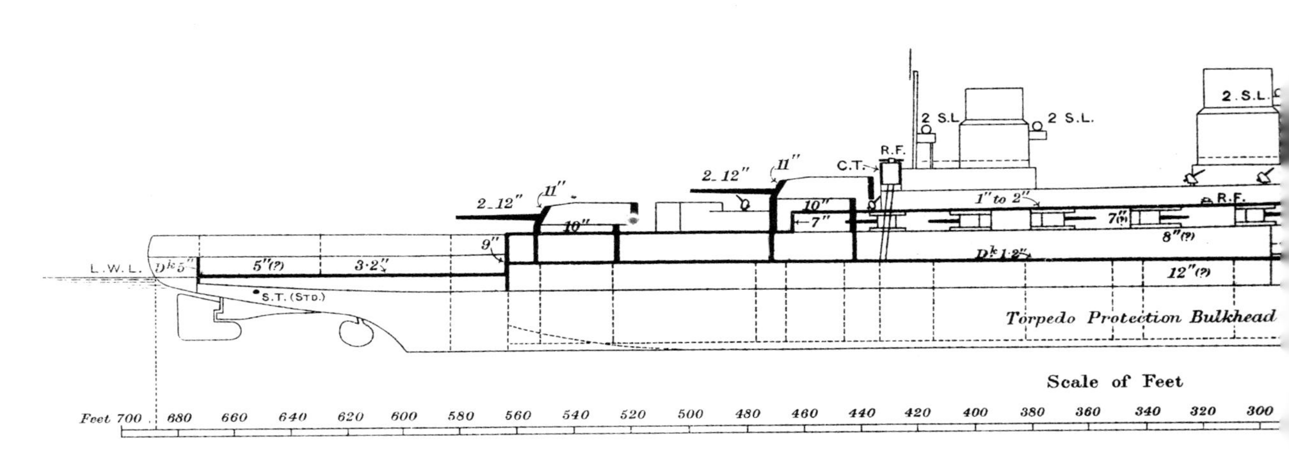

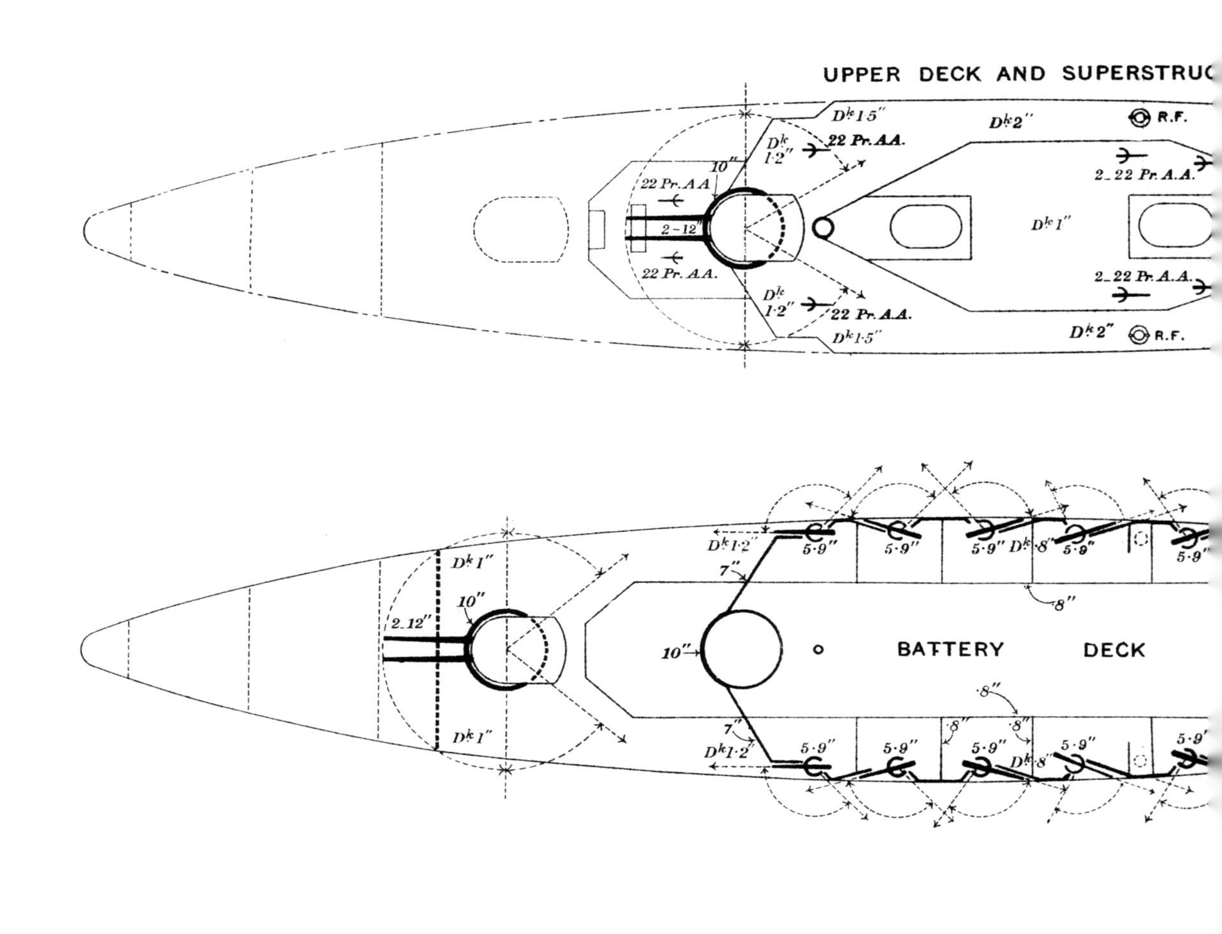

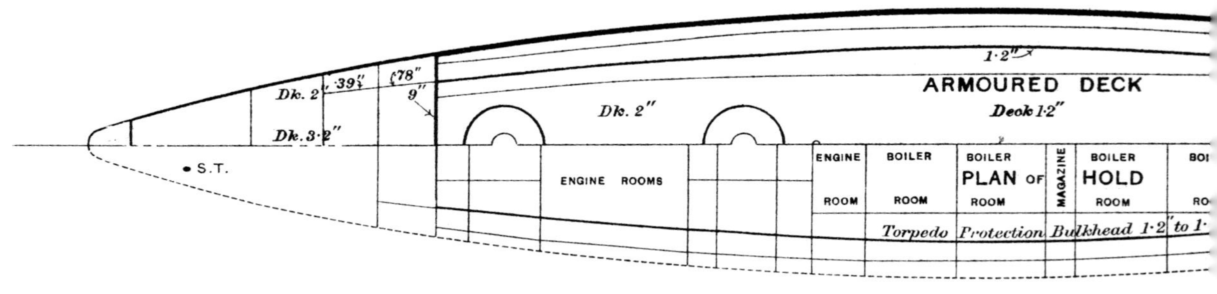

Plate 18.
C.B. 1182. P.
October, 1918.

S.L.

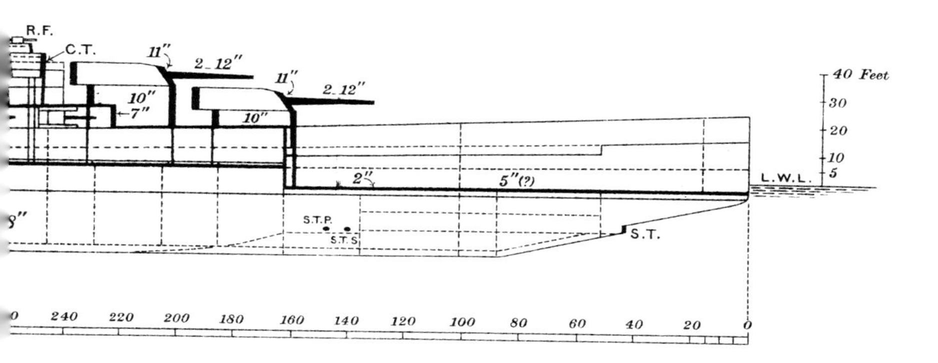

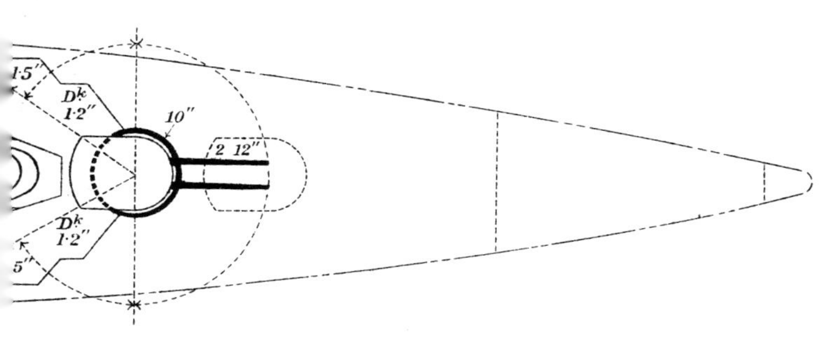

MIDSHIP SECTION.

Scale = Twice that of Plan

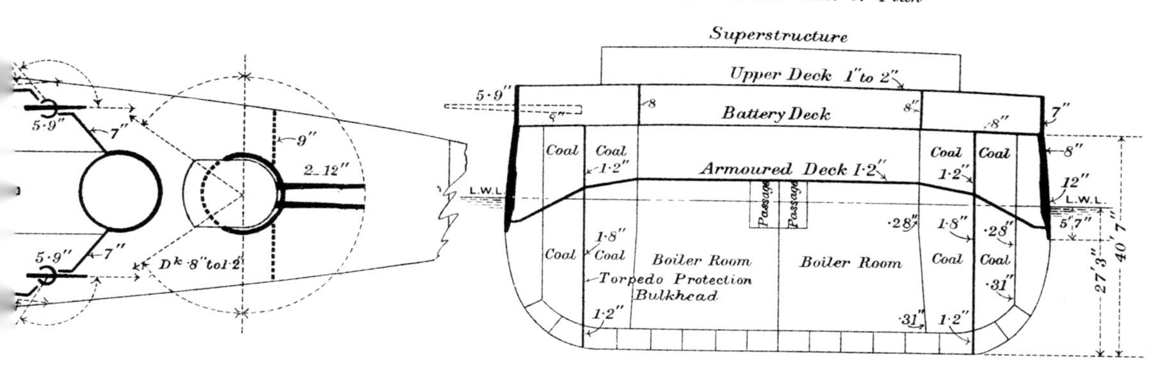

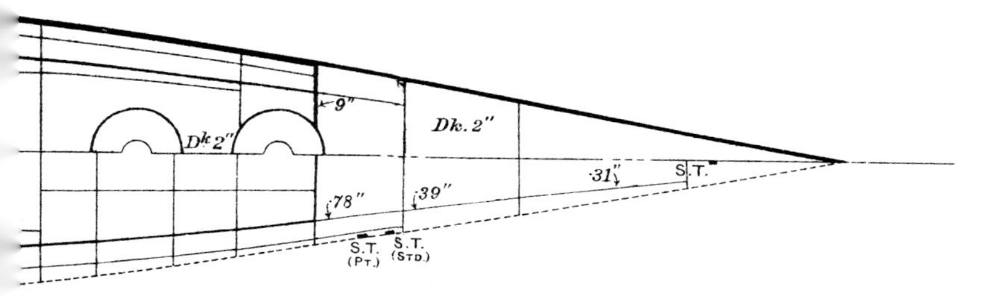

Ordnance Survey, November, 1918.

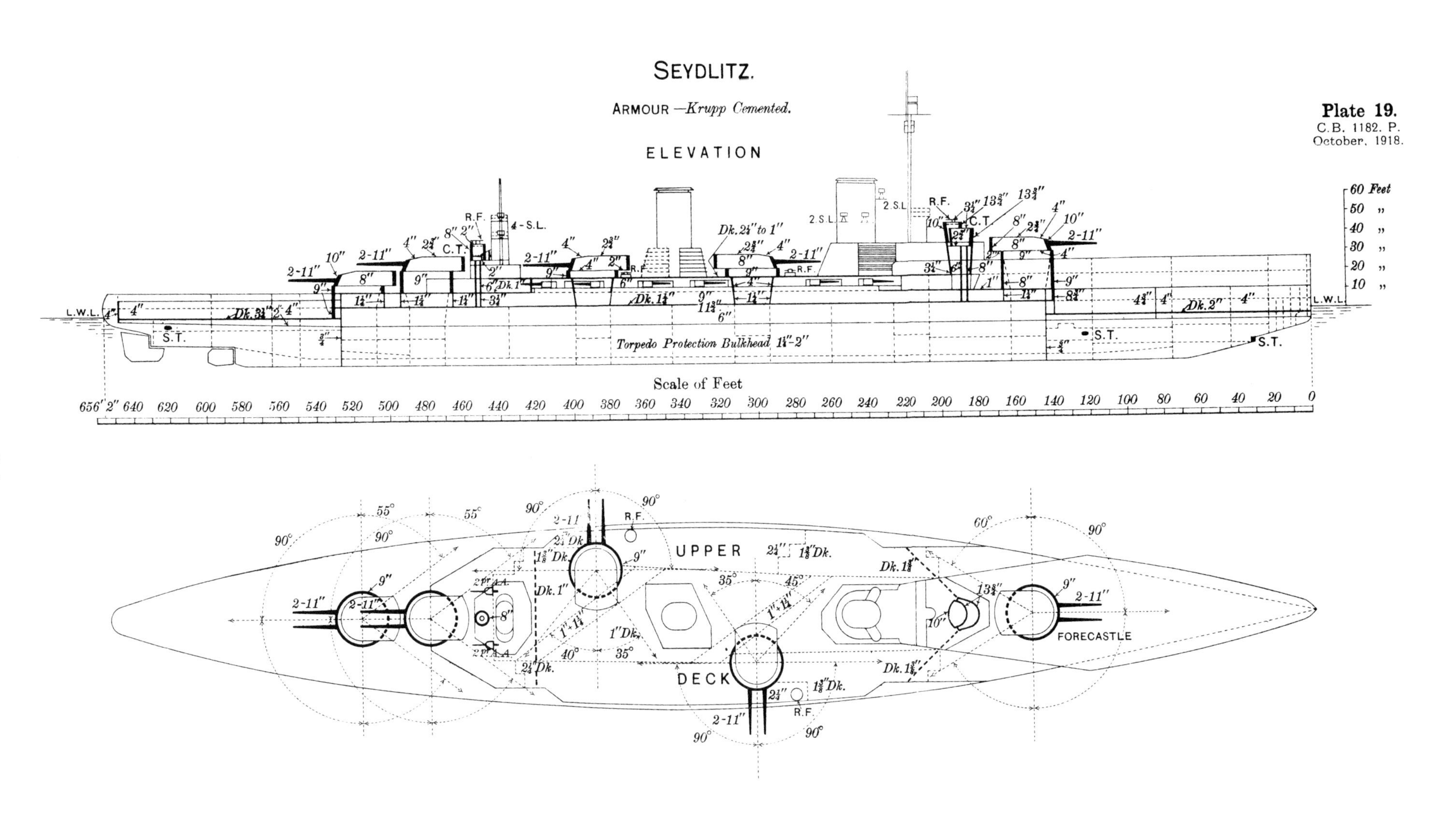
SEYDLITZ.
ARMOUR —Krupp Cemented.
ELEVATION
Plate 19.
C.B. 1182. P.
October, 1918.
Torpedo Protection Bulkhead 1¼"-2"
Scale of Feet
UPPER
DECK
FORECASTLE
L.W.L.
S.T.
C.T.
R.F.

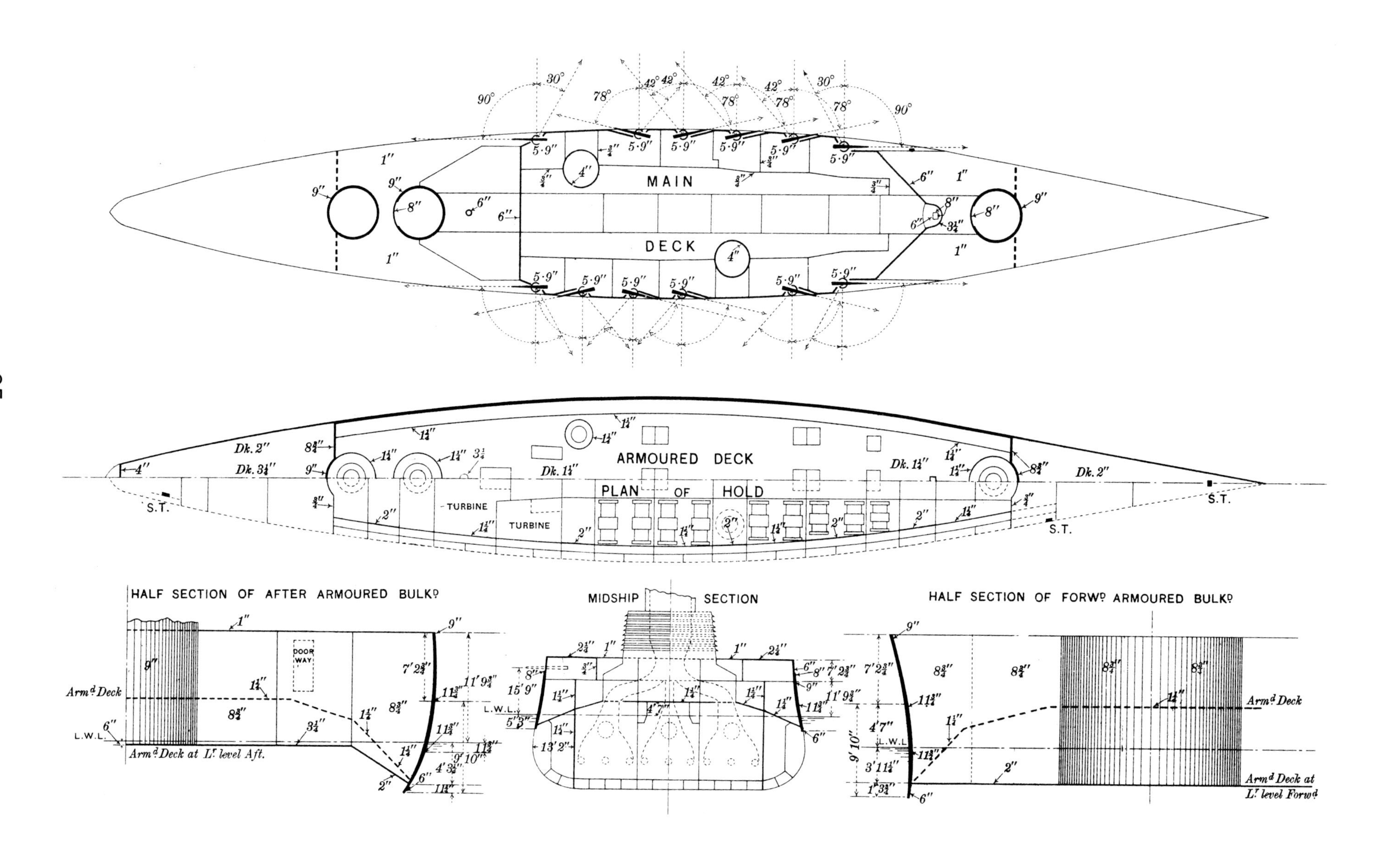

MAIN DECK
ARMOURED DECK
PLAN OF HOLD
TURBINE
TURBINE
S.T.
S.T.
S.T.
HALF SECTION OF AFTER ARMOURED BULKD
MIDSHIP SECTION
HALF SECTION OF FORWD ARMOURED BULKD
DOOR WAY
Armd Deck
Armd Deck at Lr level Aft.
Armd Deck
Armd Deck at Lr level Forwd
L.W.L.

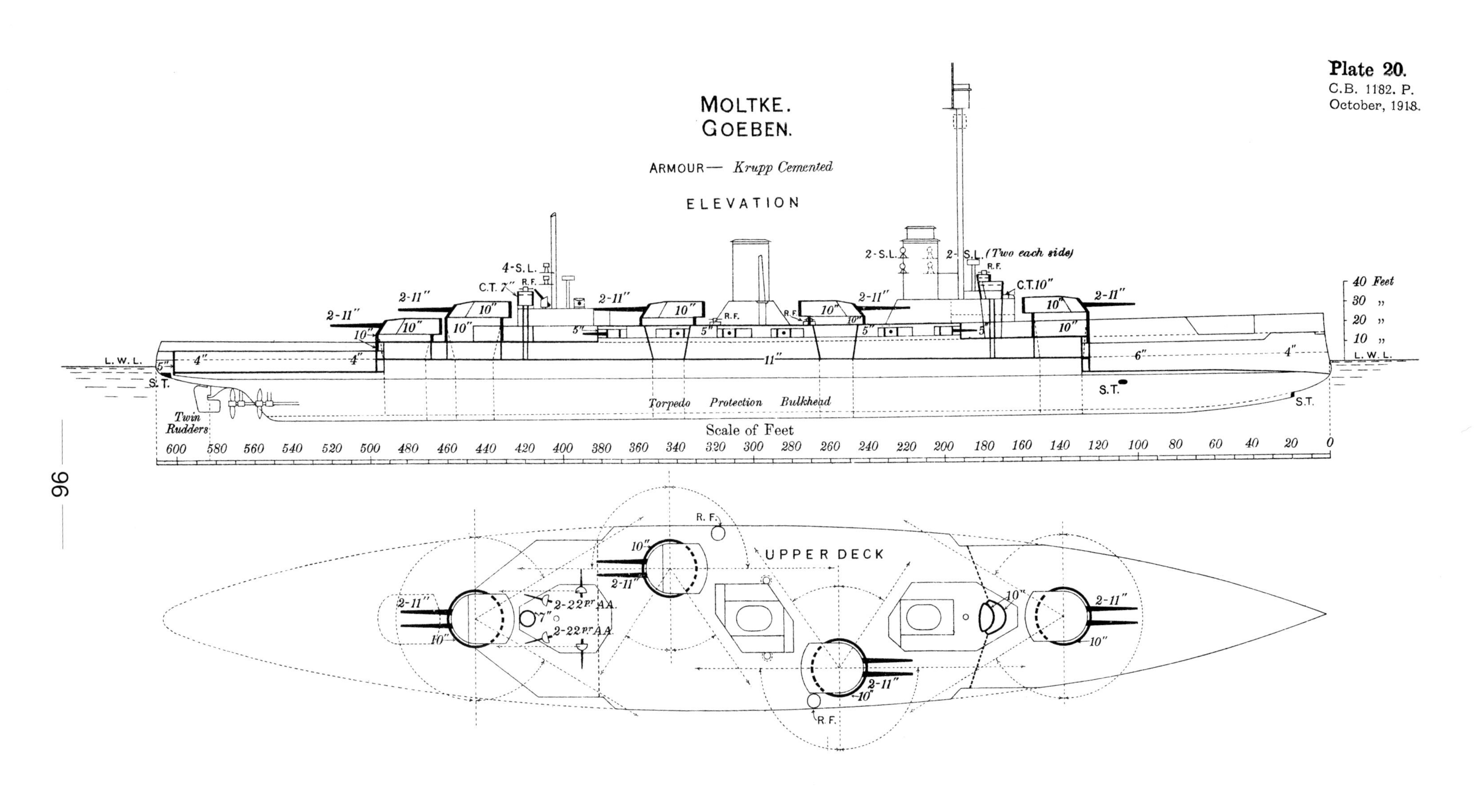
Plate 20.
C.B. 1182. P.
October, 1918.
MOLTKE.
GOEBEN.
ARMOUR— Krupp Cemented
ELEVATION
2-S.L.
2-S.L. (Two each side)
R.F.
C.T.10"
4-S.L.
C.T. 7"
2-11"
10"
5"
11"
6"
4"
L.W.L.
S.T.
40 Feet
30 "
20 "
10 "
Torpedo Protection Bulkhead
Twin Rudders
Scale of Feet
600 580 560 540 520 500 480 460 440 420 400 380 360 340 320 300 280 260 240 220 200 180 160 140 120 100 80 60 40 20 0
UPPER DECK
2-22pr AA.
7"

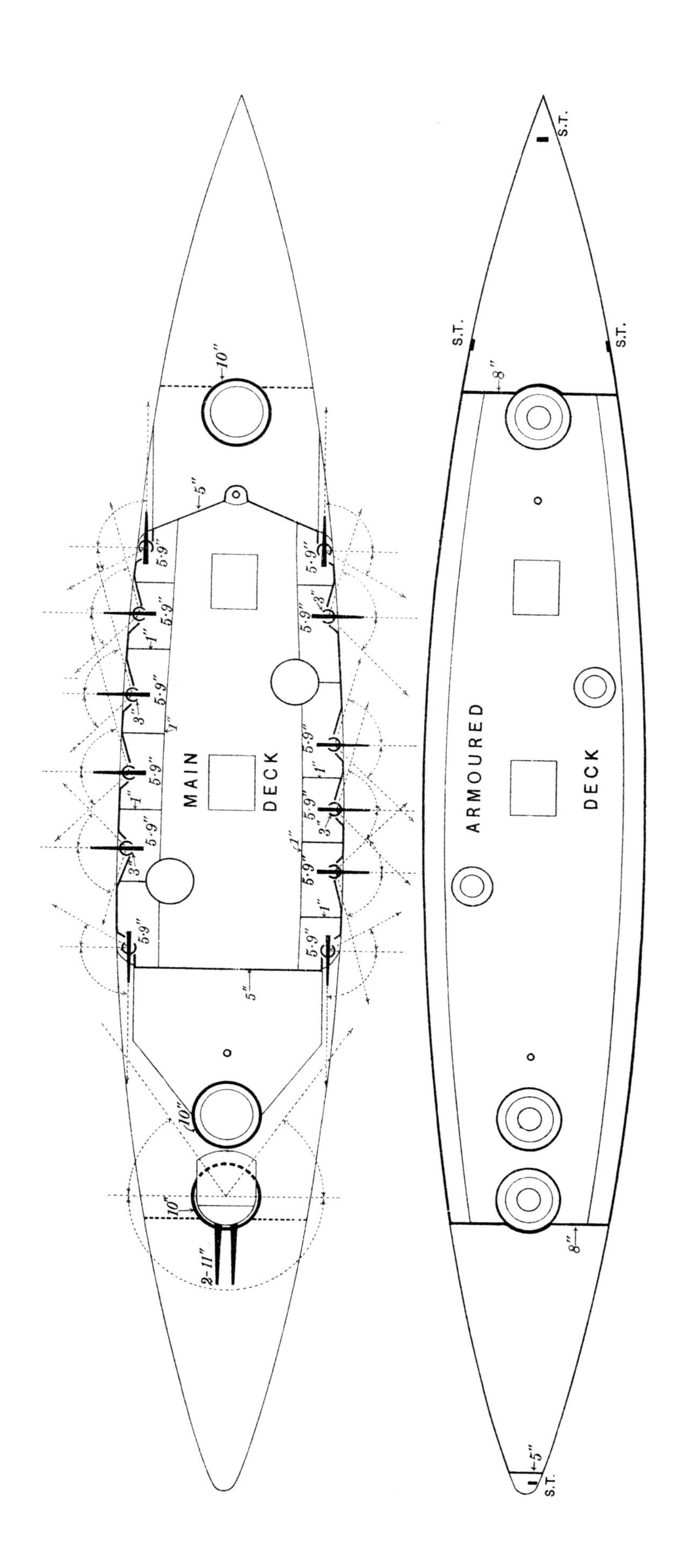

Ordnance Survey, November, 1918.

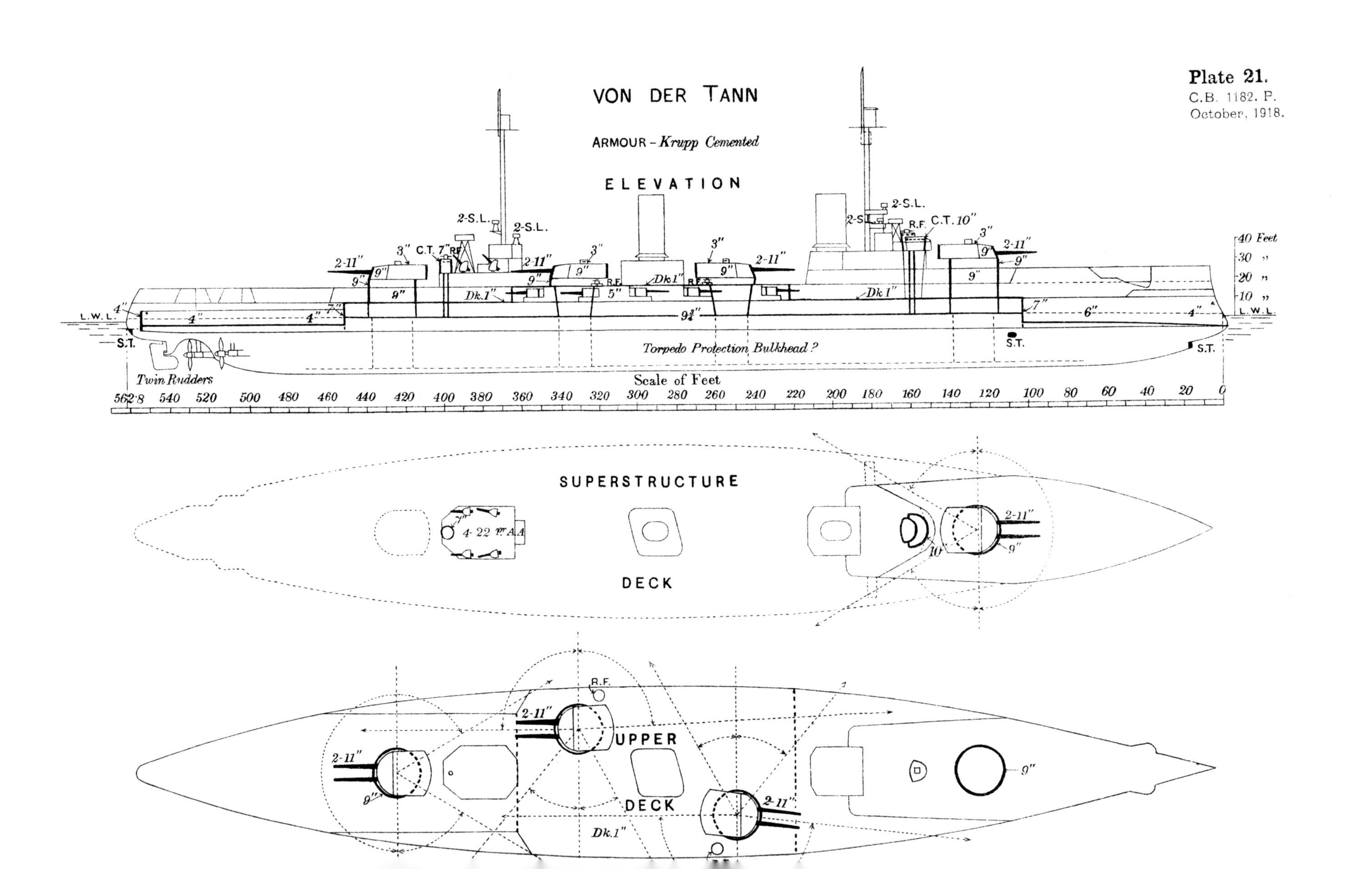

Plate 21.
C.B. 1182. P.
October, 1918.
VON DER TANN
ARMOUR - Krupp Cemented
ELEVATION
SUPERSTRUCTURE
DECK
UPPER
DECK
Torpedo Protection Bulkhead ?
Scale of Feet
Twin Rudders
L.W.L.
S.T.
C.T. 10"
C.T. 7"
2-S.L.
R.F.
Dk.1"
2-11"
4-22 ᵖʳ A.A.
40 Feet
30 "
20 "
10 "

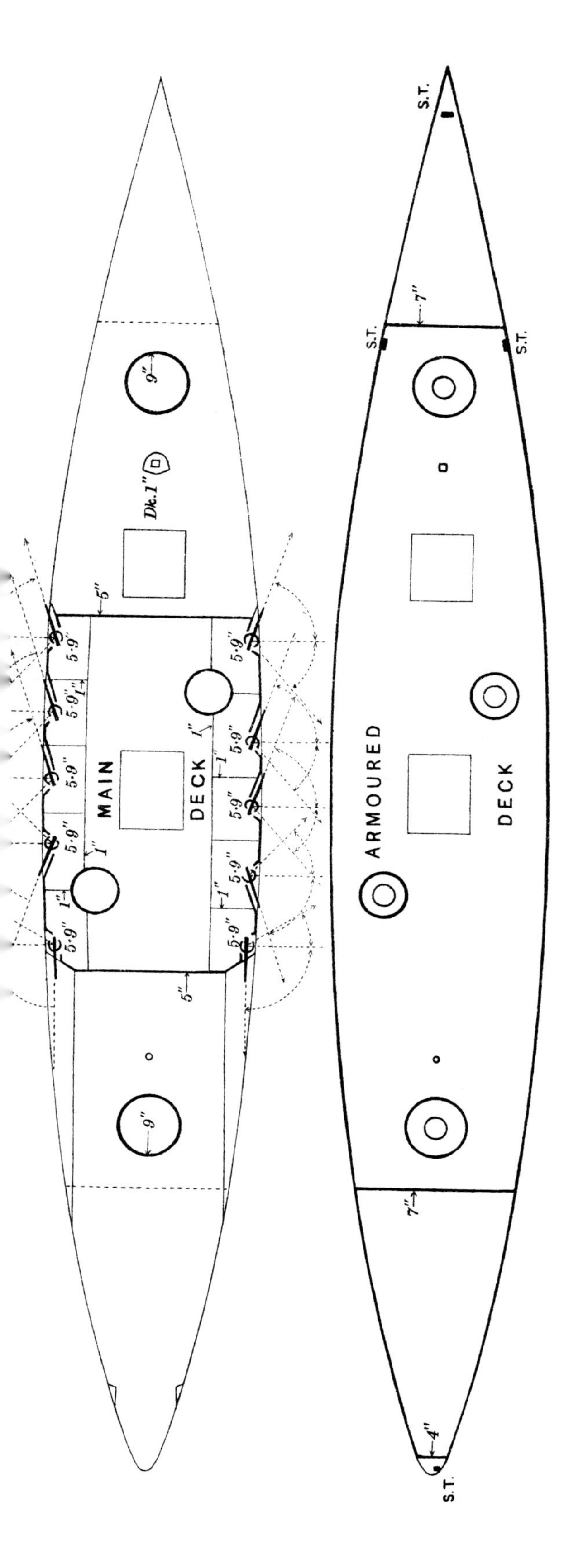

Ordnance Survey, November 1918.

Plate 22.
C.B. 1182. P.
October, 1918.

HANNOVER
SCHLESWIG-HOLSTEIN. SCHLESIEN
DEUTSCHLAND.

ARMOUR – *Krupp Cemented*

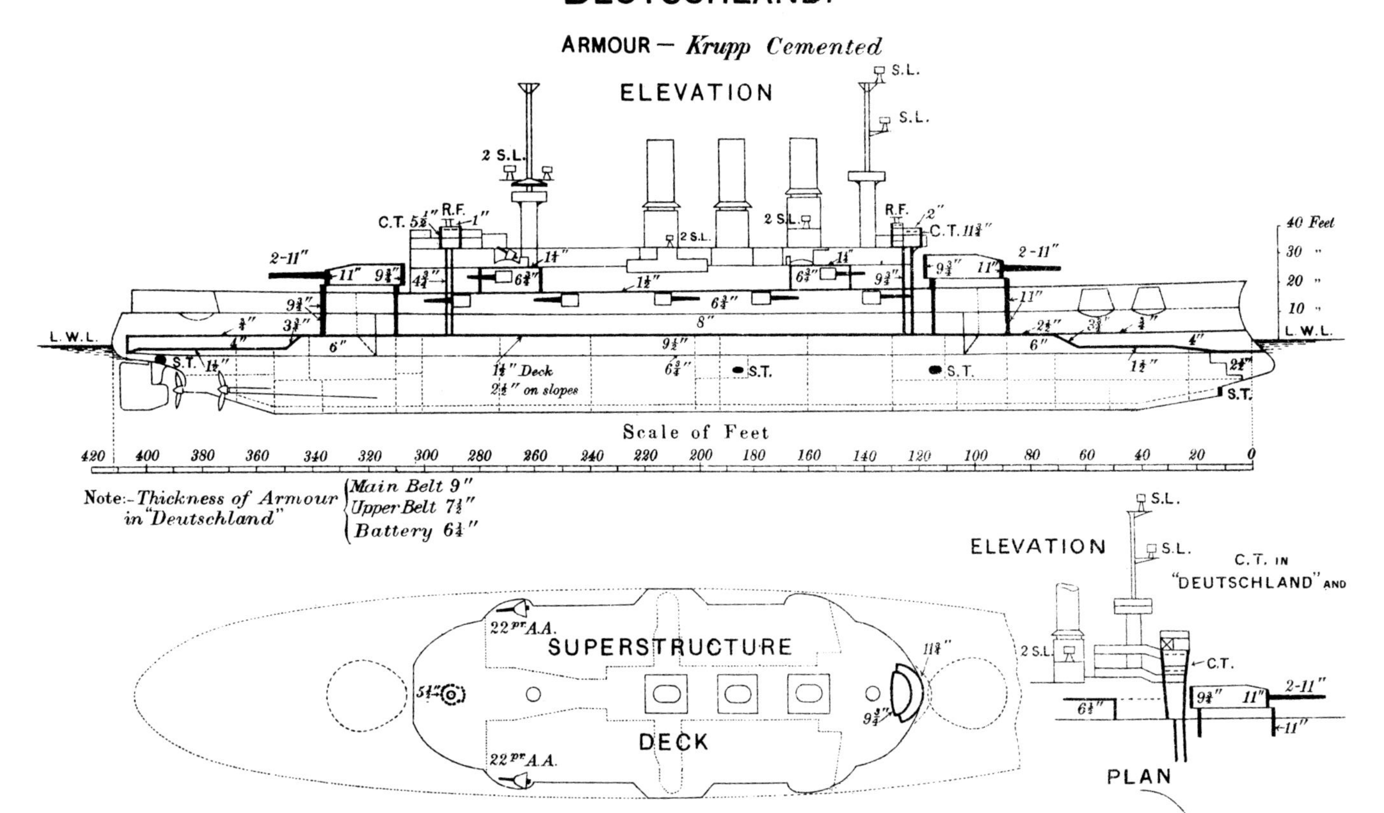

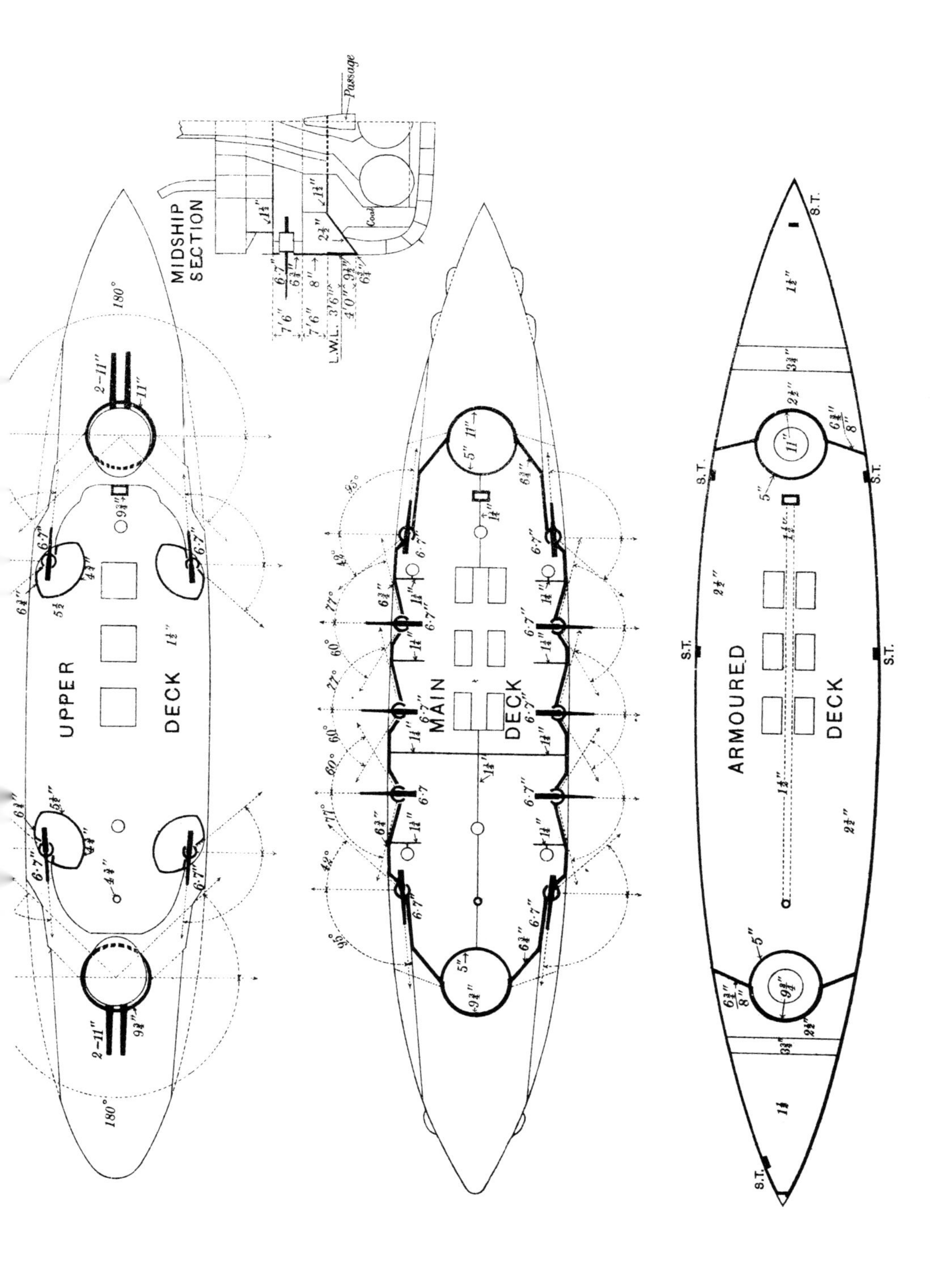

Ordnance Survey, November, 1918.

Plate 23.
C.B. 1182. P.
October, 1918.

Braunschweig.

Armour - *Krupp Cemented*

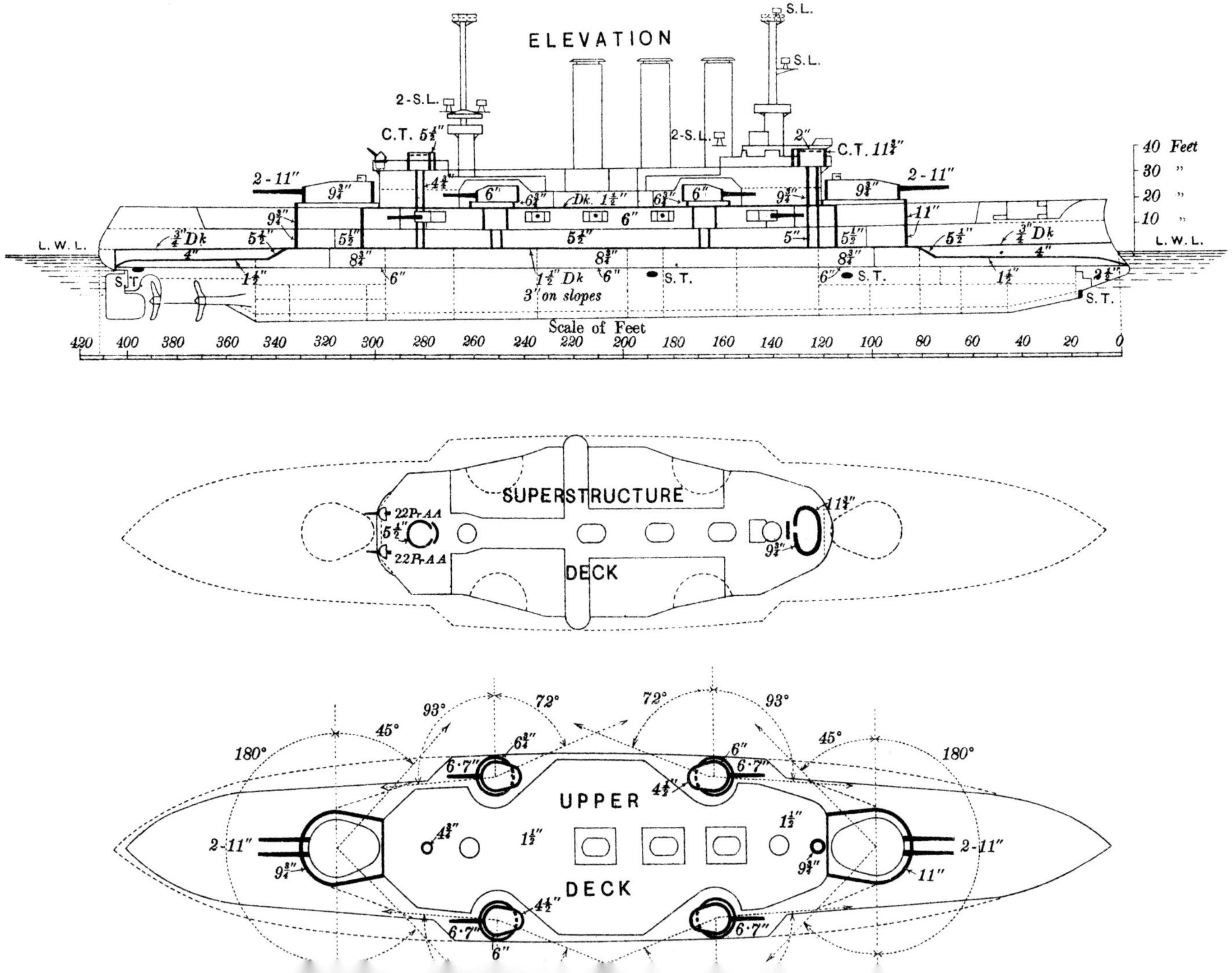
ELEVATION
SUPERSTRUCTURE DECK
UPPER DECK
Scale of Feet
L. W. L.
S.L.
2-S.L.
C.T. 5½"
C.T. 11¾"
2-11"
S.T.
3" on slopes
22Pr AA

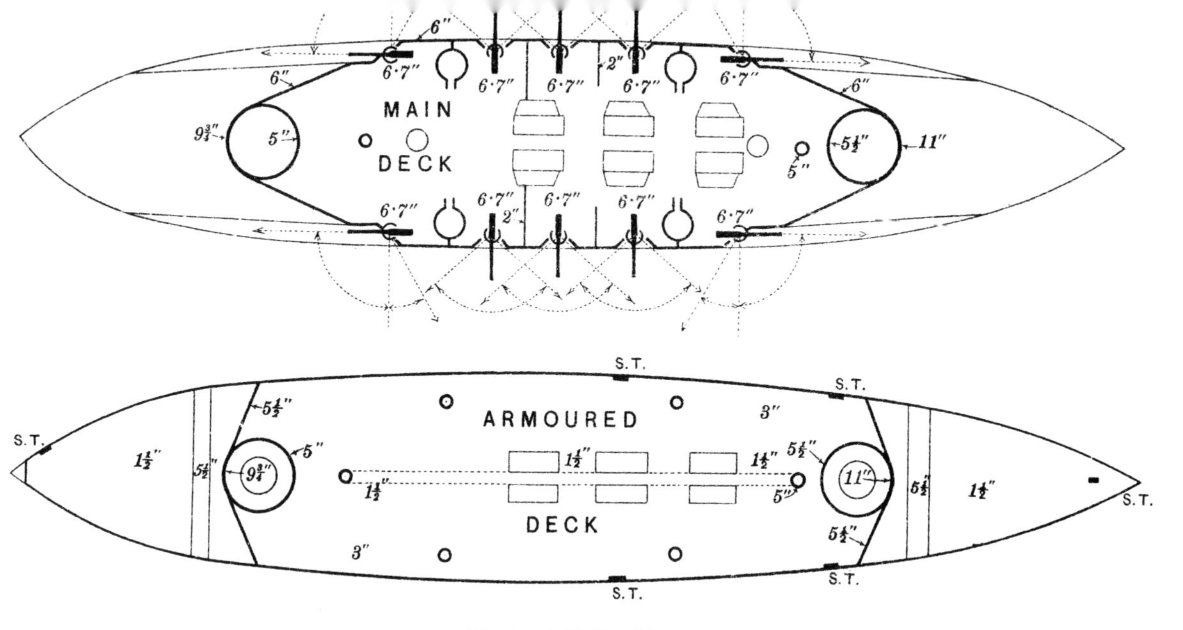

Vessels of Similar Type,

ELSASS * HESSEN

* PREUSSEN * LOTHRINGEN

* *Thickness of Turret Armour differs slightly. See text.*

Ordnance Survey, November. 1918.

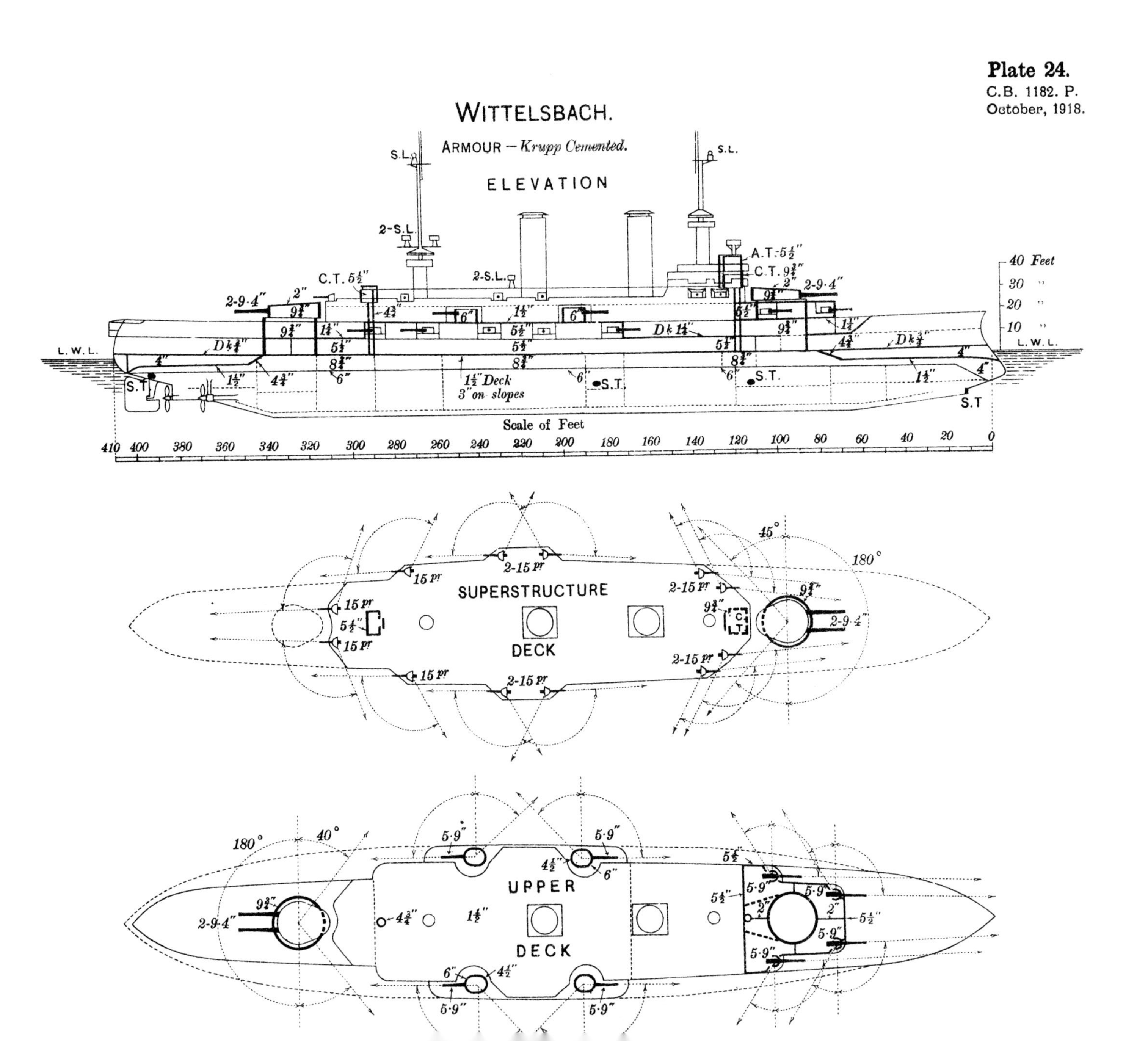

Plate 24.
C.B. 1182. P.
October, 1918.
WITTELSBACH.
ARMOUR – Krupp Cemented.
ELEVATION
Scale of Feet
L.W.L.
SUPERSTRUCTURE
DECK
UPPER
DECK

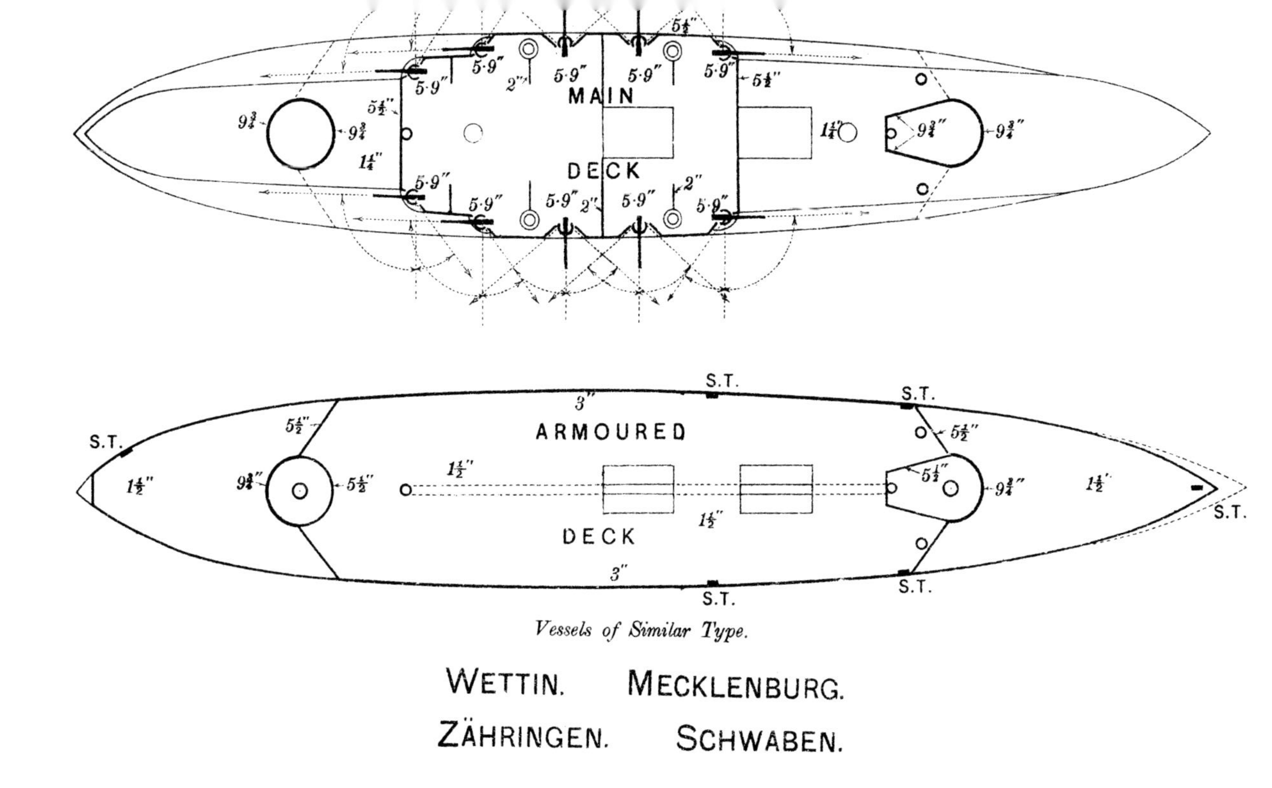

Vessels of Similar Type.

WETTIN. MECKLENBURG.
ZÄHRINGEN. SCHWABEN.

Ordnance Survey, November, 1918.

SECTION III
CRUISERS AND LIGHT CRUISERS

GERMAN NAVY.

PART III.

SECTION 3.

CRUISERS AND LIGHT CRUISERS (including Minelaying Cruisers).

Name.	Classification.	Page.	Plate No.
Amazone	Light Cruiser	21	16
Arcona	„	20	5 & 16
Augsburg	„	13	4 & 13
Berlin	„	17	15
Bremen	„	4	—
Bremse	Minelaying Cruiser.	3	1 & 6
Brummer	„	3	1 & 6
Cöln	Light Cruiser	4	—
Danzig	„	17	15
Dresden	„	4	—
Emden	„	4	1 & 7
Ersatz Amazone	„	4	—
Ersatz Medusa	„	4	—
Frankfurt	„	5	2 & 8
Freya	Obsolete Light Cruiser.	29	20
Fürst Bismarck	Obsolete Cruiser.	27	19
Gazelle	Light Cruiser	21	16
Graudenz	Light Cruiser	8	3 & 10
Hamburg	„	17	15
Hansa	Obsolete Light Cruiser.	29	20
Hertha	„	29	20
Kaiserin Augusta	„	31	21
Karlsruhe	Light Cruiser	4	1 & 7
Kolberg	„	13	4 & 13
Königsberg	„	4	1 & 7
Leipzig (?)	„	4	—
Lübeck	„	17	15
Magdeburg (?)	„	4	—
Medusa	„	21	16
München	„	17	15
Niobe	„	21	16
Nürnberg	„	4	1 & 7
Nymphe	„	21	16
P.	„	4	—
Pillau	„	6	2 & 9
Prinz Heinrich	Obsolete Cruiser.	25	18

O (33) AS 4924—2 Pk 2972 1000 8/18 E & S

Cruisers and Light Cruisers—*continued.*

Name.	Classification.	Page.	Plate No.	Name.	Classification.	Page.	Plate No.
Q. - - - -	Light Cruiser	4	—	**Strassburg** - -	Light Cruiser	10	4 & 12
R. - - - -	,,	4	—	**Stuttgart** - -	,,	15	5 & 14
Regensburg - -	,,	8	3 & 11	**Thetis** - - -	,,	21	16
Roon - - -	Obsolete Cruiser.	23	17	**Victoria Louise** -	Obsolete Light Cruiser.	29	20
Stettin - - -	Light Cruiser	15	5 & 14	**Vineta** - - -	,,	29	20
Stralsund - -	,,	10	4 & 12	**Wiesbaden** - -	Light Cruiser	4	—

LIST OF PLATES.

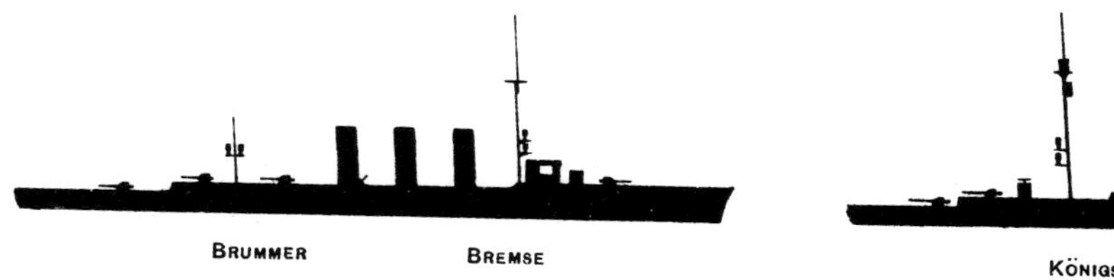

Brummer Bremse

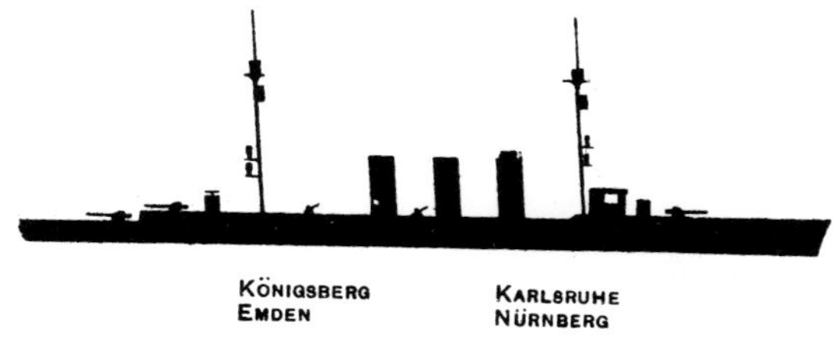

Königsberg
Emden

Karlsruhe
Nürnberg

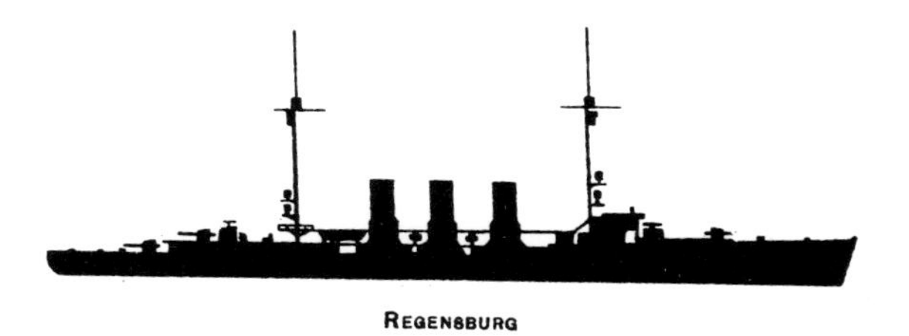

Regensburg

Strassburg

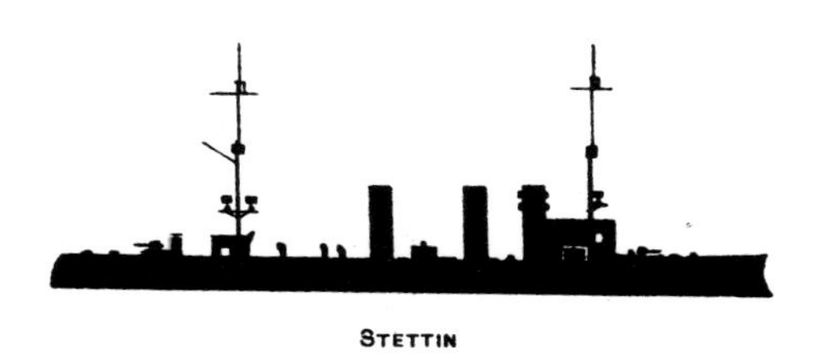

Stettin

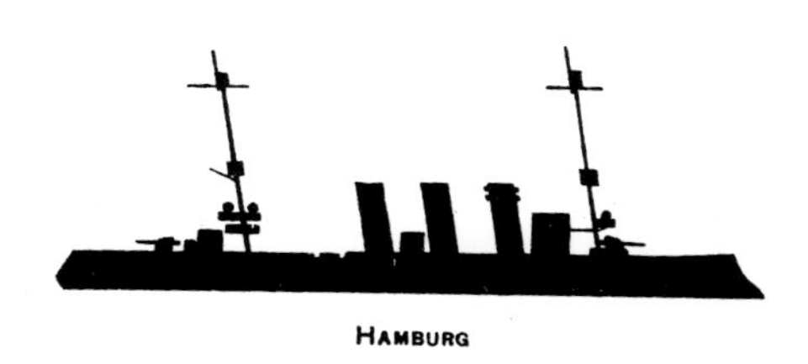

Hamburg

Arcona

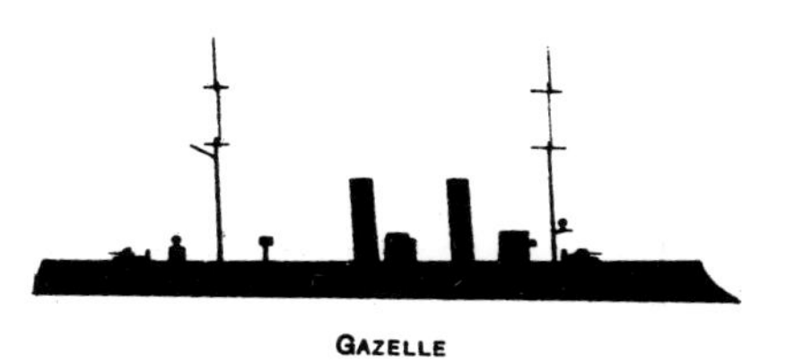

Gazelle

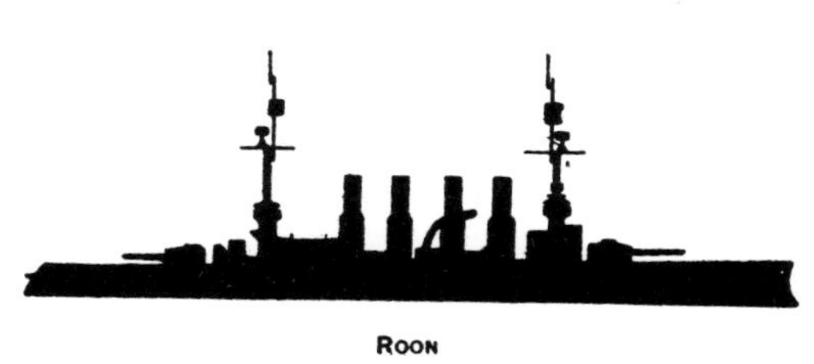

Roon

Prinz Heinrich

CRUISERS AND LIGHT CRUISERS.

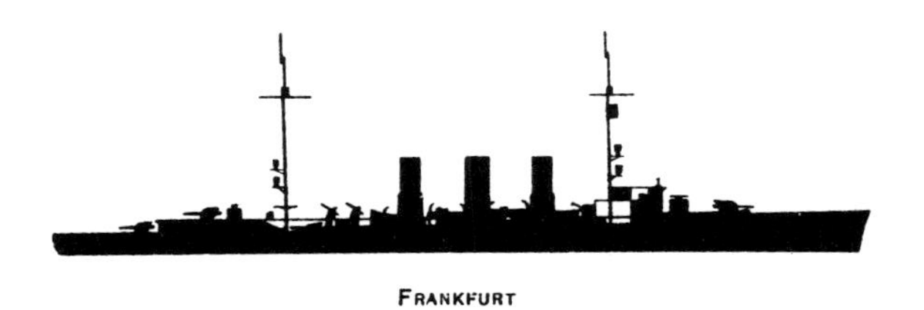

FRANKFURT

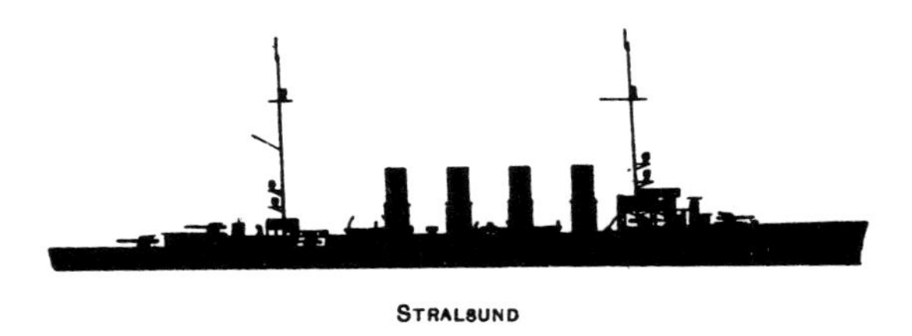

STRALSUND

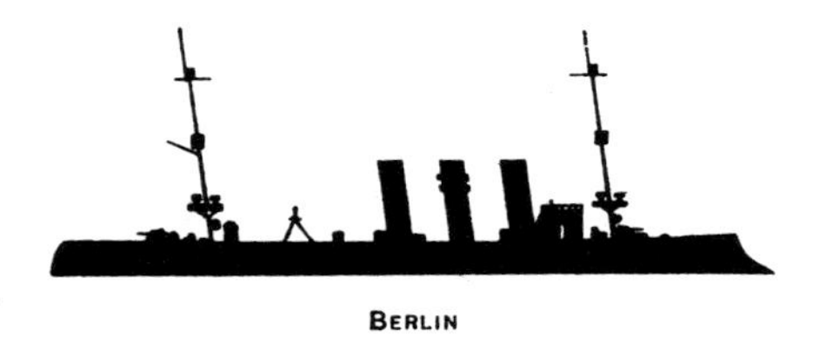

BERLIN

MÜNCHEN

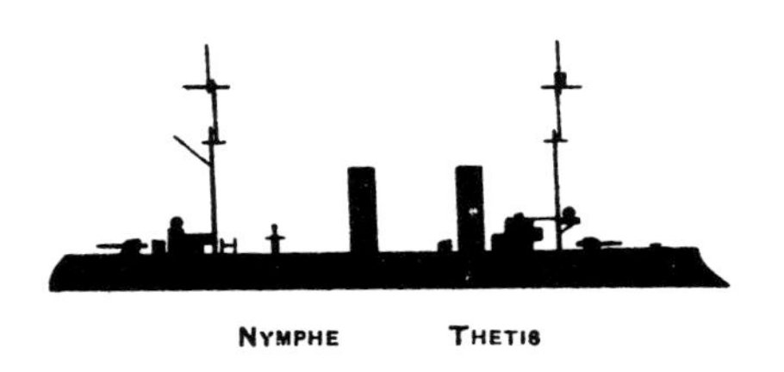

NYMPHE THETIS

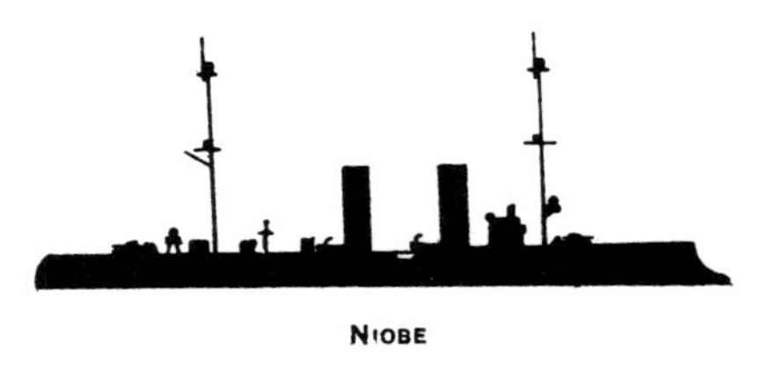

NIOBE

FÜRST BISMARCK

HERTHA VICTORIA LOUISE
VINETA HANSA

GERMANY.

C.B. 1182Q.
I.D. 1159
June, 1918.

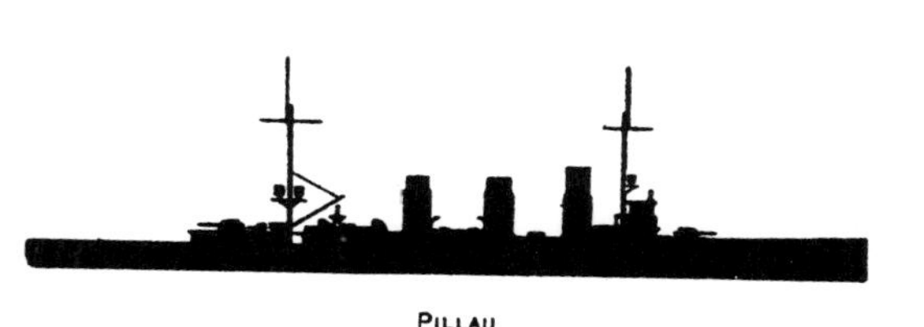

Pillau

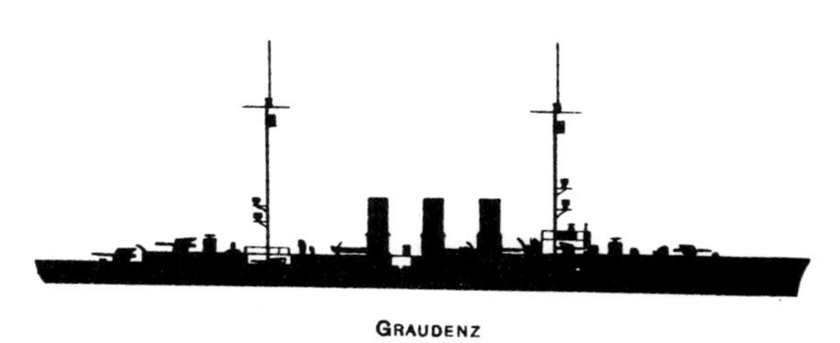

Graudenz

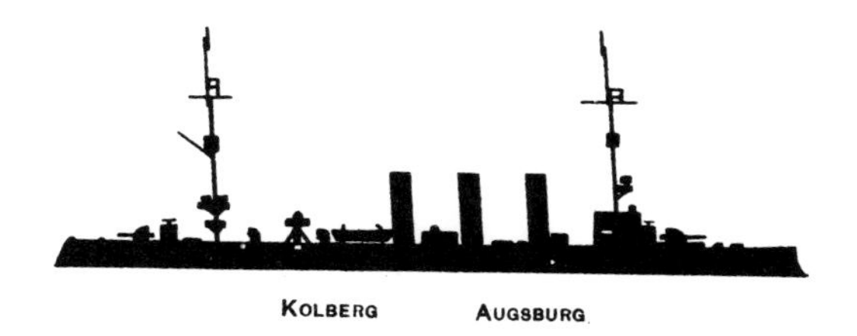

Kolberg Augsburg

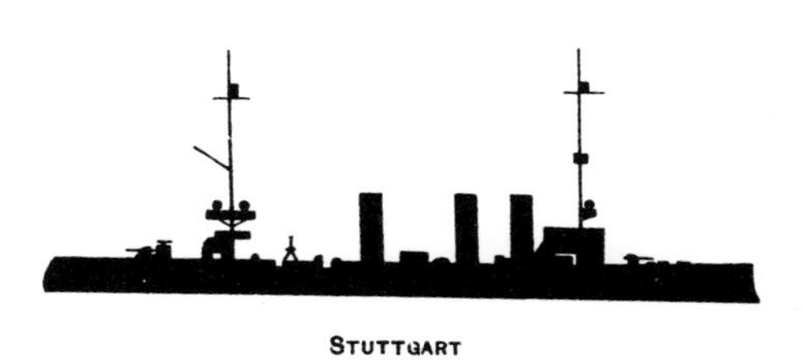

Stuttgart

Lübeck

Danzig

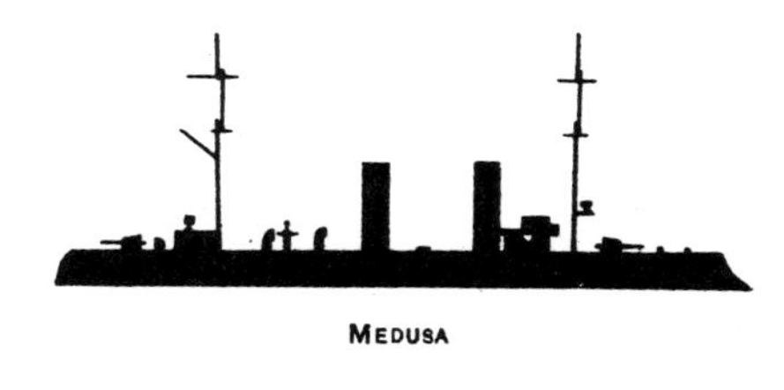

Medusa

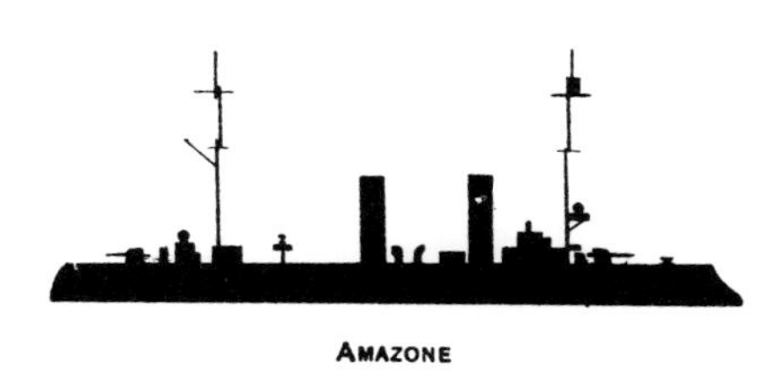

Amazone

Freya

Kaiserin Augusta

Ordnance Survey, July, 1918.

BRUMMER.
(From a Sketch.)
BREMSE similar.

KÖNIGSBERG.
(From a Sketch.)
KARLSRUHE, EMDEN and NÜRNBERG similar.

o (33) AS 4924—20 Pk 2972 1000 8/18 E & S

MINELAYING CRUISERS.

Brummer *and* Bremse.

(*See Plates* 1 *and* 6.)

Name.	Where Built or Building.	Laid Down.	Launched.	Commissioned for Trials.	Completed Trials.
Brummer - - -	Vulcan Yard, Stettin - - - - -	1914	1915	–.3.16	—
Bremse - - -	" " " - - - - -	1914	1915	1.7.16	—

General Remarks.—These are vessels of a new type, evolved during the war. They burn coal and oil (*not* oil only, as previously reported).

Though in all essentials light cruisers, they are here classed by themselves as "minelaying cruisers," because, though no details are available, it is known that they are adapted for minelaying to an extent which differentiates them entirely from previous types. Up to the present, however, the *Brummer* and *Bremse* have been employed almost exclusively on scouting and screening duties, working in a Scouting Group with ordinary light cruisers. The only offensive operation of importance which they are known to have performed is the raid of 17th October 1917 on the Scandinavian Convoy.

Further vessels of this class have been repeatedly reported to be under construction, but there is no evidence (June 1918) that any of them are nearing completion.

Complement.—War complement, about 480.

General Appearance.—*See* Plate 1 and Silhouette.

General Dimensions.—

Length, L.W.L. - - - - - -	*About* 430 ft. 0 ins.
Breadth - - - - - - - - -	" 41 " 0 "
Draught - - - - - - - - -	" 15 " 6 "
Displacement - - - - - - -	" 4,000 tons.

Armour.—There is no side armour and only a thin protective deck is fitted. The conning tower is before the fore bridge and entirely clear of it.

Armament.—

Guns.—

(For reported disposition and approximate arcs of training, *see* Plate 6.)

4—5·9-in. (15-cm.) Q.F. on the centre line.

2—22-pr. (8·8-cm.) semi-automatic anti-aircraft guns.

Torpedo Tubes.—

2—19·7-in. (50-cm.) above-water, one on either side amidships.

It has been reported that two further tubes were added in 1917, one on either side just abaft the foremast.

Mines.—

Rails are fitted on both sides and on the centre line on the upper deck and main deck for a total of about 360 mines, but mines are not stowed permanently on board.

Searchlights.—

4—two on fore side of foremast arranged vertically, two on mainmast placed in fore and aft line.

Machinery and Boilers.—

Main Engines.—Turbines.

Boilers.—Eight ; two burning coal, six burning oil.

A 2

Brummer *and* **Bremse**—*cont.*

Speed, Horse-Power, and Fuel.—

Speed (designed).	Horse-Power (designed).	Fuel.
Knots. *About* 34·0	*About* 46,000	Coal and oil.

LIGHT CRUISERS.

Details of light cruisers later than the "Strassburg" and "Stralsund" are incomplete in several respects, but those two vessels (see p. 10) may be taken generally as typical of modern German light cruisers.

Cöln, Dresden, Wiesbaden, (?) Leipzig, (?) Magdeburg, *and* Bremen.

Name.	Designation before Launch.	Programme Year.	Where Built or Building.	Ordered.	Laid Down.	Launched.	Commissioned for Trials.	Completed Trials.
Cöln -	Ersatz *Ariadne*	1916-17	Blohm & Voss Yard, Hamburg.	—	1915	*–.10.16	*–.11.17	*–.4.18
Dresden -	Ersatz *Amazone*	1916–17	Howaldt's Yard, Kiel.	—	1915	*–.4.16	*–.12.17	*–.6.18
Wiesbaden	Ersatz *Medusa*	1917–18	Weser Yard, Bremen.	—	1916	*1917	—	—
(?)*Leipzig*	*P.* - - -	1917–18	" "	—	1916	—	—	—
(?)*Magdeburg.*	*Q.* - - -	†	Blohm & Voss Yard, Hamburg.	—	1916	—	—	—
Bremen -	*R.* - - -	†	Weser Yard, Bremen.	—	1916	—	—	—

* Approximate dates.
† Under the pre-war programme of construction, these two vessels might be laid down at any time between 1914 and 1917.

General Remarks.—These vessels are being given the names of light cruisers lost during the war. It is believed that they will prove to be generally similar to the new *Königsberg* class (*see* below), but no details are yet available.

Cöln and *Dresden* are believed to have recently joined the Fleet (June 1918), and two further vessels of the class should be completed during the latter half of 1918.

Königsberg, Karlsruhe, Emden, *and* Nürnberg.

(*See Plates* 1 *and* 7.)

Name.	Designation before Launch.	Programme Year.	Where Built.	Ordered.	Laid Down.	Launched.	Commissioned for Trials.	Completed Trials.
Königsberg	Ersatz *Gazelle*	1914–15	Weser Yard, Bremen.	*11.6.14	1914	*28.12.15	*13.8.16	*–.10.16
Karlsruhe	Ersatz *Niobe* -	1914–15	Imperial Dockyard, Kiel.	*11.6.14	1914	*–.6.16	*–.11.16	*–.2.17
Emden -	Ersatz *Nymphe*	1915–16	Weser Yard, Bremen.	—	1915	*–.2.16	*–.8.16	*–.1.17
Nürnberg	Ersatz *Thetis*	1915–16.	Howaldt's Yard, Kiel.	—	1915	*20.1.16	*–.11.16	*–.5.17

* Approximate dates.

Königsberg, Karlsruhe, Emden, *and* Nürnberg—*cont.*

General Remarks.—These vessels, which have been given the names of light cruisers lost during the war, are believed to be somewhat smaller than the *Frankfurt* (*see* below), but in other respects to represent an improvement on her. They carry one 5·9-in. gun less, but can bring the same number of guns to bear on the broadside.

Complement.—War complement, about 500.

Accommodation.—Stated to be exceptionally good.

General Appearance.—*See* Plate 1 and Silhouette.

General Dimensions.—

Length, L.W.L. - - - - - -	*About* 450 ft. 0 ins.
Breadth - - - - -	„ 45 „ 0 „
Draught, designed load - - - - -	„ 16 „ 0 „
Displacement, designed load - - -	„ 4,200 tons.

Armour.—No details are yet available, but it may be assumed that a practically complete belt and armoured deck are fitted.

There are two combined conning and control towers, one on forecastle before fore bridge, one on after superstructure, abaft mainmast.

Shields are fitted to all guns.

Armament.—

Guns.—

(For reported disposition and approximate arcs of training, *see* Plate **7**.)

7—5·9-in. (15-cm.) Q.F.
3—22-pr. (8·8-cm.) semi-automatic anti-aircraft guns.
2—Machine guns.

Torpedo Tubes.—

2—19·7-in. (50-cm.) above water, one on either side amidships.
2—19·7-in. (50-cm.) submerged, fitted with gyro angling gear which can be set for every 15° from 30° before to 60° abaft the direction of the tube.

Mines.—There is no permanent stowage for mines.

Searchlights.—

4—two on fore side of foremast, two on after side of mainmast, arranged vertically in each case.

Speed, Horse-Power, and Fuel.—

Speed (designed).	Horse-Power (designed).	Fuel: (*a*) Coal. (*b*) Oil.
Knots. *About* 28·5	—	Coal and oil.

Frankfurt.

(*See Plates* 2 *and* 8.)

Programme year - - - - - -	1913–14.
Designation before launch - - - -	Ersatz *Hela*.
Where built - - - - - - -	Imperial Dockyard, Kiel.
Laid down - - - - - - -	1st December 1913.
Launched - - - - - - -	*About* March 1915.
First commissioned - - - - - -	„ August 1915.

General Remarks.—The *Frankfurt's* sister ship, the *Wiesbaden*, was sunk in action on 31st May–1st June 1916.

These two vessels were the first German light cruisers actually designed to carry 5·9-in. guns, although earlier classes have now been rearmed with guns of that calibre. The increase in armament was accompanied by a marked increase in beam. The armour protection also is believed to represent a considerable advance on *Graudenz*.

A 3

Frankfurt—*cont.*

Complement.—War complement, about 500.

General Appearance.—
See Plate 2 and Silhouette.

General Dimensions.—

Length, L.W.L. - - - -	*about* 465 ft. 0 ins.
Breadth - - - - -	51 „ 4 „
Draught, designed load - - -	17 „ 0 „
Displacement, designed load - - -	5,120 tons
Height of fore funnel to L.W.L. - -	59 ft.
„ „ lower masthead to L.W.L. -	96 „
„ „ truck to L.W.L. - -	*about* 135 „

Armour (doubtful).—

Belt.—Nearly complete, about 4″ thick amidships, 2·5″ at ends, and about 10 ft. wide.

Conning and Control Towers.—A combined conning and control tower is fitted on the forecastle, just before fore bridge, and a second one on the after superstructure, abaft mainmast; thickness of armour in each case probably about 4″.

Gun Shields.—About 5″ in front.

Horizontal.—Armoured deck, about 1″ on flat and 1·5″ on slopes.

Armament.

Guns.—
(For disposition and approximate arcs of training, *see* Plate 8.)
8—5·9-in. (15-cm.) Q.F.
2—22-pr. (8·8 cm.) semi-automatic anti-aircraft guns.

Torpedo Tubes.—
2—19·7-in. (50-cm.) above-water, one on either side between the first and second funnels.
2—19·7-in. (50-cm.), submerged; fitted with gyro angling gear which can be set for every 15° from 30° before to 60° abaft the direction of the tube.

Mines.—There is no permanent stowage for mines, but 120 can be carried if required.

Searchlights.—
4—two on fore side of foremast, two on after side of mainmast, arranged vertically in each case.

Auxiliary Machinery.—
Turbo-fans are fitted.

Speed, Horse-Power, and Fuel.

Ship.	Speed (designed).	Horse-Power (designed).	Fuel: (*a*) Coal. (*b*) Oil.
Frankfurt - - - - - - - -	Knots. 28·0	—	(*a*) 1,500 (?) (*b*) Some.

Pillau.

(*See Plates* 2 *and* 9.)

Name.	Name under which building for Russia.	Where Built.	Laid Down.	Launched.	First Commissioned.
Pillau - -	Muravev Amurski -	Schichau Works, Danzig.	12.4.13	11.4.14	*14.12.14

* Approximate date.

FRANKFURT.

PILLAU.

GRAUDENZ.

REGENSBURG

Pillau—*cont.*

General Remarks.—The *Pillau* was one of two ships originally building for Russia but commandeered by Germany on the outbreak of war in August 1914.

Her sister ship, the *Elbing*, was sunk in action on 31st May–1st June 1916.

These two vessels were originally designed to carry 8—5·1-in. guns, but 5·9-in. guns were substituted after they had been taken over by Germany. At the same time the funnels were made considerably taller, the foremost funnel being raised more than the other two.

The *Pillau* has proved a very successful ship, always steaming well and being a good sea boat.

Complement.—Peace complement, 372; war complement, about 450.

Accommodation.—The officers' accommodation is excellent, the men's very poor.

General Appearance.—*See* Plate 2 and Silhouette.

General Dimensions, &c.—

Length, L.W.L. - - - - -	- 440 ft. 11 ins.
Breadth - - - - - -	- 46 „ 0 „
Draught, designed load - - -	- 18 „ 11 „
Displacement, designed load - -	- 4,320 tons.
Height of fore funnel to L.W.L. - -	- 60 ft.
„ „ lower masthead to L.W.L. -	- 85 „
„ „ truck to L.W.L. - -	*about* 115 ft.

Armour.—

Material.—Krupp cemented.

Belt.—None.

Conning Tower.—3″ thick. Situated under fore bridge.

Horizontal.—Armoured deck, ·8″ on flat and 1·6″ on slopes, except over tiller compartment, where it is 3·1″ thick.

Transverse bulkhead forward and glacis aft, where deck drops to a lower level, 3·1″ thick.

Armament.—

Guns.—

(For disposition and approximate arcs of training *see* Plate 9.)

8—5·9-in. (15-cm.) Q.F.

2—22-pr. (8·8-cm.) semi-automatic anti-aircraft guns.

Torpedo Tubes.—

2—19·7-in. (50-cm.) above-water, one on either side amidships, training 60° before to 60° abaft the beam.

Mines.—

Rails for about 200 mines were originally fitted on upper deck, but were removed shortly before commissioning. The vessel does not appear ever to have been used for minelaying, and no mines are stowed permanently on board.

Searchlights.—

4—two on foremast abreast, two on mainmast placed in fore-and-aft line.*

Machinery and Boilers.—

Main Engines.—

Two sets, in four watertight compartments.

Type—Schichau turbines.

Boilers.—Ten in number; four for oil-firing only, six for coal and oil.

Propellers.—Two.

Speed, Horse-Power, and Fuel.—

Ship.	Speed (designed).	Horse-Power (designed).	Fuel: (*a*) Coal. (*b*) Oil.
Pillau - - - - - - - - -	Knots. 27·5	27,400	(*a*) 984 (*b*) 250

* Very possibly this disposition may now (June 1918) have been altered to that adopted in the *Frankfurt* (*see* page 6) and other modern light cruisers.

A 4

Pillau—*cont.*

Endurance.—

—	Speed.	Horse-Power.	Daily Consumption.	Radius of Action.	Remarks.
	Knots.		Tons.	Miles.	
At $\frac{4}{5}$ths designed power - -	25·6	21,920	446	1,750	
At maximum continuous seagoing speed.	25·0	19,180	400	1,900	
At $\frac{3}{5}$ths designed power - -	24·2	16,440	352	2,100	
At $\frac{2}{5}$ths designed power - -	22·0	10,960	258	2,600	
At $\frac{1}{5}$th designed power - -	18·0	5,480	164	3,340	
At 10 knots - - - - -	10·0	950	46	6,600	

Graudenz *and* Regensburg.

(*See Plates* 3, 10, *and* 11.)

Name.	Designation before Launch.	Pro-gramme Year.	Where Built.	Ordered.	Laid Down.	Launched.	Com-missioned for Trials.	Completed Trials.
Graudenz -	Ersatz *Prinzess Wilhelm.*	1912–13	Imperial Dockyard, Kiel.	*–.4.12	23.10.12	25.10.13	*–.8.14	*–.10.14
Regensburg -	Ersatz *Irene* -	1912–13	Weser Yard, Bremen	*–.5.12	1912	25.4.14	*–.12.14	*–.3.15

* Approximate dates.

Cost.—

	£
Hull, machinery, &c. - - - - - -	318,003
Gun armament - - - - - -	73,386
Torpedo armament - - - - -	26,418
Total - - - - -	£417,807

General Remarks.—These vessels generally resemble *Strassburg* and *Stralsund* (*see* p. 10), but have three instead of four funnels. They were rearmed with 5·9-in. in place of 4·1-in. guns and fitted with deck torpedo tubes in 1915–16.

Complement.—Peace complement, 364 ; war complement, about 480.

Accommodation.—The men are berthed on two mess decks extending from the bow to the bridge. The officers' quarters are in the after superstructure.

General Appearance.—*See* Plate 3 and Silhouettes. It should be noted that in both ships the three funnels are equally spaced, but in *Regensburg* they are midway between the masts, whereas in *Graudenz* they are nearer the foremast. In *Graudenz* the two foremost broadside guns are mounted on the forecastle, abreast the bridge, whilst in *Regensburg*, when last reported, they were down in the waist (where the 4·1-in. guns used to be).

General Dimensions, &c.—

Length, L.W.L. - - - - - -	456 ft. 0 ins.
Breadth, extreme - - - -	*about* 45 ,, 0 ,,
Draught, designed load - - - -	17 ,, 0 ,,
Displacement, designed load - - -	4,842 tons.
Height of fore funnel to L.W.L. - -	59 ft.
,, ,, lower masthead to L.W.L.	115 ,,
,, ,, truck to L.W.L. { *Graudenz* -	*about* 155 ,,
,, ,, truck to L.W.L. { *Regensburg* -	,, 150 ,,

Armour (doubtful).—

Belt.—Nearly complete, about 4″ thick amidships, 2·5″ at ends, and about 10 ft. wide amidships.

Conning and Control Towers.—A combined conning and control tower is fitted on the forecastle, just before fore bridge, and a second one on the after superstructure, abaft mainmast ; thickness of armour in each case probably about 4″.

Horizontal.—Armoured deck about 1″ on flat, 2″ on slopes.

Graudenz *and* Regensburg—*cont.*

Armament.—

Guns.—

(For disposition and approximate arcs of training *see* Plates 10 and 11.)

*__7__—5·9-in. (15-cm.) Q.F. (disposition differs slightly in the two vessels, as mentioned under General Appearance).

2—22-pr. (8·8-cm.) semi-automatic anti-aircraft guns.

2—Machine guns.

Torpedo Tubes.—

†**2**—19·7-in. (50-cm.) above-water, one on either side amidships, probably training 60° before to 60° abaft the beam.

Mines.—There is no permanent stowage for mines.

Searchlights.—

4—two on fore side of foremast, two on after side of mainmast arranged vertically in each case.

Compasses.—The master gyro compass is placed below the armour deck forward. The standard magnetic compass is fitted on a platform abaft the funnels.

Machinery and Boilers.—

Main Engines.—Turbines, 2 sets, Navy type. In separate watertight compartments divided by an unpierced longitudinal bulkhead.

Boilers.—12 Schulz, Navy type: two oil-burning in foremost boiler room; 10 mixed fuel in two rooms containing four boilers each and one containing two. Working pressure, 256 lbs. per sq. in.

Propellers.—Two, three-bladed.

Auxiliary Machinery.—

Includes three dynamos, one in auxiliary engine room and one on deck over main engine room.

The two auxiliary engine rooms are abaft the main engine rooms.

Speed, Horse-Power, and Fuel.—

Ship.	Speed (designed).	Horse-Power (designed).	Fuel: (*a*) Coal. (*b*) Oil.
	Knots.		
Graudenz	27·5	26,000	(*a*) 1,279
Regensburg			(*b*) 220

Endurance.—

—	Speed.	Horse-Power.	Daily Consumption.	Radius of Action.	Remarks.
	Knots.		Tons.	Miles.	
At ⅘ths designed power	25·5	20,800	420	2,300	
At maximum continuous seagoing speed.	24·8	18,200	380	2,480	
At ⅗ths designed power	23·9	15,600	334	2,720	
At ⅖ths designed power	21·7	10,400	245	3,360	
At ⅕th designed power	18·0	5,200	156	4,400	
At 10 knots	10·0	1,000	48	7,900	

* The original primary armament of these vessels was 12—4·1-in.

† Two 19·7-in. (50-cm.) submerged tubes were originally fitted, but subsequently removed.

O AS 4924—2 B

Strassburg *and* Stralsund.

(*See Plates 4 and 12.*)

Name.	Designation before Launch.	Pro-gramme Year.	Where Built.	Ordered.	Laid Down.	Launched.	Com-missioned for Trials.	Completed Trials.
Strassburg -	Ersatz *Condor* -	1910–11	Imperial Dockyard, Wilhelmshaven.	-.4.10	—	24.8.11	1.10.12	23.12.12
Stralsund -	Ersatz *Cormoran*	1910–11	Weser Yard, Bremen -	-.4.10	-.10.10	4.11.11	10.12.12	15.2.13

Cost.—

	£
Hull, machinery, &c. - - - - - -	318,003
Gun armament - - - - - - -	73,386
Torpedo armament - - - - - -	24,950
Total - - - - - -	£416,339

General Remarks.—These were the first German war vessels to be built on the Isherwood (longitudinal) system. They are now the only German light cruisers with four funnels. The lines of the hull are much finer than in previous classes, and there is no poop. The special feature of the class, however, was the introduction of vertical side armour (for a portion of the ship's length).

In 1916 the *Stralsund* was rearmed and fitted with deck torpedo tubes. The *Strassburg* was similarly altered in 1917.

These vessels have been fitted during the war with a new form of fire control; a special control tower has been added on the top of the original fore conning tower and an after combined conning and control tower has also been added.

Of the other two vessels of this class, the *Breslau* was sold to Turkey on 11th August 1914 and re-named *Midilli*, and was sunk by mine off Imbros 20th January 1918; whilst the *Magdeburg* was destroyed in the Gulf of Finland on 20th August 1914.

Complement.—Peace complement: *Strassburg*, 370; *Stralsund*, 373. War complement, about 450.

General Appearance.—*See* Plate 4 and Silhouettes.

General Dimensions.—

Length, L.W.L. - - - - - -	446 ft. 2 ins.
Breadth, extreme - - - - - -	43 ,, 7 ,,
Draught, designed load - - - - -	15 ,, 9 ,,
Displacement, designed load - - -	4,480 tons.
Height of fore funnel to L.W.L. - - *about*	59 ft.
Height of fore lower masthead to L.W.L. - -	96 ,,
Height of fore truck to L.W.L. - - -	138 ,,

Armour.—

Material.—Krupp.

Belt.—3·9″ thick amidships, 2·4″ at ends. The belt is narrow—probably about 6 ft. in width amidships. It extends from about 20 ft. from bow to within about 65 ft. of stern.

Fore Conning Tower.—Oval in shape, about 13 ft. by 9 ft.; 3·9″ thick, except in rear, where it is 3·1″. Observation slits, about 12 ins. by 2 ins., are cut in the armour. The fore conning tower is on forecastle, just before fore-bridge. Access is obtained either through an armoured door on the after side, or from the control tower above, or through the armoured tube leading to the transmitting station below.

Fore Control Tower.—Above fore conning tower. Circular in shape, about 8 ft. in diameter, with a flat roof; sides probably 3·9″, roof ·8″ thick. Observation slits

Strassburg *and* Stralsund—*cont.*

Armour—*continued.*

as in fore conning tower. In the floor of the control tower there is a circular hole, giving access to the fore conning tower by means of a ladder. This hole is fitted with two light semi-circular armoured doors and is ordinarily closed by a grating. Direct access to the control tower is given by a hinged armoured door, reached by a light gangway from fore-bridge.

After Conning and Control Tower.—On superstructure deck, abaft mainmast Circular in shape, about 8 ft. in diameter and 3·9″ thick. Access is given by a hinged armoured door.

Communication Tube.—A circular armoured tube, about 2 ft. in diameter and 3″ thick, leads from the fore conning tower to the transmitting station.

Horizontal.—Armoured deck, ·8″ on flat and 2″ on slopes.

Armament.—

Guns and Ammunition Supply.—

(For disposition and approximate arcs of training, *see* Plate 12.)

***7**—5·9-in. (15-cm.) Q.F., L/45, model 1916.

The ammunition supply is by whip through tubular hoists from the handing rooms and shell rooms. Electric bollards are fitted for working the whips. A ready supply of six complete rounds is kept at each gun.

2—22-pr. (8·8-cm.) semi-automatic anti-aircraft guns.

2—Machine guns.

Magazines and Stowage.—There are three magazines, which are situated *below* the shell rooms. For the 5·9-in. guns about 200 rounds per gun are carried, more than half the supply of shell being H.E.A.P. and the remainder H.E. and star shell. Both natural and artificial flooding arrangements are fitted to the magazines.

Torpedo Tubes.—

2—19·7-in. (50-cm.) above-water, one on either side amidships, probably training 60° before to 60° abaft the beam.

†**2**—19·7-in. (50-cm.) submerged, fitted with gyro angling gear which can be set for every 15° from 30° before to 60° abaft the direction of the tube.

Mines.—There is no permanent stowage for mines, but 120 can be carried if required. A small mine shute is fitted on either quarter.

Searchlights.—

4—probably 43·3-in., two on fore side of foremast and two on after side of mainmast, arranged vertically in each case. Mechanical control is fitted.

Range-finders.—Range-finders are permanently fitted on fore control tower, on after conning and control tower, and on engine room casing between the third and fourth funnels. They are believed to be 10-ft. (3-m.) instruments, but the fore control tower range-finder may be larger.

Fire Control.—The system in use is believed to consist of (1) a training director, and (2) sights kept in adjustment by follow-the-pointer gear; both (1) and (2) being operated through the transmitting station from either control tower.

Compasses.—A gyro installation was fitted to these vessels in 1916, the master compass being in the transmitting station below the armoured deck and receivers being fitted in both conning towers.

Machinery and Boilers.—

Main Engines.—

Strassburg.—Parsons turbines (modified) on two shafts.

Stralsund.—Bergmann turbines on three shafts.

Boilers.—16 Schulz, Navy type.

Propellers.—*Strassburg.*—Two. *Stralsund.*—Three.

* The original primary armament of these vessels was 12—4·1-in.

† These two submerged tubes may possibly have been removed (June 1918)

Speed, Horse-Power, and Fuel.—

Ship.	Speed (designed).	Horse-Power (designed).	Fuel : (*a*) Coal. (*b*) Oil.
	Knots.		
Strassburg - - - - - - - - } *Stralsund* - - - - - - - - }	26·75	24,200	(*a*) 1,181. (*b*) Some.

Steam Trials.—

Ship.	Nature of Trial.	Date.	Speed.	Horse-Power.	Revolutions.	Remarks.
Strassburg - -	Measured mile -	—.12.12	28·28	33,742	—	
	6 hours forced draught.	—.12.12	26·9	25,647	305	
Stralsund - -	Measured mile -	—.1.13	28·27	35,515	—	
	6 hours forced draught.	—.1.13	26·95	27,032	397	

Endurance.—

—	Speed.	Horse-Power.	Daily Consumption.	Radius of Action.	Remarks.
	Knots.		Tons.	Miles.	
At $\frac{4}{5}$ths designed power - -	25·1	19,360	394	1,715	
At maximum continuous seagoing speed.	24·4	16,940	354	1,850	
At $\frac{3}{5}$ths designed power - -	23·5	14,520	311	2,030	
At $\frac{2}{5}$ths designed power - -	21·3	9,680	228	2,515	
At $\frac{1}{5}$th designed power - -	17·6	4,840	145	3,290	
At 10 knots - - - - -	10·0	950	46	5,850	

N.B.—Radius of action calculated on coal only.

STRASSBURG.
STRALSUND similar.

KOLBERG.
AUGSBURG similar.

C.B. 1182Q.
July, 1918.

STUTTGART.
STETTIN similar.

ARCONA.

o AS 4924--20

Kolberg *and* Augsburg.

(*See Plates* 4 *and* 13.)

Name.	Designation before Launch.	Programme Year.	Where Built.	Ordered.	Laid Down.	Launched.	Commissioned for Trials.	Completed Trials.
Kolberg -	Ersatz *Greif* -	1907–08	Schichau Works, Danzig	–.10.07	–.–.07	14.11.08	21.6.10	?
Augsburg -	Ersatz *Sperber* -	1908–09	Imperial Dockyard, Kiel.	–.4.08	22.8.08	10.7.09	1.11.10	?

Cost. —

	Kolberg.	*Augsburg.*
	£	£
Hull, machinery, &c. - -	293,541	318,003
Gun armament - -	68,492	73,386
Torpedo armament - -	18,835	26,418
Total - -	£380,868	£417,807

Complement.—Peace complement, 379; war complement, about 440.

General Appearance.—*See* Plate 4 and Silhouette.

General Dimensions.—

Length { between perpendiculars - - -		401 ft. 11 ins.
Length { L.W.L. - - - - -		426 „ 6 „
Breadth, extreme - - - - -		45 „ 11 „
Draught, designed, load - - - -		16 „ 5 „
„ maximum load - - - -		18 „ 4 „
Displacement, designed load - - - -		4,280 tons.
Height of forecastle to L.W.L. - -	about	21 ft. 0 ins.
„ poop to L.W.L. - - - -		18 „ 4 „
„ fore bridge to L.W.L. - - -		31 „ 6 „
„ „ compass position to L.W.L. - -		39 „ 4 „
„ foremost searchlights to L.W.L. - -		55 „ 9 „
„ funnels to L.W.L. - - - -		59 „ 0 „
„ fore crosstrees to L.W.L. - - -		67 „ 3 „
„ „ lower masthead to L.W.L. - -		102 „ 0 „
„ „ topmast head to L.W.L. - -		123 „ 0 „
„ „ truck to L.W.L. - -	*about*	145 „ 0 „

Armour.—

Conning and Control Towers.—Two combined conning and control towers are fitted: one on the forecastle, close before fore bridge, thickness 3·9″, except in rear, where it is 3·1″; one on the poop, just before the after gun, thickness probably similar.

Horizontal.—Armoured deck, ·8″ on flat and 2″ on slopes. Transverse bulkhead fitted forward, and glacis aft, where deck drops to a lower level.

Armament.*—

Guns and Ammunition Supply.—

(For disposition and approximate arcs of training *see* Plate 13.)

6—5·9-in. (15-cm.) Q.F.
2—22-pr. (8·8-cm.) semi-automatic anti-aircraft guns.
2—Machine guns.

* Rearmed in 1916. Original armament was 12—4·1-inch, 2—machine guns, 1—7-pr. boat and field gun, 2 submerged tubes (17·7″).

B 3

Kolberg *and* Augsburg—*cont.*

Armament—*cont.*

Torpedo Tubes.—

2 above-water tubes, probably 19·7-in. (50-cm.), one on either side amidships and probably training 60° before to 60° abaft the beam.

Mines.—There is no permanent stowage for mines, but 120 can be carried if required.

Searchlights.—

4—43·3-in. Two on foremast abreast, and two on mainmast placed in fore-and-aft line.

Machinery and Boilers.—

Main Engines.—
Kolberg—Melms and Pfenninger turbines. *Augsburg*—Parsons turbines.

Boilers.—
Fifteen—Schulz, Navy type.

Propellers.—
Kolberg, two. *Augsburg*, four.

Speed, Horse-Power, and Fuel.

Ship.	Speed (designed).	Horse-Power (designed).	Fuel: (*a*) Coal. (*b*) Oil.
	Knots.		
Kolberg / *Augsburg*	25	19,600	(*a*) 890. (*b*) Some.

Steam Trials.—

Ship.	Nature of Trial.	Date.	Speed.	Horse-Power.	Revolutions.	Remarks.
Kolberg	Full power	—.5.11	26·3	30,400	—	After repair.
Augsburg	Measured mile	—.3.11	27·01	31,340	516	Maximum.

Endurance.—

—	Speed.	Horse-Power.	Daily Consumption.	Radius of Action.	Remarks.
	Knots.		Tons.	Miles.	
At $\frac{4}{5}$ths designed power	23·6	16,000	324	1,470	
At maximum continuous seagoing speed.	23·0	14,000	300	1,540	
At $\frac{3}{5}$ths designed power	22·0	12,000	258	1,690	
At $\frac{2}{5}$ths designed power	19·8	8,000	214	1,900	
At $\frac{1}{5}$th designed power	16·2	4,000	128	2,560	
At 10 knots	10·0	950	46	4,400	

N.B.—Radius of action calculated on coal only.

Stuttgart *and* **Stettin.**

(*See Plates* 5 *and* 14.)

Name.	Designation before Launch.	Programme Year.	Where Built.	Laid Down.	Launched.	First Commissioned.	Completed Trials.
Stuttgart -	O. - -	1905–06	Imperial Dockyard, Danzig	–.5.05	22.9.06	1.2.08	9.4.08
Stettin -	Ersatz *Wacht*	1905–06	Vulcan Works, Stettin -	–.12.05	7.3.07	29.10.07	1.2.08

Cost.—

	£
Hull, machinery, &c. - - - - - -	232,387
Gun armament - - - - - -	66,047
Torpedo armament - - - - -	13,699
Total - - - - - -	£312,133

General Remarks.—These ships are an improved *Hamburg* type. *Stettin* differs very slightly from *Stuttgart*. A noticeable feature of both ships is that the centre funnel is closer to the foremost than to the aftermost funnel.

Of the other two vessels of this class, the *Königsberg* was destroyed in the Rufigi river on 11th July 1915, and the *Nürnberg* was sunk off the Falkland Islands on 8th December 1914.

Complement.—Peace complement: *Stuttgart*, 322; *Stettin*, 328. War complement, about 390.

General Appearance.—*See* Plate 5 and Silhouettes.

General Dimensions, &c.—

——	*Stuttgart.*	*Stettin.*
Length between perpendiculars -	360 ft. 11 ins.	363 ft. 10 ins.
„ , L.W.L. - - -	383 „ 2 „	——
Breadth, extreme - - -	43 „ 8 „	43 ft. 8 ins.
Draught, designed load - -	15 „ 9 „	15 „ 9 „
Displacement, designed load -	3,400 tons	3,494 tons.

Height of fore funnel to L.W.L. - - - 59 ft. 0 ins.
„ „ lower masthead to L.W.L. - - 80 „ 0 „
„ „ truck to L.W.L. - - *about* 140 „ 0 „

Armour.—

Conning and Control Towers.—Two combined conning and control towers are fitted: one on the forecastle under fore bridge, thickness 3·9″ (3·1″ in rear), ·8″ on top; one on the poop, just before the two after guns, thickness probably similar.

Communication Tube.—1·2″.

Gun Shields.—1·9″.

Horizontal.—Complete armoured deck.— Amidships, ·8″ on the flat, 1·8″ on the slopes. Transverse bulkhead forward and glacis aft, where deck drops to a lower level, 3·1″. Deck forward, before transverse bulkhead, 1·4″ to ·8″. Deck aft, abaft after glacis, 1″. Glacis over engines, 3·9″.

Armament.—

Guns and Ammunition Supply.—
(For disposition and arcs of training *see* Plate 14.)
10—4·1-in. (10·5-cm.) Q.F., L/40, on C.P. Mounting C/04.
2—·31-in. (8-mm.) Maxim automatic.
Overhead rails run along the upper deck each side for transport of ammunition, which comes up a double tube under the forecastle and poop.

B 4

Stuttgart *and* **Stettin**—*cont.*

Armament—*cont.*

Magazines and Stowage.—
4·1-in.—100 rounds per gun.

Torpedo Tubes.—
2—17·7-in. submerged ; one each side between 2nd and 3rd funnels.

Torpedoes.—Four per tube.

Mines.—There is no permanent stowage for mines, but 108 can be carried if required.

Searchlights.—
4—two on foremast abreast, and two on mainmast placed in fore-and-aft line.

Machinery and Boilers.—

Main Engines.—

Stuttgart.—
Two sets, in separate engine rooms.
Type—Vertical, four cylinder, triple expansion.
Revolutions at full power, 145 (designed).

Stettin.—
Four sets.
Type—Parsons turbines.
Revolutions at full power, 540 (designed).
Working pressure at turbines, 175 lbs. per sq. in.

Boilers (both ships).—
10 Schulz, in five boiler rooms.
H.S. : *Stuttgart,* 40,000 sq. ft. ; *Stettin,* 41,000 sq. ft. G.S., 644 sq. ft.
Working pressure, 220 lbs. per sq. in.

Auxiliary Machinery includes—
Three dynamos—two in engine-room and one on deck.

Propellers.—
Stuttgart.—Two—three-bladed.
Stettin.—Four—three-bladed. Diameter, 6 ft. 3 ins. ; pitch, 5 ft. 7 ins.

Speed, Horse-Power, and Fuel.—

Ship.	Speed (designed).	Horse-Power (designed).	Fuel : (*a*) Coal. (*b*) Oil.
	Knots.		
Stuttgart - - - - - - -	23·5	13,000	(*a*) 866
Stettin - - - - - - - -	23·5	13,400	(*a*) 866

Stuttgart *and* Stettin—*cont.*

Steam Trials.—

Ship.	Nature of Trial.	Date.	Speed.	Horse-Power.	Revolutions.	Remarks.
Stuttgart - - -	—	—	23·7	13,745	—	
Stettin - - -	Measured mile -	—.1.08	25·17	21,600	584	33 fms. at Neukrug.
" - - -	6 hours full power	"	23·96	15,448	524	Coal per hour, turbines only :—
" - - -	24 hours - -	"	20·2	7,000	415	5·9 tons.
" - - -	" - -	"	12·0	1,344	243	1·788 tons.
" - - -	" - -	"	12·0	1,403	243	1·564 "
" - - -	Coal consumption trial.	"	17·0	4,195	344	3·838 "

The times and distances in which the *Stettin* was brought to a standstill with the astern turbines working at full power are given in the following table :—

Number of Boilers.	Speed.	Distance.	Time.
	Knots.	Feet.	Min. Secs.
11	24	1,410	1 7
11	20	1,190	1 6½
8	16	770	1 3
6	11	590	57
5	9	500	57½
3	5	260	56½

Endurance.

—	Speed.	Horse-Power.	Daily Consumption.	Radius of Action.	Remarks.
	Knots.		Tons.	Miles.	
At 4/5ths designed power - -	22·3	10,400	245	1,800	*Stuttgart.*
	22·6	10,800	221	2,020	*Stettin.*
At maximum continuous seagoing speed.	21·5	8,800	207	2,045	*Stuttgart.*
	21·9	9,500	204	2,130	*Stettin.*
At 3/5ths designed power - -	20·8	7,800	192	2,130	*Stuttgart.*
	21·2	8,160	175	2,220	*Stettin.*
At 2/5ths designed power - -	18·7	5,200	130	2,840	*Stuttgart.*
	18·8	5,440	146	2,550	*Stettin.*
At 1/5th designed power - -	14·7	2,600	75	3,860	*Stuttgart.*
	14·8	2,720	87	3,340	*Stettin.*
At 10 knots - - - - -	10·0	845	33	6,100	*Stuttgart.*
	10·0	860	41	4,750	*Stettin.*

Hamburg, Berlin, München, Lübeck, *and* Danzig.

(*See Plate* 15.)

Name.	Designation before Launch.	Programme Year.	Where Built.	Laid Down.	Launched.	First Commissioned.
Hamburg -	K. - -	1902–03	Vulcan Works, Stettin -	1.8.02	25.7.03	8.3.04
Berlin -	Ersatz *Zieten*	1902–03	Imperial Dockyard, Danzig	26.8.02	22.9.03	4.4.05
München -	M. - -	1903–04	Weser Yard, Bremen -	18.8.03	30.4.04	10.1.05
Lübeck -	Ersatz *Merkur*	1903–04	Vulcan Works, Stettin -	12.5.03	26.3.04	26.4.05
Danzig -	Ersatz *Alexandrine.*	1904–05	Imperial Dockyard, Danzig	–.8.04	23.9.05	1.2.07

Hamburg, Berlin, München, Lübeck, *and* Danzig—*cont.*

General Remarks.—*Lübeck* and *Danzig* are now (June 1918) attached to the Submarine School as target ships; *Hamburg* is used as flagship of the Commodore (S); *München* was torpedoed on 19th October 1916, but reached port, and, although a considerable time under repair, does not appear to have been recommissioned.

Lübeck is fitted with turbines, whereas the remaining ships have reciprocating engines. *Hamburg* was re-boilered in 1910, and underwent alterations to fit her as parent ship of submarines. *München* is fitted with submarine sound signalling apparatus.

Of the two other vessels of this type, the *Leipzig* was sunk off the Falkland Islands on 8th December 1914 and the *Bremen* was sunk in the Baltic on 17th December 1915.

Complement.—Peace complement, 303; war complement, about 350.

General Appearance.—*See* Silhouettes. The *Danzig* differs from the other vessels of the type in having her conning tower before and her bridge round foremast. The other vessels have both conning tower and bridge abaft foremast.

General Dimensions, &c.—

Length between perpendiculars	340 ft. 6 ins.
Length, L.W.L.	362 „ 7 „
Breadth, extreme	43 „ 4 „
Draught, designed load	16 „ 5 „
Displacement, designed load	3,200 tons.
Height of fore funnel to L.W.L.	59 ft. 0 ins.
„ „ crosstrees to L.W.L.	71 „ 0 „
„ „ truck	*about* 135 „ 0 „

Armour.—

Conning Tower.—3·9″ thick. Dimensions about 5 ft. 6 ins. by 9 ft.

Communication Tube.—1·2″.

Gun Shields.—1·9″.

Horizontal.—Complete armoured deck. Amidships, ·8″ on the flat, 1·9″ on the slopes. Transverse bulkhead forward and glacis aft, 3·1″. Deck forward before transverse bulkhead, 1·4″ to ·8″. Deck aft abaft after glacis, 1″. Glacis over engines, 3·9″.

Armament.—

Guns and Ammunition Supply.—

(For disposition and arcs of training, *see* Plate 15.)

10—4·1-in. (10·5-cm.) Q.F., L/40, on C.P. Mounting C/04.
Hoists for 4·1-in. guns, electric, alternative hand.

2—·31-in. (8-mm.), Maxim automatic.

Magazines and Stowage.—
There are two magazines, one forward and one aft.

Torpedo Tubes.—
2—17·7-in. (45-cm.) submerged; before conning tower.

Mines.—There is no permanent stowage for mines, but 108 can be carried if required.

Searchlights.—

Hamburg, München, and *Danzig.*—One on foremast, two on mainmast in fore-and-aft line.

Lübeck.—One on each mast.

Berlin.—Two on foremast and two on mainmast, in fore-and-aft line.

Anchors.—Stockless. Hawse-pipes rather far aft.

Hamburg, Berlin, München, Lübeck, *and* Danzig—*cont.*

Machinery and Boilers.—

Main Engines, except *Lübeck*.—

Two sets, vertical, four cylinder, triple expansion, in separate engine-rooms abreast of one another, separated by a longitudinal watertight bulkhead.

Lübeck—

Main Engines.—Four sets, Parsons turbines, in two engine rooms divided by a longitudinal watertight bulkhead.

Two shafts on each side of the ship. Port and starboard outer shafts, H.P. main turbine, and astern turbine. Port inner shaft—L.P. cruising, L.P. main with astern turbine. Starboard inner shaft—H.P. cruising turbine, L.P. main turbine with astern turbine.

Saving of weight on reciprocating engines of sister ships, 52 tons.

Boilers (all ships).—

Ten Schulz, Navy type, in three boiler rooms, two in fore, four in middle, and four in after boiler room. Working pressure, 215 lbs. per sq. in.

Total heating surface, 32,650 sq. ft. Total grate surface, 650 sq. ft.

Auxiliary Machinery includes—

Three dynamos, two in engine-room and one on deck.

Propellers.—Two, four-bladed ; except *Lübeck*, uncertain.

Speed, Horse-Power, and Fuel.—

Ship.	Speed (designed).	Horse-Power (designed).	Fuel : (*a*) Coal. (*b*) Oil.
	Knots.		
Class except - - - - -	22	10,000	(*a*) 846
Lübeck - - - - - -	22	11,200	

Steam Trials.—

Ship.	Nature of Trial.	Speed.	H.P.	Revolutions.	Remarks.
		Knots.		No.	Coal per I.H.P. per hour. Lbs.
Hamburg - -	Measured mile trials - -	23·1	11,848	147	—
München - -	,, ,, - -	23·45	12,388	143	—
Lübeck - - -	,, ,, - -	23·1	14,158(T)	—	—
Hamburg - -	6 hours' forced draught -	22·1	10,746	141	2·2
Berlin - - -	,, ,, -	22·6	10,857	138	2·1
München - -	,, ,, -	22·0	10,580	135	2·3
Lübeck - - -	,, ,, -	22·6	13,879(T)	663	—
Danzig - - -	,, ,, -	22·7	12,110	—	—
Hamburg - -	24 hours' coal consumption -	20·2	7,244	124	1·82
Berlin - - -	,, ,, -	20·3	7,121	121	1·76
München - -	,, ,, -	20·4	7,165	121	1·96

Steam Trials—*cont.*

Comparative Stopping Trials.—

Speed ahead on Reversing.	Distance traversed.				
	Lübeck.				*Hamburg*.
	Eight Small Propellers.	Four Large Propellers.	Four Large and Four Small Propellers.	Four Extra Propellers.	
Knots.	Ft.	Ft.	Ft.	Ft.	Ft.
5	334	170	164	246	184
9	384	413	361	482	361
11	754	692	636	705	590
22	1,430	1,758	1,627	1,640	918

Arcona.

(*See Plates* 5 *and* 16.)

Name.	Designation before Launch.	Programme Year.	Where Built.	Launched.	First Commissioned.
Arcona (R. 1912) -	H.	1901–2	Weser Yard, Bremen - -	22.4.02	12.5.03

General Remarks.—This vessel is an improved *Gazelle* (*see* page 21), and the particulars, including armament, are the same, except as stated below.

Of the other two vessels of the type, the *Undine* was sunk in the Baltic on 7th November 1915 and the *Frauenlob* was sunk in the North Sea on 31st May 1916.

During the war, the *Arcona* has been stationed almost continuously at Emden, performing the duties of guardship, flagship of the Ems patrol and wireless repeater for submarines.

Complement.—In peace time, 281.

General Appearance.—*See* Plate 5 and Silhouette.

General Dimensions, &c.—

Length between perpendiculars - - -	328 ft. 1 in.
Length, L.W.L. - - - - -	342 ,, 6 ,,
Breadth, extreme - - - - -	40 ,, 4 ,,
Draught, designed load - - - -	16 ,, 5 ,,
Displacement, designed load - - - -	2,656 tons.
Height of fore funnel to L.W.L. - - -	60 ft. 0 ins.
,, ,, lower masthead to L.W.L. - -	70 ,, 0 ,,
,, ,, truck to L.W.L. - -	*about* 130 ,, 0 ,,

Armour.—

Conning Tower.—3·1″ thick; communication tube, 1·2″.

Horizontal.—Complete armoured deck, ·8″ thick on flat, and 2″ on slopes.

Armoured Coamings.—3·1″ thick.

Mines.—*Arcona* was altered in 1912 and fitted for minelaying, but she is rarely used for this purpose. According to an estimate she can carry over 400 contact mines, which are placed along the lower deck on rails, a double row being situated on each side of the ship.

Searchlights.—

3—One on foremast, two on mainmast in fore-and-aft line.

Machinery and Boilers.—

Main Engines.—

Two sets, vertical, triple expansion, in separate watertight compartments.
Diameter of cylinders—31·1 ins., 47·2 ins., and 73·2 ins. Stroke 27·6 ins.

Boilers.—

Nine, Schulz, Navy type.
Heating surface—24,973 sq. ft. Grate surface—455 sq. ft.

Propellers.—Two, three-bladed. Diameter, 11 ft. 6 ins.; pitch, 16 ft. 3 ins.

Speed, Horse-Power, and Fuel.—

Ship.	Speed (designed).	Horse-Power (designed).	Fuel: (*a*) Coal. (*b*) Oil.
Arcona - - - - - - - - -	Knots. 21	8,000	(*a*) 698

Arcona—*cont.*

Steam Trials.—

Ship.	Nature of Trial.	Date.	Speed.	Horse-Power.	Revolutions.	Remarks.
Arcona - -	6 hours' forced draught.	—	21·0	8,291	161	
	24 hours' coal consumption.	—	19·55	5,881	148	Coal per I.H.P. per Hour. 1·98 lbs.
		—	12·04	1,318	88	1·98 „

Gazelle, Nymphe, Niobe, Thetis, Medusa, *and* Amazone.

(*See Plate* 16.)

Name.	Designation before Launch.	Programme Year.	Where Built.	Launched.	First Commissioned.
Gazelle - - - -	—	1896–97	Germania Yard, Kiel - -	31.3.98	23.11.98
Nymphe - - - -	A.	1898–99	Germania Yard, Kiel - -	21.11.99	20.9.00
Niobe - - - -	B.	1898–99	Weser Yard, Bremen - -	18.7.99	25.6.00
Thetis - - - -	C.	1899–00	Imperial Dockyard, Danzig -	3.7.00	14.9.01
Medusa - - - -	E.	1900–01	Weser Yard, Bremen - -	5.12.00	26.7.01
Amazone - - - -	F.	1900–01	Germania Yard, Kiel - -	6.10.00	15.11.01

General Remarks.—These vessels are now (June 1918) practically obsolete and are used only in the Baltic, where *Thetis* and *Amazone* are attached to the Submarine School as target ships.

The *Ariadne*, which was of the same type, was sunk in action on 28th August 1914.

Complement.—In peace time: *Gazelle* 268, *Niobe* 269, remainder 275.

General Appearance.—*See* Silhouettes.

General Dimensions, &c.—

Length { between perpendiculars - - - 328 ft. 1 in.
Length { L.W.L. - - - - - - 342 „ 6 „
Length { extreme - - - - - - 344 „ 6 „
Breadth, extreme - - - - - - 38 „ 9 „
Draught { designed load - - - - 16 „ 5 „
Draught { maximum load - - - - 17 „ 6 „
Displacement, designed, 2,608 tons (*Gazelle* and *Niobe* 2,558 tons.)

Armour.—

Conning Tower.—*Gazelle*, *Niobe*, and *Nymphe*, 2·8″ thick, the remainder 3·1″.

Communication Tube.—1·2″.

Gun Shields.—1″.

Horizontal.—There is a complete curved armoured deck, ¾″ on the flat and 2″ on the slopes, 3 ft. above the load water-line amidships over the engines and boilers, and 4 ft. below load water-line at the sides. At the ends of the machinery spaces the deck dips sharply down to 1 ft. below load water-line. All openings, except for funnels, are protected by 3·1″ armoured coamings, 1 ft. high.

C 3

Gazelle, Nymphe, Niobe, Thetis, Medusa, *and* Amazone—*cont*

Armament.—

Guns and Ammunition Supply.—

(For disposition and arcs of training, *see* Plate 16.)

10—4·1 in. (10·5-cm.) Q.F., L/40, on C.P. Mounting C/97.

2—·31-in. (8-mm.) Maxim automatic.

All hoists electric, alternative hand.

Magazines and Stowage.—

Three magazines on each side. Stowage, 4·1-in.—100 rounds per gun.

Torpedo Tubes.—

2—17·7-in., submerged; one on each broadside. (*Gazelle* has one submerged, under ram, and two above-water tubes.)

Torpedoes.—Gazelle carries three for the bow and five for the broadside tubes.

Mines.—There is no permanent stowage for mines, but 70 can be carried if required.

Searchlights.—

2—One on foremast, one on raised platform on poop.

Machinery and Boilers.—

Main Engines.—

Two sets, vertical, four-cylinder, triple expansion, balanced cranks, in separate watertight compartments.

Diameter of cylinders—30·7 ins., 44·5 ins., two of 51 ins. Stroke 28·3 ins.

Boilers.—

Gazelle has eight, and the remaining ships nine. In two separate watertight compartments.

Type—Thornycroft in *Niobe*, Schulz in the remaining ships.

H.S. and G.S.—*Nymphe*, H.S. 25,000 sq. ft.; G.S. 488 sq. ft.

Propellers.—Two, three-bladed. Diameter, 11 ft. 5 ins.; pitch, 16 ft. 5 ins.

Speed, Horse-Power, and Fuel.—

Ship.	Speed (designed).	Horse-Power (designed).	Fuel: (*a*) Coal. (*b*) Oil.
	Knots.		
Class except *Niobe* and *Gazelle* - - -	22	8,000	(*a*) 571
Niobe - - - - - - - - -	22	8,000	(*a*) 590
Gazelle - - - - - - - - -	20	6,000	(*a*) 590

Steam Trials.—

Ship.	Nature of Trial.	Speed.	Horse-Power.	Revolutions.	Remarks.
Gazelle - - -	Full power - -	20·2	6,366	—	
Nymphe - - -	" - -	20·8	8,400	—	
Niobe - - -	" - -	21·6	8,631	166	
Thetis - - -	" - -	21·75	8,888	172	
Medusa - - -	" - -	22·2	8,500	—	
Amazone - - -	" - -	21·5	8,603	161	

OBSOLETE CRUISERS.

It is not thought that these vessels will again be employed on active service, but details of them are given on the chance of their being used for some special purpose, such as blocking.

Roon.

(*See Plate* 17.)

Programme year - - - - -	1902–03.
Designation before launch - - -	Ersatz *Kaiser.*
Where built - - - - -	Imperial Dockyard, Kiel.
Laid down - - - - -	1st August 1902.
Launched - - - - -	27th June 1903.
First commissioned - - - -	5th April 1906.

General Remarks.—Now (June 1918) used as Torpedo Experiments Ship and believed to be entirely disarmed. The *Yorck*, the *Roon*'s sister ship, struck a mine and sank off the Jade river on 4th November 1914.

Complement.—Peace complement, 633 (as private ship); war complement, about 725.

General Appearance.—*See* Silhouette.

General Dimensions, &c.—

Length between perpendiculars - - -	403 ft. 6 ins.
,, L.W.L. - - - - -	417 ,, 8 ,,
,, extreme - - - - -	419 ,, 4 ,,
Breadth, extreme - - - - -	66 ,, 3 ,,
Draught, designed load - - - -	23 ,, 11 ,,
Displacement, designed load - - -	9,348 tons.
Height of fore funnel to L.W.L. - - -	72 ft. 0 ins.
,, ,, lower masthead to L.W.L. - -	92 ,, 0 ,,

Constructive Details.—The ship is well subdivided.

There is a central watertight passage, under the armoured deck, extending for about 185 ft. of the middle section and passing through the funnel casing; it contains electric leads, voice pipes, steering, and engine-room communications, &c., and is entered by a watertight door at each end. At the forward end it opens on to a compartment 23 ft. square, which serves as the transmitting station, communicating with the conning tower by an armoured tube, of oblong section, 2½ ft. by 1½ ft.

Armour.—

Material.—Krupp non-cemented.

Belt.—Complete, 7 ft. 6 ins. wide, lower edge about 4 ft. below L.W.L., 3·9″ thick amidships, 3·1″ forward and aft.

Citadel.—7 ft. wide, 200 ft. long, 3·9″ thick, closed at the ends by diagonal bulkheads of same thickness, which embrace the armoured tubes to turrets.

There are also three transverse bulkheads in connection with belt, 3·1″ thick, one just before ammunition tube to fore turret, one just abaft tube to after turret, and one right aft.

Battery.—3·9″ thick, 86 ft. long, containing six 5·9-in. guns, separated by splinter screens 2″ thick.

The whole battery is divided in two, by fore and aft armoured bulkheads 1¼″ thick fitted between the funnel uptakes, and is enclosed by bulkheads 3·9″ thick.

Turrets.—Turrets for 8·2-in. guns—5·9″ thick.
,, 5·9-in. ,, —3·9″ ,,

C 4

Roon—*cont.*

Armour—*cont.*

Ammunition Tubes.—For 8·2-in. gun turrets—3·9″ thick. For 5·9-in. gun turrets, above battery armour—3·9″.

Conning Towers.—Fore—5·9″ thick; after—3·2″ thick.

Communication Tubes.—To fore conning tower—3·2″ thick; to after—·8″ thick.

Horizontal.—There are two protective decks, the principal of which extends from stem to stern, and is 1½″ thick on the flat, and 2″ on the slopes. The second is formed by the lower deck outside citadel, 8″ thick, and the roof of citadel outside battery and roof of battery, 1·2″ thick.

***Former Armament.**—

Guns and Ammunition Supply.—

(For approximate arcs of training, *see* Plate 17.)

4—8·2-in. (21-cm.) Q.F., L/40; in pairs in turrets forward and aft on Drehscheiben-Lafette C/01.

10—5·9-in. (15-cm.) Q.F., L/40, on C.P. Mounting C/02: four in single turrets on upper deck, six in armoured battery on main deck.

14—15-pr. (8·8-cm.) Q.F., L/35.

2—·31-in. (8-mm.) Maxim automatic.

Magazines and Stowage.—

8·2-in. magazines and shell-rooms surround base of turret trunk.

5·9-in. magazines and shell-rooms are situated before and abaft the main deck battery.

Torpedo Tubes.—

4—17·7-in. (45-cm.), submerged; one under ram, one on either broadside just abaft fore turret, and one in stern on port quarter.

It is believed that the broadside tubes are fixed about 17° before the beam.

Torpedoes.—

12 were carried in peace, 15 in war time.

Searchlights.—

6—35·4-in. (90-cm.); one in fore upper top, two on hood of fore fighting top, one in maintop, and two on fore bridge.

Machinery and Boilers.—

Main Engines.—

Three sets, vertical, four cylinder, triple expansion, in separate watertight compartments.

Boilers.—

Sixteen in number. Dürr, water tube, fitted with superheaters, 8 fitted to burn coal and oil.

Auxiliary Machinery includes—

Dynamos—four in number, 110 volts. (Accumulators are fitted for lighting ship, &c., for half an hour in case of sudden breakdown of dynamo.)

Propellers.—Three; outer screws, 3-bladed, diameter 15 ft. 3 ins.; centre screw, 4-bladed, diameter 14 ft., pitch 19 ft. 9 ins.

Speed, Horse-Power, and Fuel.

Ship.	Speed (designed).	Horse-Power (designed).	Fuel: (*a*) Coal. (*b*) Oil.
Roon - - - - - - - - -	Knots. 21	19,000	(*a*) 1,545 (*b*) 197

* All the guns are believed to have been removed (June 1918).

Roon—*cont.*

Steam Trials.—

Ship.	Nature of Trial.	Date.	Speed.	Horse-Power.	Revolutions.	Remarks.
						Coal per I.H.P. per Hour.
Roon - - - -	6 hours full power	—.6.06	21·17	20,625	118	2·1 lbs.
,, - - - -	24 hours coal consumption.	—.6.06	13·64	3,800	76	1·97 ,,

---

Prinz Heinrich.

(*See Plate* 18.)

Programme year - - - - - 1898–99.
Designation before launch - - - "A."
Where built - - - - - Imperial Dockyard, Kiel.
Laid down - - - - - 1st December 1898.
Launched - - - - - 22nd March 1900.
First commissioned - - - - 11th March 1902.

General Remarks.—Now (June 1918) used as a harbour training ship and believed to be entirely disarmed.

Complement.—In peace time, 567 (as private ship).

General Appearance.—*See* Silhouette.

General Dimensions, &c.—

Length between perpendiculars - - - - 393 ft. 8 ins.
Length, L.W.L. - - - - - - 409 ,, 9 ,,
Breadth, extreme - - - - - - 64 ,, 3 ,,
Draught, designed load - - - - - 23 ,, 11 ,,
Displacement, designed load - - - - - 8,756 tons.
Height of fore funnel to L.W.L. - - - - 70 ft. 0 ins.
,, ,, ,, lower masthead to L.W.L. - - 95 ,, 0 ,,

Constructive Details.—The ship is well subdivided. There is a central watertight passage under the armoured deck as in *Roon* (*see* p. 23).

Armour.—

Material.—Krupp, non-cemented.

Belt.—Complete, 7 ft. 6 ins. wide, extending to about 4 ft. 6 ins. below L.W.L., 3·9″ thick amidships, 3·1″ forward and aft.

Citadel.—140 ft. long, 7 ft. wide, 4″ thick, closed at the ends by diagonal bulkheads of the same thickness, which embrace the armoured tubes to conning towers.

Battery.—Above the citadel is the battery armour, 3·9″ thick. The battery is closed by transverse bulkheads of the same thickness. Splinter bulkheads 1·2″ thick are fitted between the guns, extending about one-third of the way across the ship. There is no centre line protection.

Turrets.—
For 9·4-in. guns—5·9″ thick.
For 5·9-in. guns—3·9″ thick.

Ammunition Tubes.—
For 9·4-in. turrets—3·9″ thick.
For 5·9-in. turrets, above side armour—3·9″ thick.

Conning Towers.—
Fore—5·9″ thick; communication tube—3·2″.
After—·8″ thick; ,, ,, ·5″.

Control Tower.—3·9″.

O AS 4924—2 D

Prinz Heinrich—*cont.*

Armour—*cont.*

Horizontal.—Two protective decks. The principal one extends from stem to stern and is 1½″ on the flat, 2″ on the slopes. The second is formed by the lower deck, outside citadel, ·8″ thick, and the roof of citadel outside battery and roof of battery, 1·2″ thick.

*__Former Armament.__—

Guns and Ammunition Supply.—

(For arcs of training, *see* Plate 18.)

2—9·4-in. (24-cm.) Q.F., L/40; in turrets forward and aft on Drehscheiben-Lafette C/99.

10—5·9-in. (15-cm.) Q.F., L/40; four in single turrets on Turm-Lafette C. 97/99; six in armoured battery on main deck on C.P. Mounting C/97.

10—15-pr. (8·8-cm.) Q.F., L/30.

2—·31-in. (8-mm.) Maxim automatic.

Torpedo Tubes.—

3—17·7-in. (45-cm.), submerged; one under ram and one each side in line with after side of foremost turret.

1—17·7-in. (45-cm.), above water (armoured), in stern.

Torpedoes.—12 carried in peace, 15 in war time.

Searchlights.—

5—35·4-in. (90-cm.): on foremast—one on crosstrees, two below on S.L. platform abreast; on mainmast—two on S.L. platform abreast.

Machinery and Boilers.—

Main Engines.—

Three sets, vertical, four-cylinder, triple expansion.

Diameters of cylinders, 35 ins.; 52·4 ins.; two of 60·2 ins. Stroke, 37·4 ins.

Boilers.—

Fourteen; Dürr, with super-heaters, eight of which are fitted to burn coal and oil.

H.S., 42,820 sq. ft.; superheated surface, 2,357 sq. ft.; G.S., 1,107 sq. ft.

Auxiliary Machinery includes—

Dynamos—Four, 110 volts.

Propellers.—Three; outer screws, three-bladed; diameter, 15 ft. 3 ins.; centre screw, four-bladed; diameter, 14 ft.; pitch, variable, 17 ft. 8 ins. to 21 ft. 7 ins.

At trials, pitch of 18 ft. 8 ins. gave best results.

Surface area of the blades of each propeller, 49·1 sq. ft.

Speed, Horse-Power, and Fuel.—

Ship.	Speed (designed).	Horse-Power (designed).	Fuel: (*a*) Coal. (*b*) Oil.
Prinz Heinrich - - - - - - -	Knots. 20	15,000	(*a*) 1,564 (*b*) 197

* All of these guns are believed to have been removed (June 1918).

Prinz Heinrich—*cont.*

Steam Trials.—

Ship.	Nature of Trial.	Date.	Speed.	Horse-Power.	Revolutions.	Remarks.
Prinz Heinrich - -	6 hours full power	—	20·0	15,703	127	Coal per 1 h.p. per hour, 2·1 lbs.
„ „ - -	24 hours - -	—	18·16	10,355	111	1·91 lbs., air pressure, ·39 ins.

Fürst Bismarck.

(*See Plate* 19.)

Programme year - - - - - 1895–96.
Designation before launch - - - *Ersatz* Leipzig.
Where built - - - - - Imperial Dockyard, Kiel.
Laid down - - - - - 1st April 1896.
Launched - - - - - 25th September 1897.
First commissioned - - - - 1st April 1900.

General Remarks.—Underwent alterations, including removal of military masts, 1912–1914, to adapt her for service as Torpedo School Ship, but is now (March 1918) used as a harbour training ship for seaman ratings, and is believed to be entirely disarmed.

Complement.—In peace time, 594 (as private ship).

General Appearance.—*See* Silhouette.

General Dimensions, &c.—

Length between perpendiculars - - - - - 393 ft. 8 ins.
„ L.W.L. - - - - - - 412 „ 5 „
„ extreme - - - - - - 416 „ 8 „
Breadth, extreme - - - - - - - 66 „ 11 „
Height of axis of 9·4-in. guns, forward, at mean draught About 34 ft.
Height of axis of 9·4-in. guns, aft, at mean draught - „ 23 „
Draught, designed load - - - - - - 26 ft.
Displacement, designed load - - - - 10,520 tons.

Constructive Details.—

A central passage of small dimensions, for communications, extends almost the whole length of the ship.
There is a complete belt of cork cofferdams above the armour belt.

Armour.—

Material.—Krupp steel.

Belt.—Complete, 7 ft. 6 ins. wide, 7·9″ thick amidships, tapering to 3·9″ at lower edge; 3·9″ at ends.

Turrets.—
For 9·4-in. guns—7·9″ thick.
For 5·9-in. guns—3·9″ thick.

Casemates.—3·9″ thick.

Ammunition Tubes.—
9·4-in. guns, 7·9″ thick.
5·9-in. guns, 3·1″ thick.

Conning Towers.—Fore, 7·9″; after, 3·9″ thick.

Communication Tube.—Forward, 7·9″; aft, 3″ thick.

Horizontal.—Deck at upper edge of belt, 2″ over machinery, 1¼″ forward and aft; underwater deck at ends, 2″. Splinter decks, ·8″.

D 2

Fürst Bismarck—*cont.*

***Former Armament.**—

Guns and Ammunition Supply.—

(For disposition and arcs of training, *see* Plate 19.)

4—9·4-in. (24-cm.) Q.F., L/40; in pairs in turrets, on Drehscheiben-Lafette C/97.

12—5·9-in. (15-cm.) Q.F., L/40; six in single turrets, on Turm-Lafette C/97; six in casemates on C.P. Mounting C/97. Separate electric, alternative hand, dredger hoists to each upper deck casemate; in main deck double gun casemates two hoists in one armoured tube.

10—15-pr. (8·8-cm.) Q.F., L/30.

4—·31-in. (8-mm.) Maxim automatic.

Magazines and Stowage.—The after submerged torpedo room and the central casemate magazine are between the engine and boiler rooms.

Torpedo Tubes.—

5—17·7-in. (45-cm.), submerged. One under ram, fixed ahead; two on each broadside, fixed about 17° before the beam.

1—17·7-in. (45-cm.), armoured, above water, in stern.

Searchlights.—

7—One on a platform on foremast, four on two platforms on fore side of forward funnel, two on a platform above after conning tower.

Machinery and Boilers.—

Main Engines.—

Three sets, vertical, four-cylinder, triple expansion.

Diameter of cylinders, 37 ins.; 57·5 ins.; two of 63 ins. Length of stroke, 33·8 ins.

Boilers.—

12 in number, eight cylindrical, four Schulz, Navy type, fitted to burn coal and oil.

Total heating surface	Cylindrical - -	23,400 sq. ft.
	Schulz - - -	17,450 „ „
Total grate surface -	Cylindrical - -	655 „ „
	Schulz - - -	327 „ „

Auxiliary Machinery includes—

Five dynamos, each 490 ampères, 110 volts.

Propellers.—Three.

Speed, Horse-Power, and Fuel.—

Ship.	Speed (designed).	Horse-Power (designed).	Fuel: (*a*) Coal. (*b*) Oil.
Fürst Bismarck - - - - - - -	Knots. 19	13,000	(*a*) 1,377 (*b*) 98

Steam Trials.

Ship.	Nature of Trial.	Date.	Speed.	Horse-Power.	Revolutions.	Remarks.
Fürst Bismarck -	Full power - -	—	18·51	12,976	118	

* All of the guns are believed to have been removed (June 1918).

OBSOLETE LIGHT CRUISERS.

It is not thought that these vessels will again be employed on active service, but details of them are given on the chance of their being used for some special purpose, such as blocking.

Hertha, Victoria Louise, Freya, Vineta, *and* Hansa.

(*See Plate* 20.)

Name.	Name before Launch.	Programme Year.	Where Built.	Laid Down.	Launched.	First Commissioned.	Reconstructed.
Hertha - -	K.	1895–96	Vulcan Works, Stettin.	–.9.95	14.4.97	23.7.98	1908
Victoria Louise	L.	1895–96	Weser Yard, Bremen	–.9.95	29.3.97	20.2.99	1908
*Freya** - -	*Ersatz* Freya	1895–96	Imperial Dockyard, Danzig.	–.10.95	27.4.97	20.10.00	1913*
Vineta - -	M.	1896–97	Imperial Dockyard, Danzig.	–.9.96	9.12.97	13.9.99	1911
Hansa - -	N.	1896–97	Vulcan Works, Stettin.	–.3.96	12.3.98	20.4.99	1909

* Partially reconstructed in 1907.

General Remarks.—Between 1908 and 1913 all these ships underwent extensive alterations to adapt them for service as seagoing training ships for cadets and boys, in lieu of the former old masted training ships.

Of the five vessels, only the *Freya* is now (June 1918) used as a seagoing training ship; of the others, all of which are believed to be completely disarmed, *Vineta* and *Hansa* serve as accommodation ships for submarines' crews; *Hertha* serves as a seaplane depôt ship; *Victoria Louise* is used as an accommodation ship in Danzig Dockyard.

Complement.—As training ships these vessels carried a total complement of 637, including about 75 cadets and 250 boys under training; they were fully manned as regards engine-room ratings, but only partially manned as regards deck ratings.

As light cruisers their complement was 570.

General Appearance.—*See* Silhouettes.

General Dimensions, &c.—

	Victoria Louise, Hertha, Freya.	*Vineta, Hansa.*
Length between perpendiculars - -	344 ft. 6 ins.	345 ft. 6 ins.
Length, L.W.L. - - - -	357 „ 11 „	360 „ 3 „
Breadth, extreme - - -	57 „ 1 „	57 „ 9 „
Draught, designed load - - -	20 „ 4 „	21 „ 8 „
Displacement, designed load - -	5,575 tons.	5,790 tons.
Displacement, full load - - -	6,214 „	5,982 „

D 3

Hertha, Victoria Louise, Freya, Vineta, *and* Hansa—*cont.*

Armour.—

Turrets.—For 8·2-in. and 5·9-in. guns—3·9″ thick.

Casemates.—For 5·9-in. guns—3·9″ thick.

Ammunition Tubes.—3·1″.

Conning Towers.—Fore, 5·9″; after, 4″.

Horizontal.—Complete armoured deck, 1·5″ on flat, 3·9″ on the slopes amidships (the lower edge being 5 ft. below water-line), 3″ and 2·5″ forward and aft.

Glacis.—4″, round bases of funnels.

*Former Armament.**—

Guns.—

(For arcs of training, *see* Plate 20.)

2—8·2-in. (21-cm.) Q.F., L/40, mounted singly in hooded barbettes forward and aft on Turm-Lafette, C/97.
6—5·9-in. (15-cm.) Q.F., L/40; four in single turrets on Turm-Lafette C/97, on the upper deck; two in casemates on main deck.
3—15-pr. (8·8-cm.) Q.F., L/35, with shields.
11—15-pr. (8·8-cm.) Q.F., L/30, with shields.
4—machine guns.

Magazines and Stowage.—Number of rounds per gun carried, 8·2-in.—80; 5·9-in.—100; 15-pr.—250. Magazines for after 5·9-in. built into the fore part of central engine room.

Torpedo Tubes.—

3—17·7-in., submerged; one under ram, one each side in line with conning tower, fixed 20° before the beam. (*Vineta* has only 2 tubes, the port tube having been removed.)

Searchlights.—

4—one in fore top, one each side on fore bridge, and one on platform above after conning tower.

Machinery and Boilers.—

Main Engines.—

Three sets, vertical, four-cylinder, triple expansion, in separate engine rooms, the centre set abaft the other two.
Diameters of cylinders, 28·4 ins.; 43·7 ins.; and two of 51·2 ins.
Length of stroke, 29·5 ins.

Boilers.—The original boilers have probably been replaced by 8 Schulz, Navy type, in each of these vessels.

Auxiliary Machinery includes—

Hydraulic.—Pumping engine. one in centre of each engine room.
Electric.—Dynamos, four (500 ampères at 110 volts).

Propellers.—Three, three-bladed; diameter, 12 ft. 9 ins.; starboard and centre propellers right- and port left-handed in *Vineta.*

Speed, Horse-Power, and Fuel.

Speed (designed)	Horse-Power (designed).	Fuel: (*a*) Coal. (*b*) Oil.
Knots. 19	10,000	(*a*) 885, except *Vineta* and *Hansa*, (*a*) 905

* With the exception of the *Freya*, all these vessels are believed to be completely disarmed; the *Freya* only retains four or five 15-pr. guns.

Hertha, Victoria Louise, Freya, Vineta, *and* Hansa—*cont.*

Steam Trials.—

Ship.	Nature of Trial.	Speed.	Horse-Power.	Revolutions.	Remarks.
Hertha - - -	Full power - -	19·2	10,173	—	With boilers originally fitted.
Victoria Louise - -	,, - -	19·1	10,000	—	
Freya - - -	,, - -	18·5	10,919	—	
Vineta - - -	,, - -	19·2	10,000	—	
Hansa - - -	,, - -	19·2	10,000	—	

Endurance.—

—	Speed.	Horse-Power.	Daily Consumption.	Radius of Action.	Remarks.
	Knots.		Tons.	Miles.	*Vineta* and *Hansa.*
At $\frac{4}{5}$ths designed power - -	18·1	8,000	186	1,950	1,990.
At maximum continuous seagoing speed.	17·2	6,400	153	2,240	2,285.
At $\frac{3}{5}$ths designed power - -	16·8	6,000	136	2,500	2,550.
At $\frac{2}{5}$ths designed power - -	15·1	4,000	98	3,140	3,210.
At $\frac{1}{5}$th designed power - -	11·8	2,000	57	4,500	4,600.
At 10 knots - - - - -	10·0	1,450	42	4,800	4,910.

Kaiserin Augusta.

(See Plate 21.)

Programme Year - - - - - - 1888–89.
Designation before launch - - - - H.
Built at - - - - - - Germania Yard, Kiel.
Laid down - - - - - - 1890.
Launched - - - - - - 15th Jan. 1892.
First commissioned - - - - 21st Jan. 1893.
(Modernised in 1905.)

General Remarks.—
Has been used throughout the war as a Gunnery Training Ship.

Complement.—In peace time, 439.

General Appearance.—*See* Silhouette.

General Dimensions, &c.—
Length between perpendiculars - - - - - 387 ft. 10 ins.
Length, L.W.L. - - - - - - - 400 ,, 11 ,,
Breadth, extreme - - - - - - - 51 ,, 2 ,,
Draught, designed load - - - - - - 22 ,, 0 ,,
,, maximum load - - - - - - 23 ,, 7 ,,
Displacement, designed - - - - - - 5,960 tons.

Armour.—

Conning Tower.—2″, with a ·8″ roof.

Communication Tube.—3·9″.

Horizontal.—Deck 2″ amidships and 3″ on the slopes. The deck slopes down to 5 ft. below water-line.

***Former Armament.**—

Guns and Ammunition Supply.—
(For arcs of training, *see* Plate 21.)
12—5·9-in. (15-cm.) Q.F., L/35, on C.P. Mounting, C/92; two under forecastle, four each side in sponsons amidships, and two under poop.

* Most of these guns have been removed (June 1918). Only a few modern guns, of various calibres, are now carried for training purposes.

D 4

Kaiserin Augusta—*cont.*

Armament—*cont.*

Guns and Ammunition Supply—cont.

8—15-pr. (8·8-cm.) Q.F., L/30; two on forecastle, two on poop, two each side in sponsons.

2—·31-in. (8-mm.) Maxim automatic.

Torpedo Tubes.—

1—13·8-in. (35-cm.), submerged; under ram.

2—13·8-in. (35-cm.), above water; on main deck, abaft centre funnel.

Searchlights.—

2—one in each top.

Machinery and Boilers.—

Main Engines.—

Three sets, vertical, triple expansion, in separate watertight compartments, centre set abaft the other two.

Boilers.—

Eight cylindrical, double-ended, in three separate boiler rooms.

H.S., 35,700 sq. ft.; G.S., 1,033 sq. ft.

Propellers.—Three. Outer screws: diameter, 14·76 ft.; pitch, 20 ft. Centre screw: diameter, 13·8 ft.; pitch, 19 ft.

Speed, Horse-Power, and Fuel.—

	Speed (designed).	Horse-Power (designed).	Fuel: (a) Coal. (b) Oil.
	Knots. 21·5	12,000	(a) 788

Steam Trials.—

Ship.	Nature of Trial.	Speed.	Horse-Power.	Revolutions.	Remarks.
Kaiserin Augusta -	Full power - - -	21·59	13,612	127	Mean draught, 21 ft. 1 in.
" " -	Coal consumption -	17·67	6,971	102	" " 21 " 8 "
" " -	Coal consumption (centre engine only).	11·7	2,337	90	" " 21 " 8 "

Endurance.—

——	Speed.	Horse-Power.	Daily Consumption.	Radius of Action.	Remarks.
	Knots.		Tons.	Miles.	
At $\frac{4}{5}$ths designed power - -	19·4	9,600	225	1,550	
At maximum continuous seagoing speed.	19·3	9,500	223	1,590	
At $\frac{3}{5}$ths designed power - -	17·9	7,200	171	1,880	
At $\frac{2}{5}$ths designed power - -	15·9	4,800	117	2,440	
At $\frac{1}{5}$th designed power - -	12·0	2,400	68	3,170	
At 10 knots - - - - -	10·0	1,600	47	3,830	

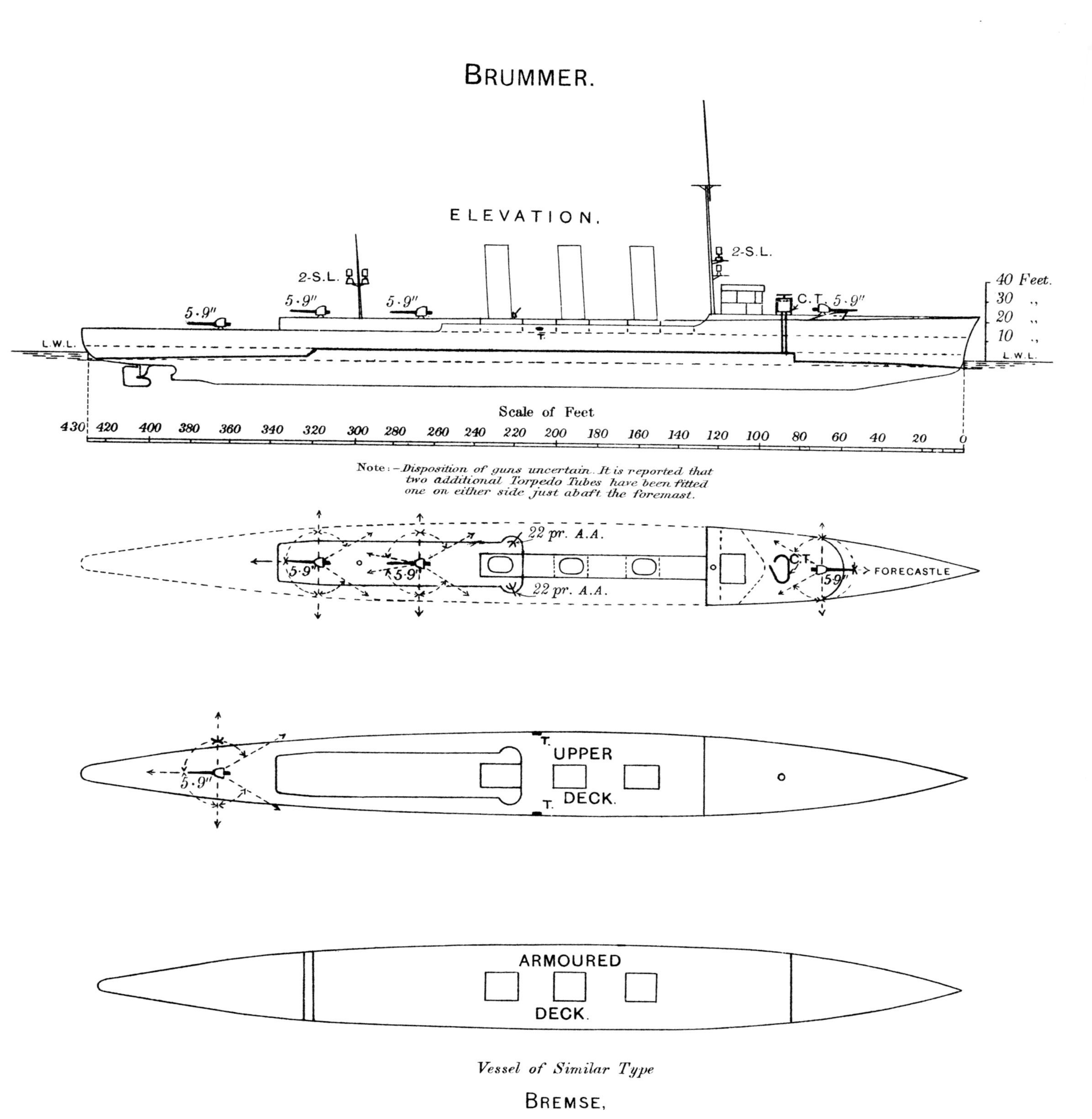

Vessel of Similar Type

BREMSE.

Ordnance Survey, July, 1918.

KÖNIGSBERG

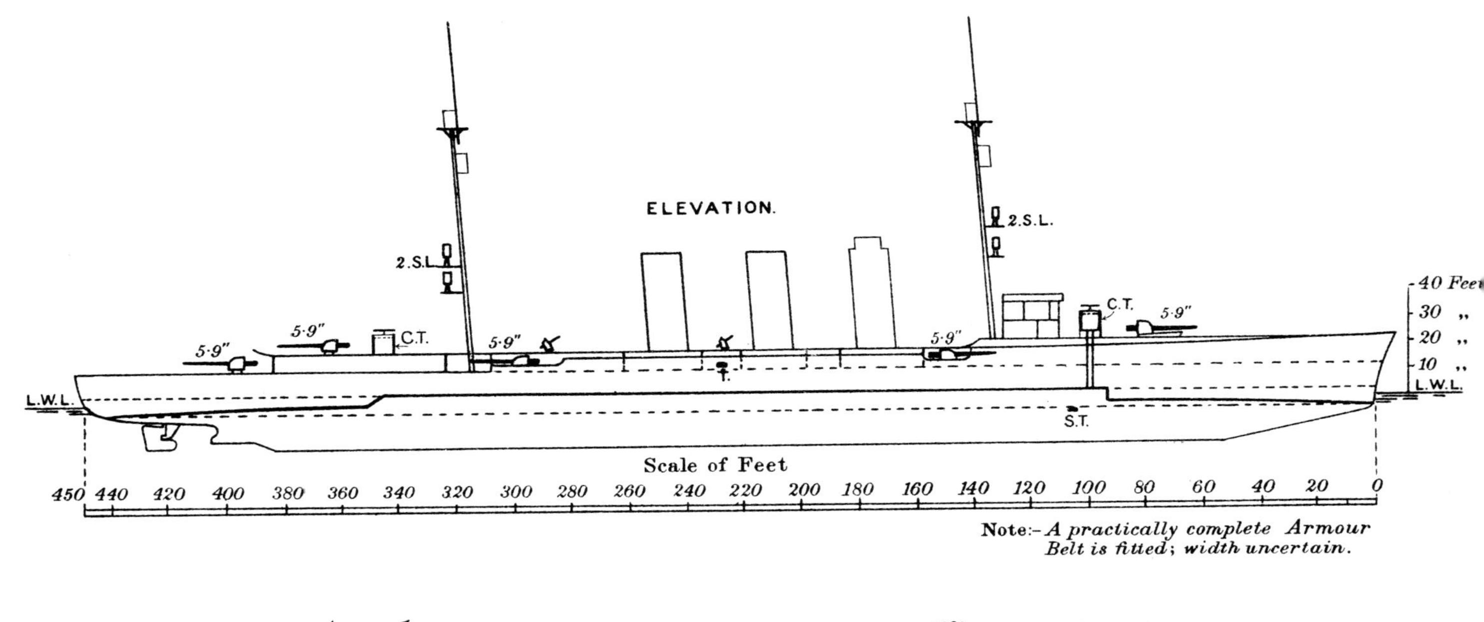

Note:—*A practically complete Armour Belt is fitted; width uncertain.*

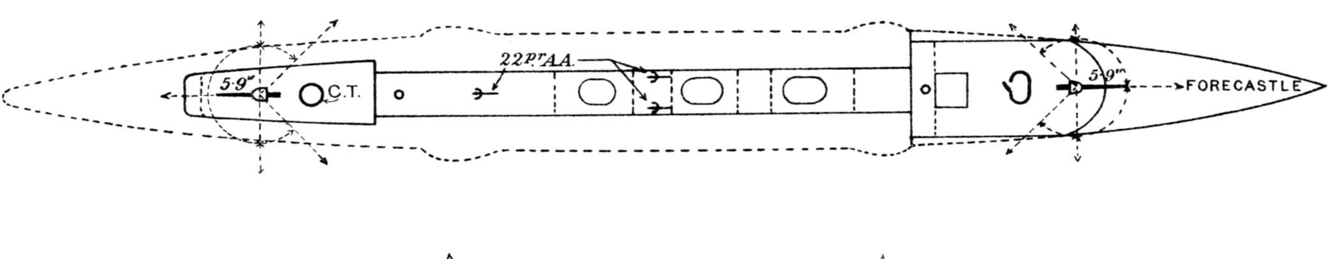

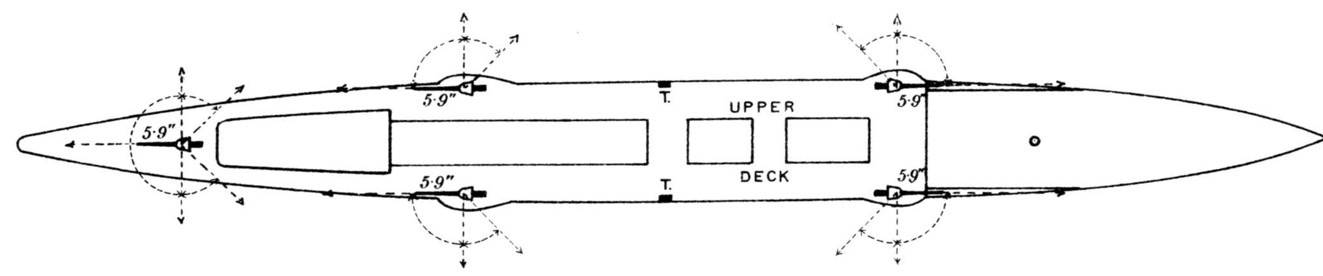

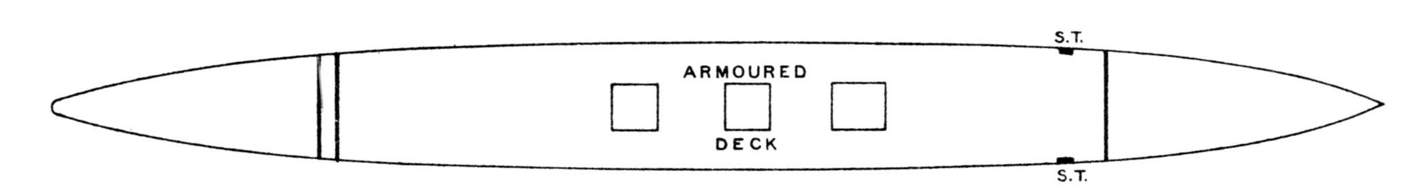

Vessels of Similar Type

KARLSRUHE, EMDEN, NÜRNBERG.

Ordnance Survey, July, 1918

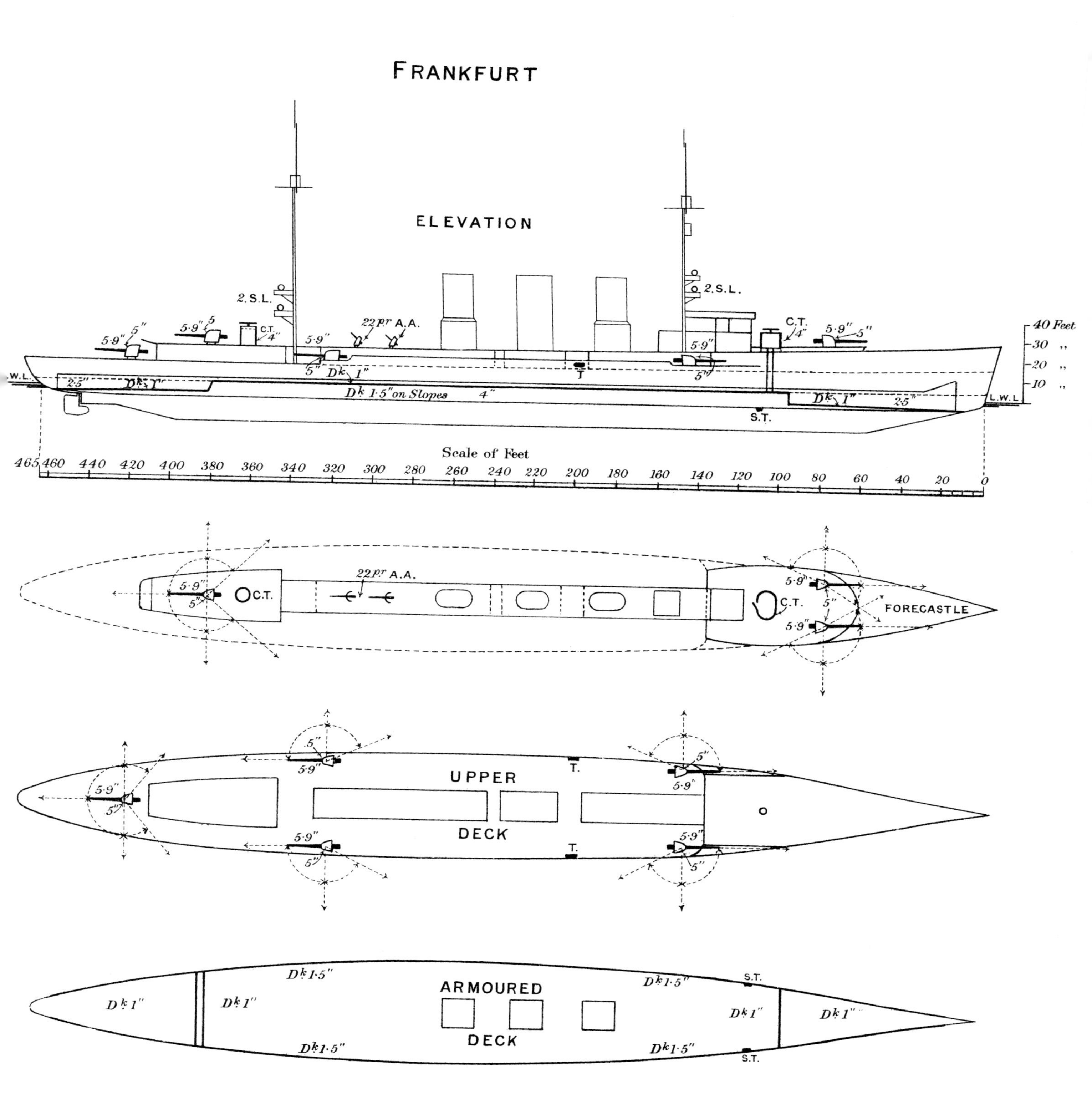

Ordnance Survey, July, 1918.

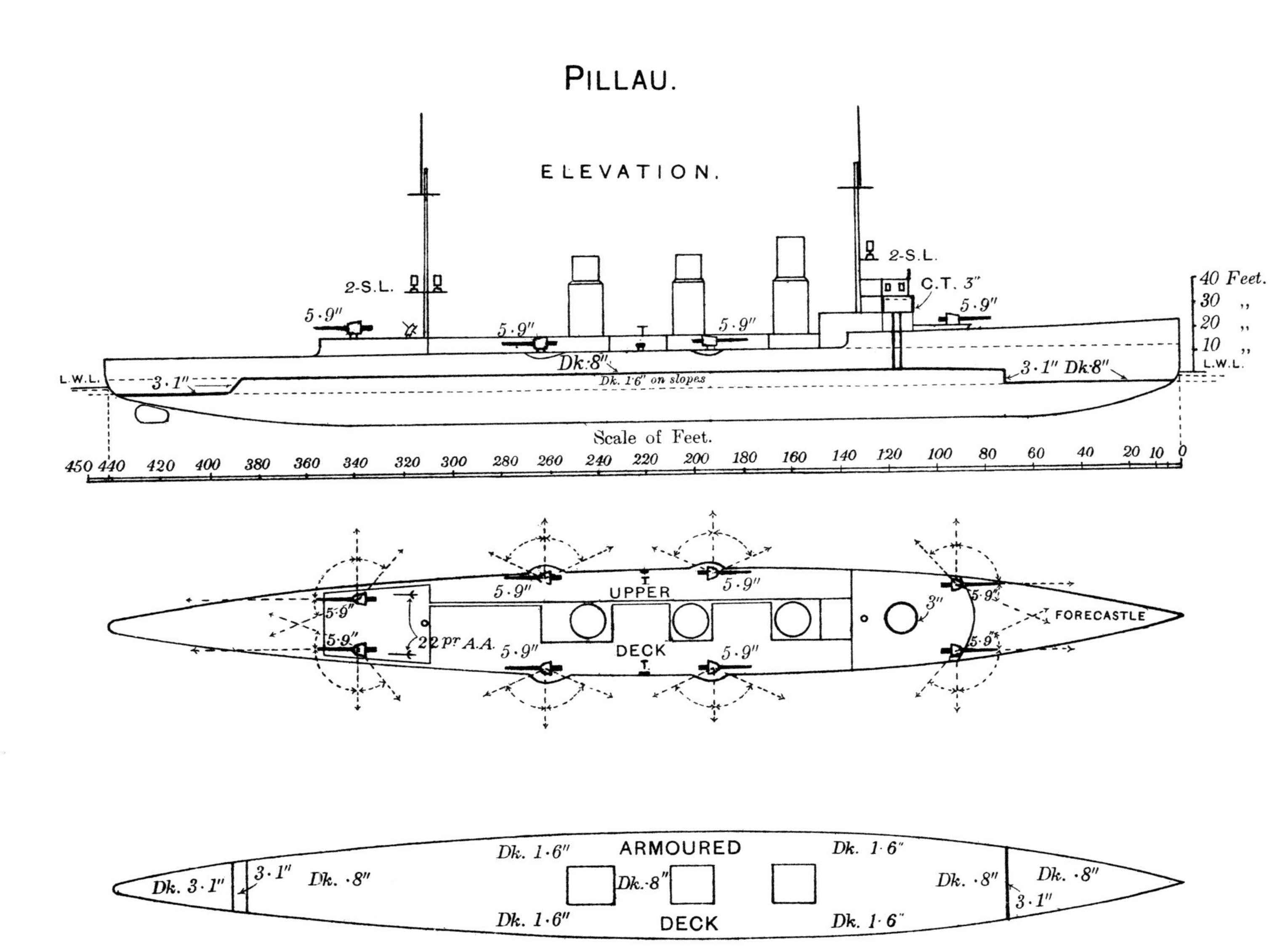

Ordnance Survey, July, 1918

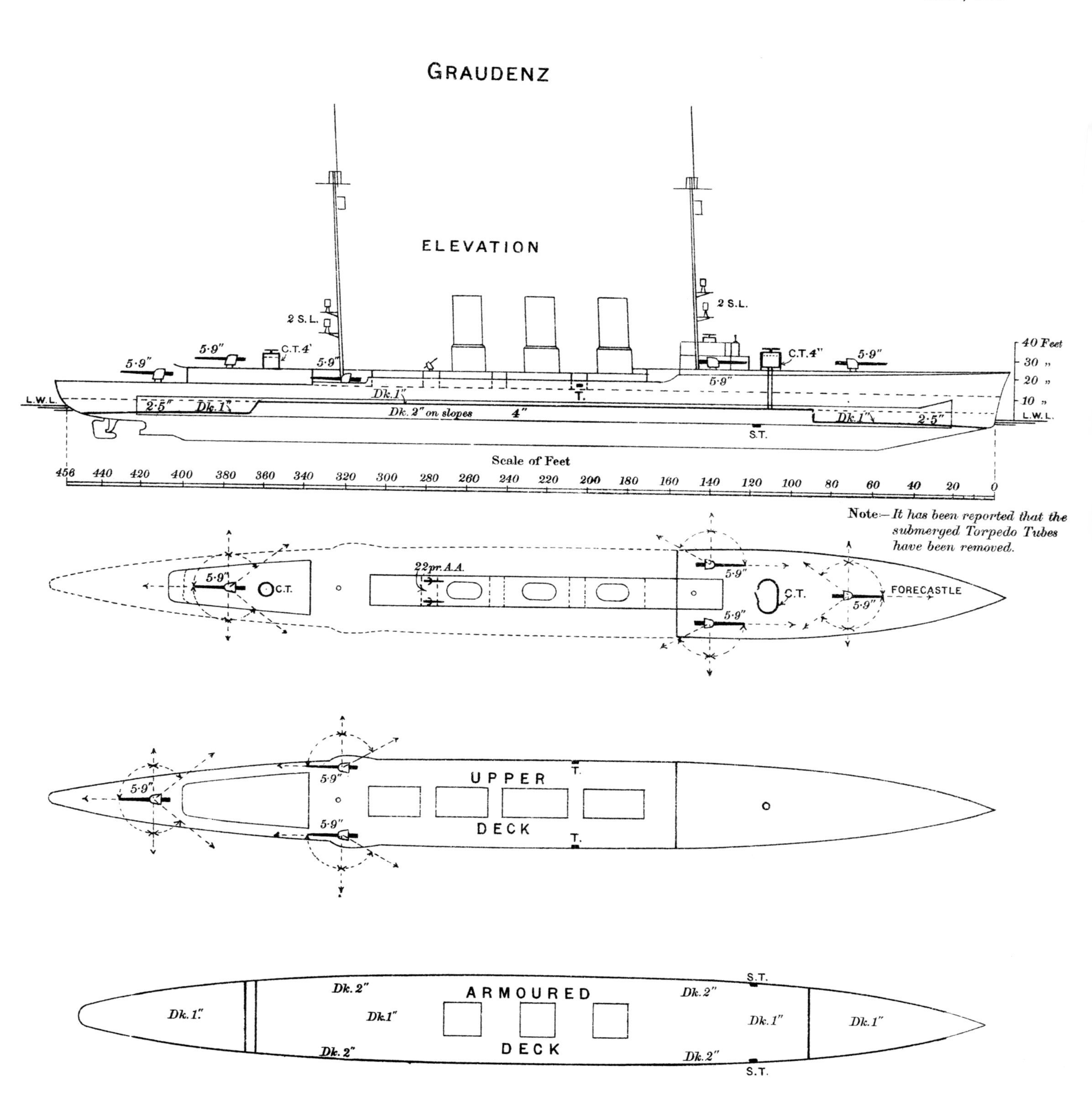

Note:—*It has been reported that the submerged Torpedo Tubes have been removed.*

Ordnance Survey, July, 1918.

REGENSBURG

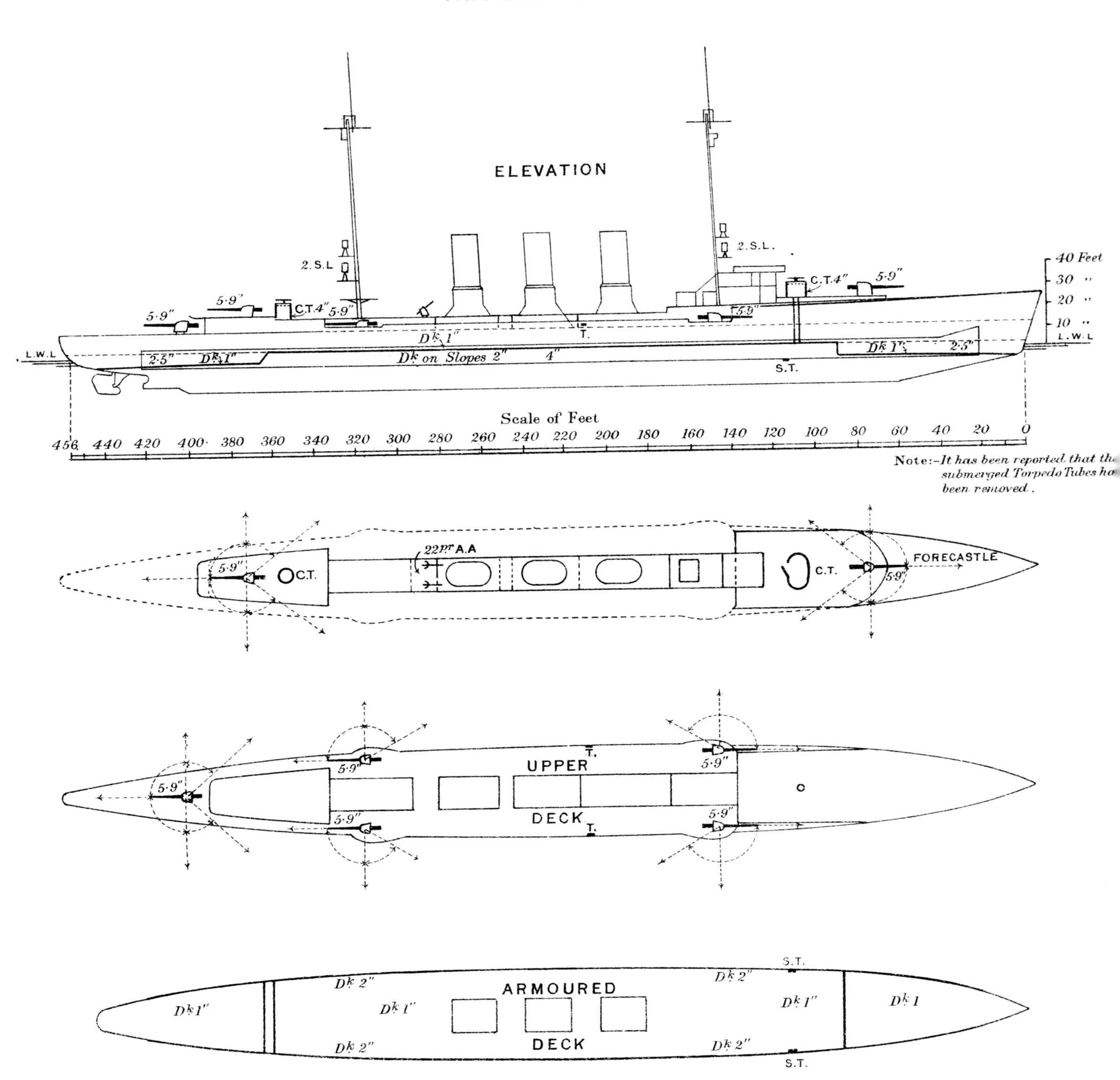

Ordnance Survey, July, 1918.

Strassburg, Stralsund.

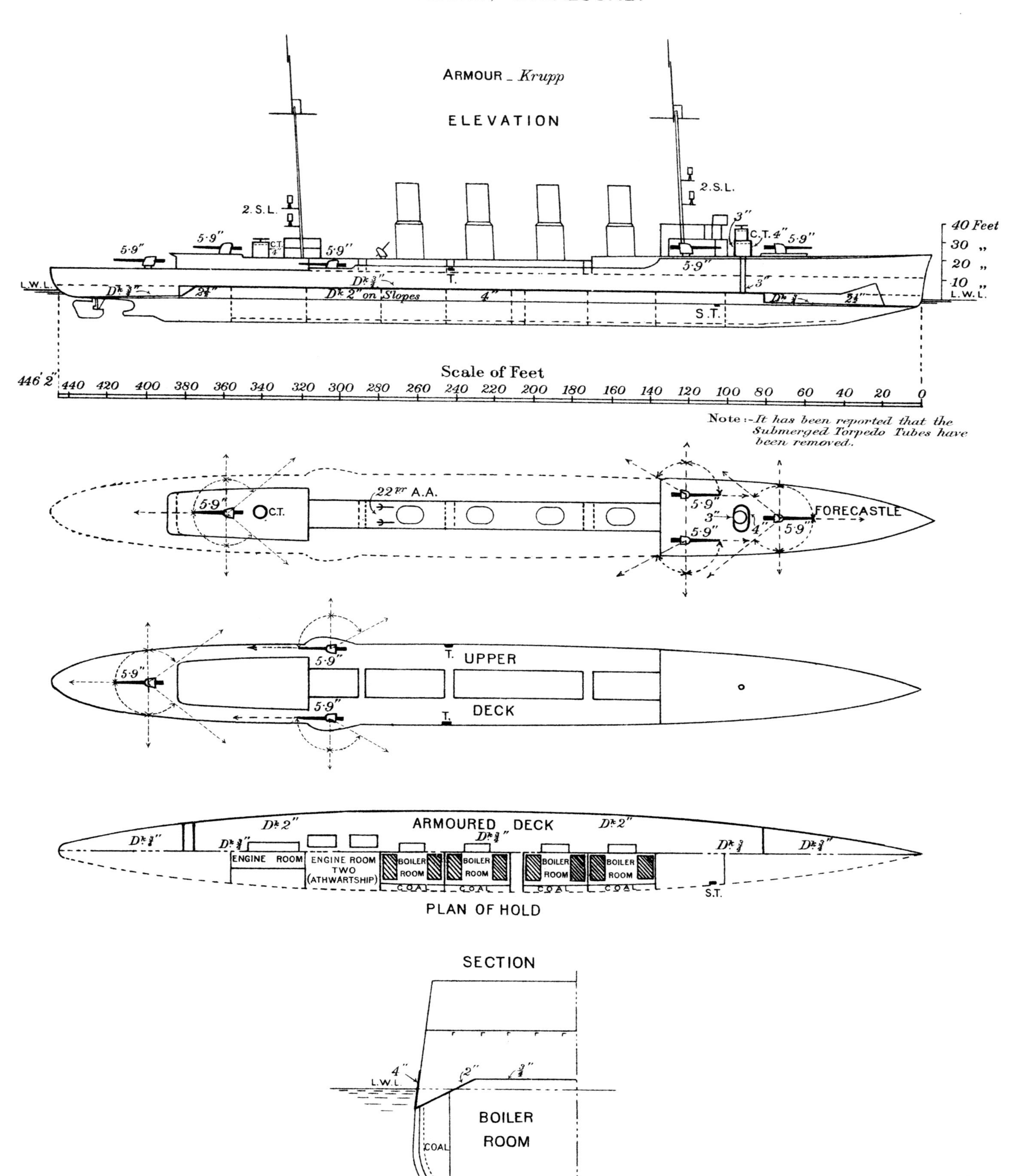

Ordnance Survey, July, 1918.

Plate 13.
C.B. 1182 Q.
June, 1918.

Kolberg and Augsburg.

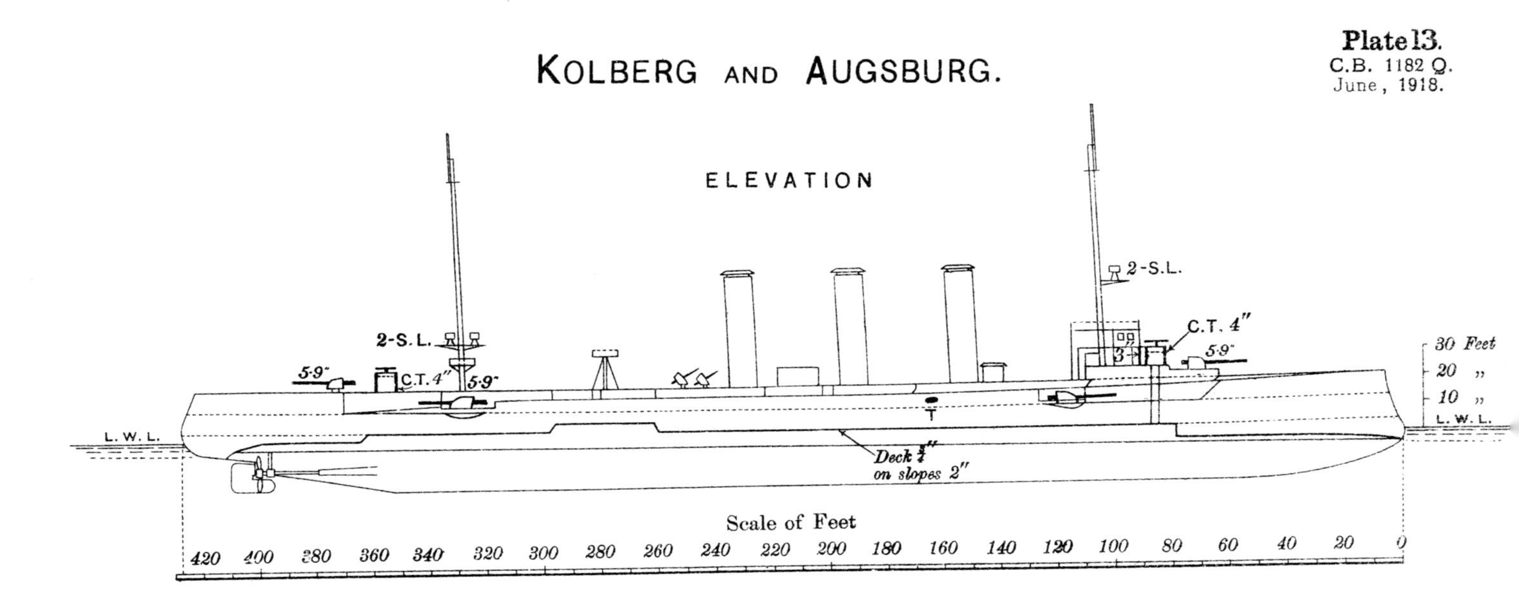

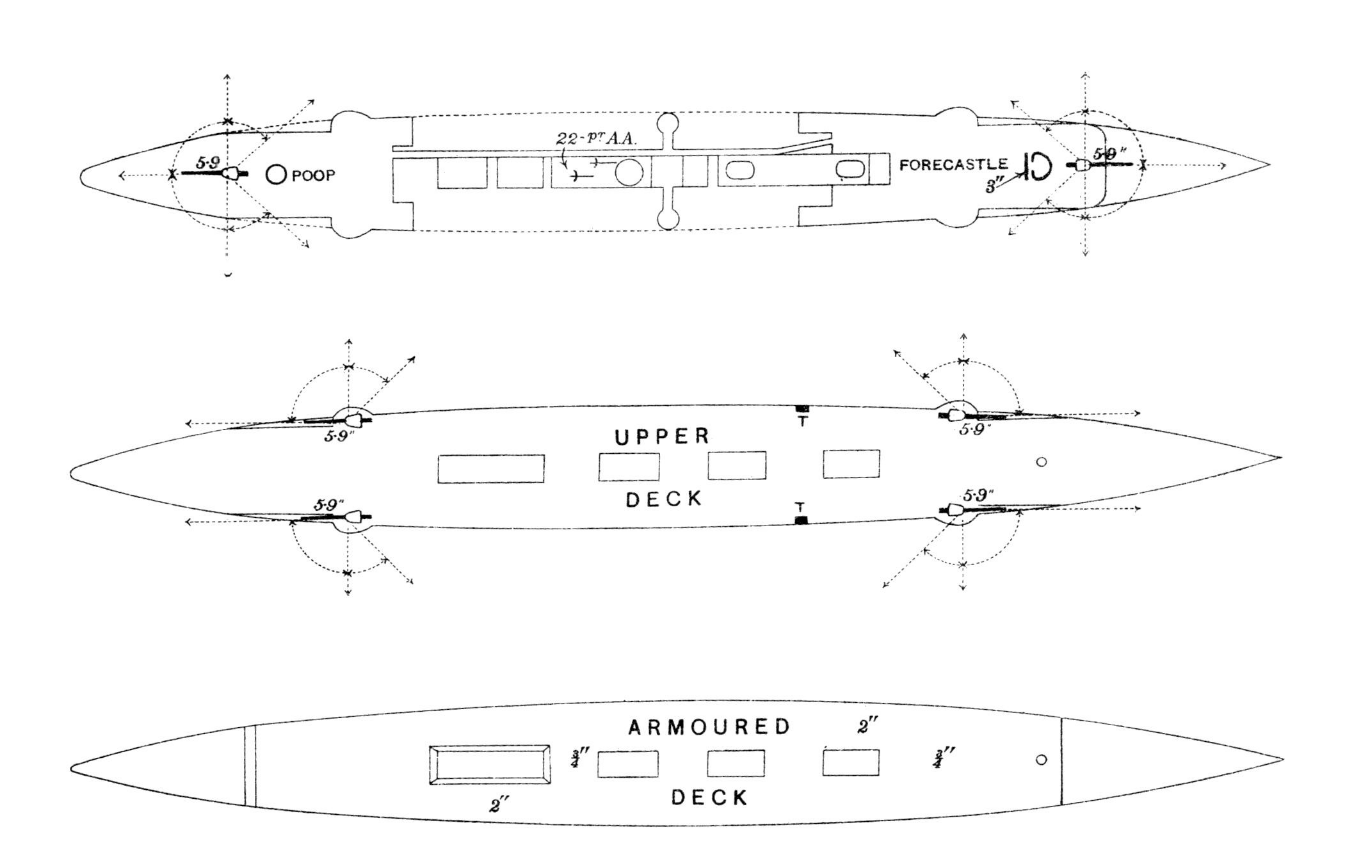

ARMOURED DECK

2"

¾"

¾"

2"

Ordnance Survey, July, 1918.

STUTTGART AND STETTIN.

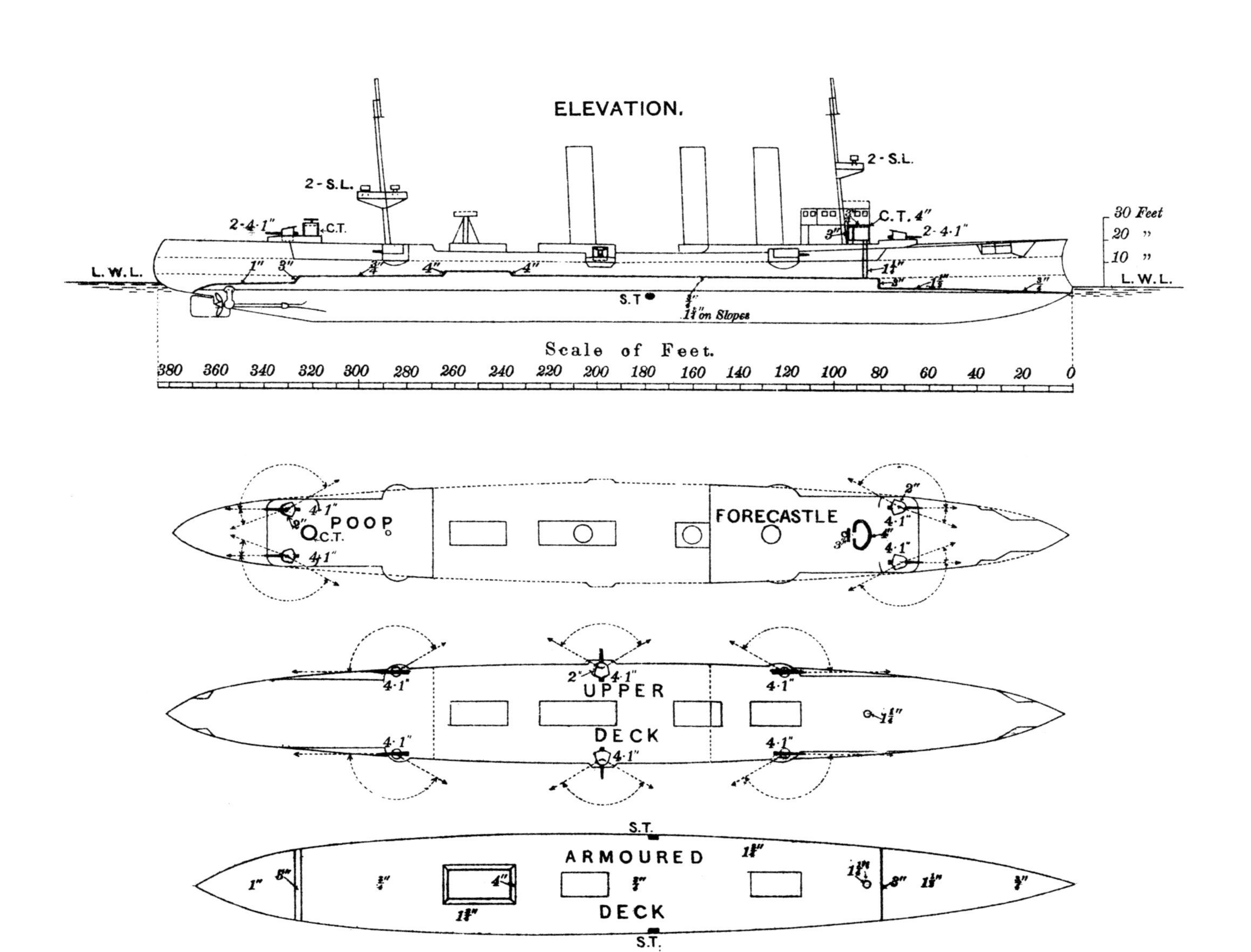

Ordnance Survey, July, 1918.

HAMBURG, BERLIN, MÜNCHEN, LÜBECK, AND DANZIG.

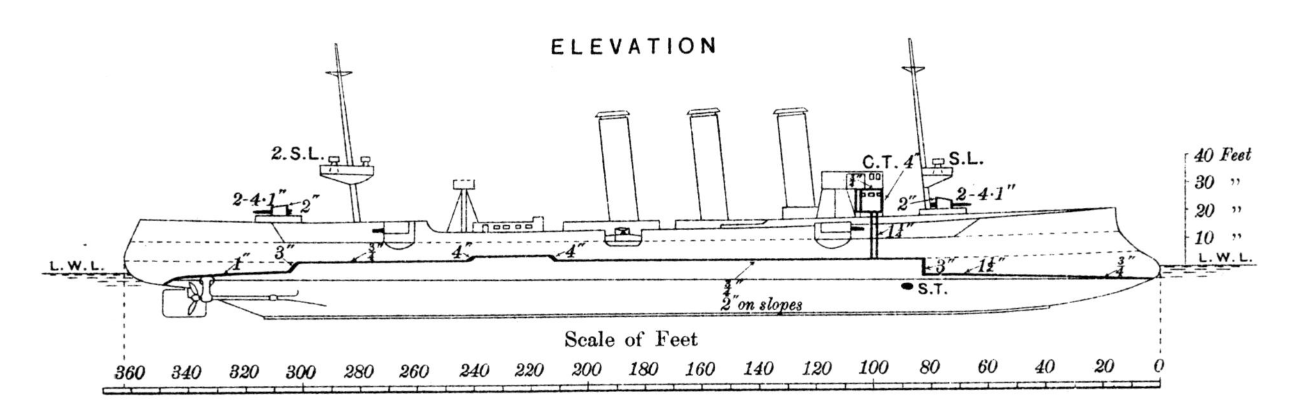

Note.—In "Danzig" the Conning Tower is before and the bridges around foremast.
There are marked differences in the funnels and masts of the different vessels of this class; for details see silhouettes.

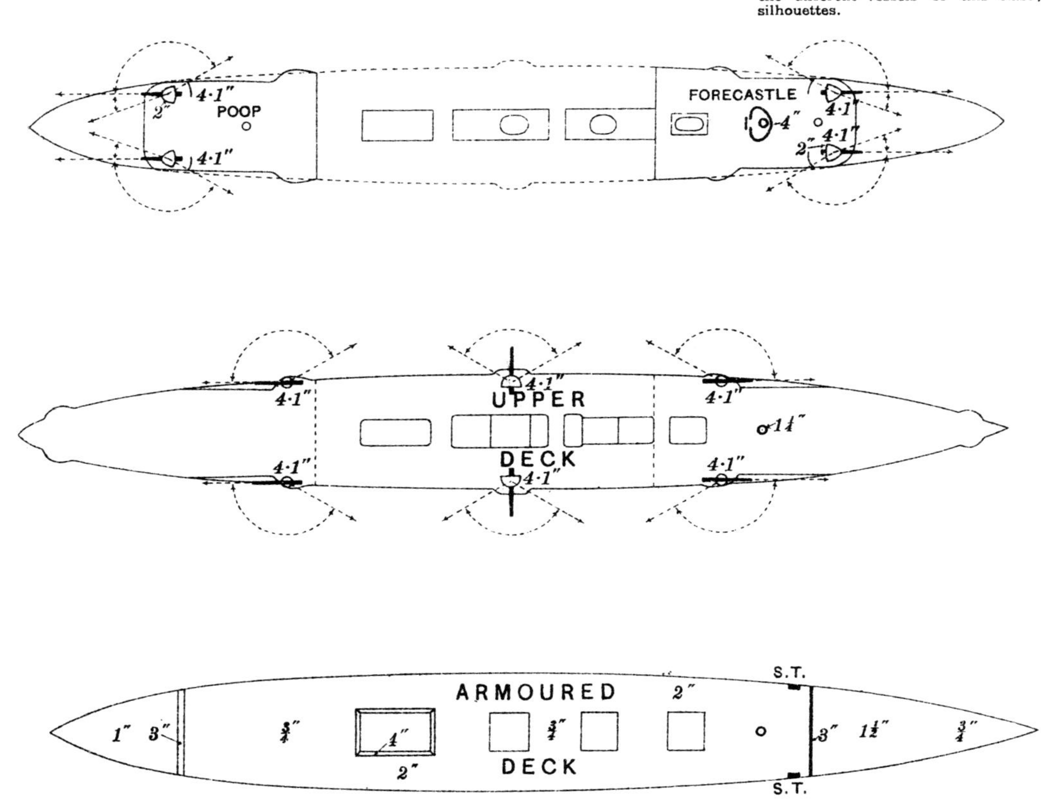

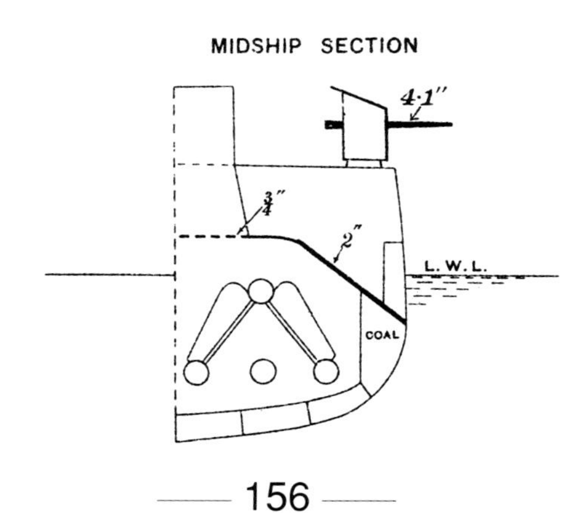

Ordnance Survey, July, 1918.

NYMPHE, NIOBE, THETIS, MEDUSA, AND AMAZONE.

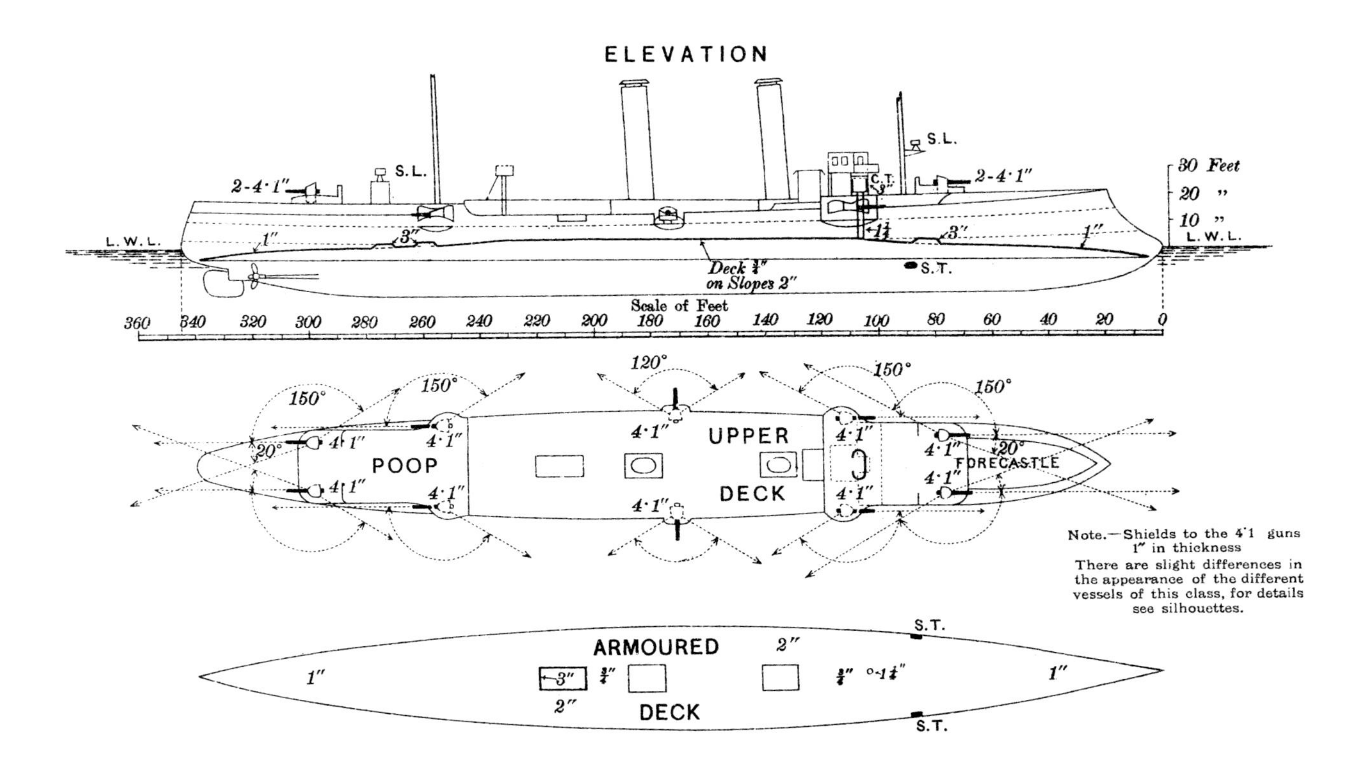

Note.—Shields to the 4·1 guns 1" in thickness
There are slight differences in the appearance of the different vessels of this class, for details see silhouettes.

Vessels of Similar Type,

GAZELLE.

The Conning Tower is just before the 2nd funnel. One submerged Tube under ram, and two above water, one either side between 2nd funnel and mainmast.

ARCONA.

Slightly broader than the above, Armour and Armament similar.

Ordnance Survey, July, 1918.

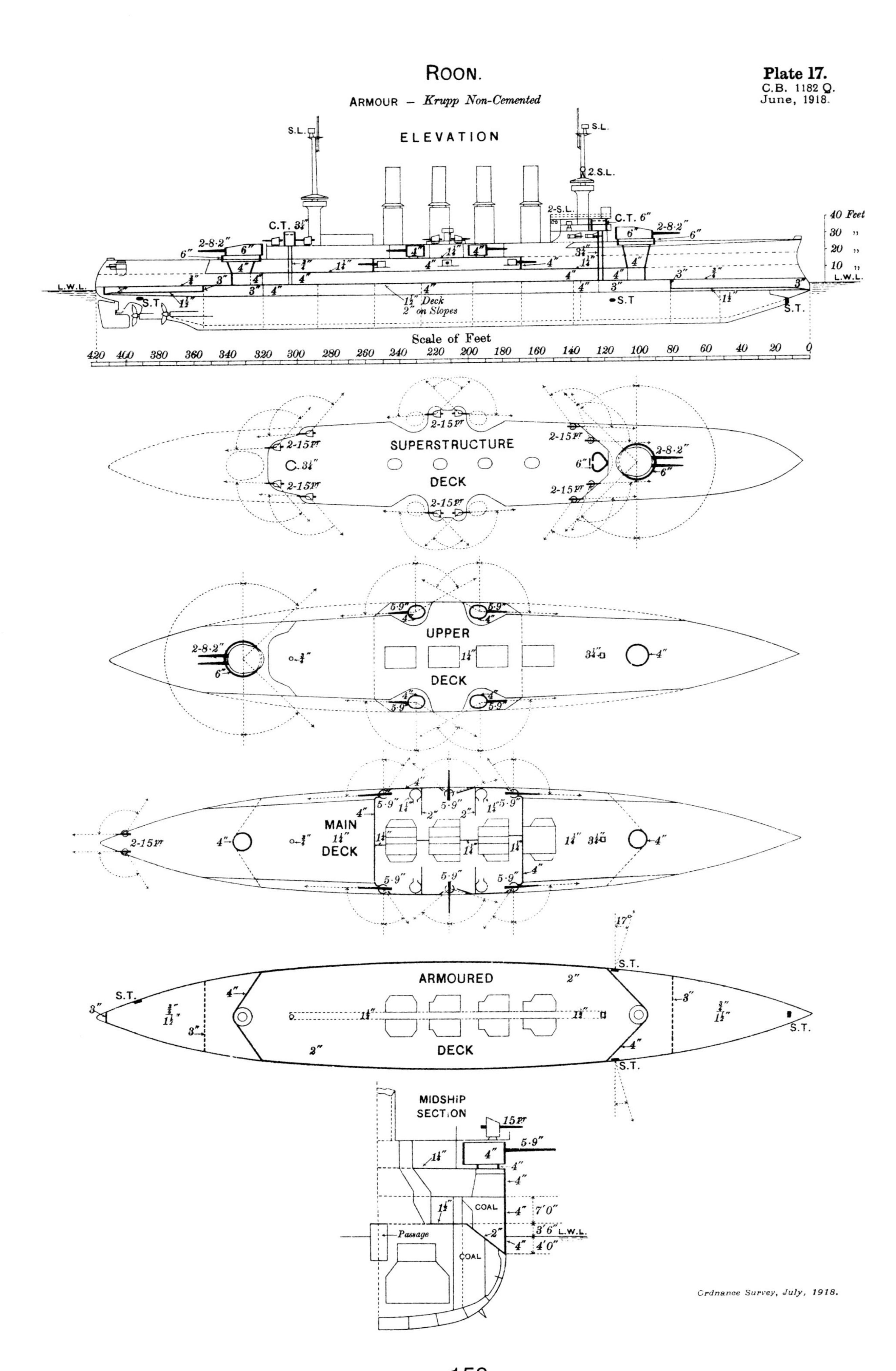
ROON.
Plate 17.
C.B. 1182 Q.
June, 1918.
ARMOUR – Krupp Non-Cemented
ELEVATION
Scale of Feet
SUPERSTRUCTURE
DECK
UPPER
DECK
MAIN
DECK
ARMOURED
DECK
MIDSHIP
SECTION
COAL
Passage
L.W.L.
Ordnance Survey, July, 1918.

PRINZ HEINRICH.

ARMOUR – *Krupp Non-Cemented.*

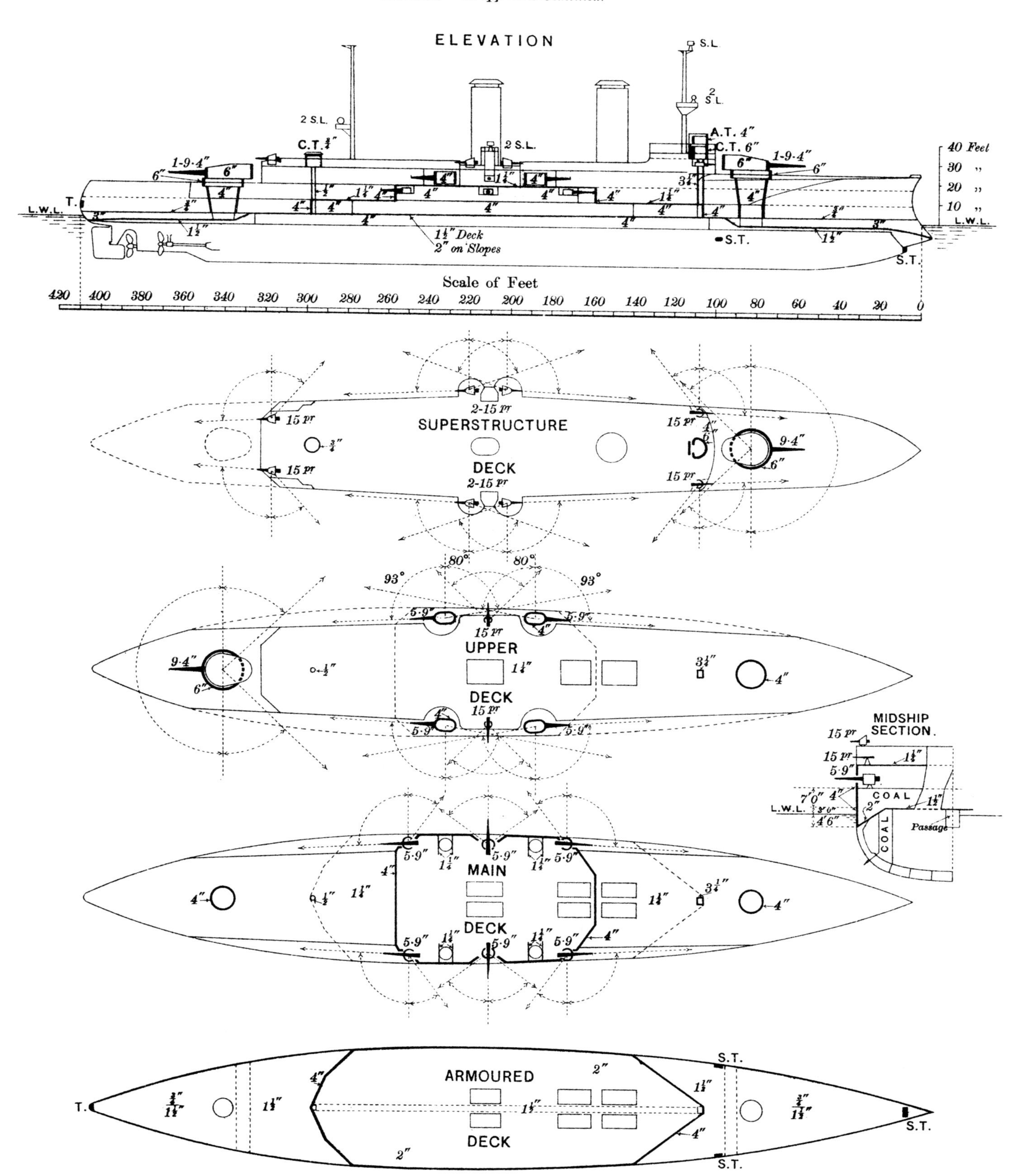

Ordnance Survey, July, 1918.

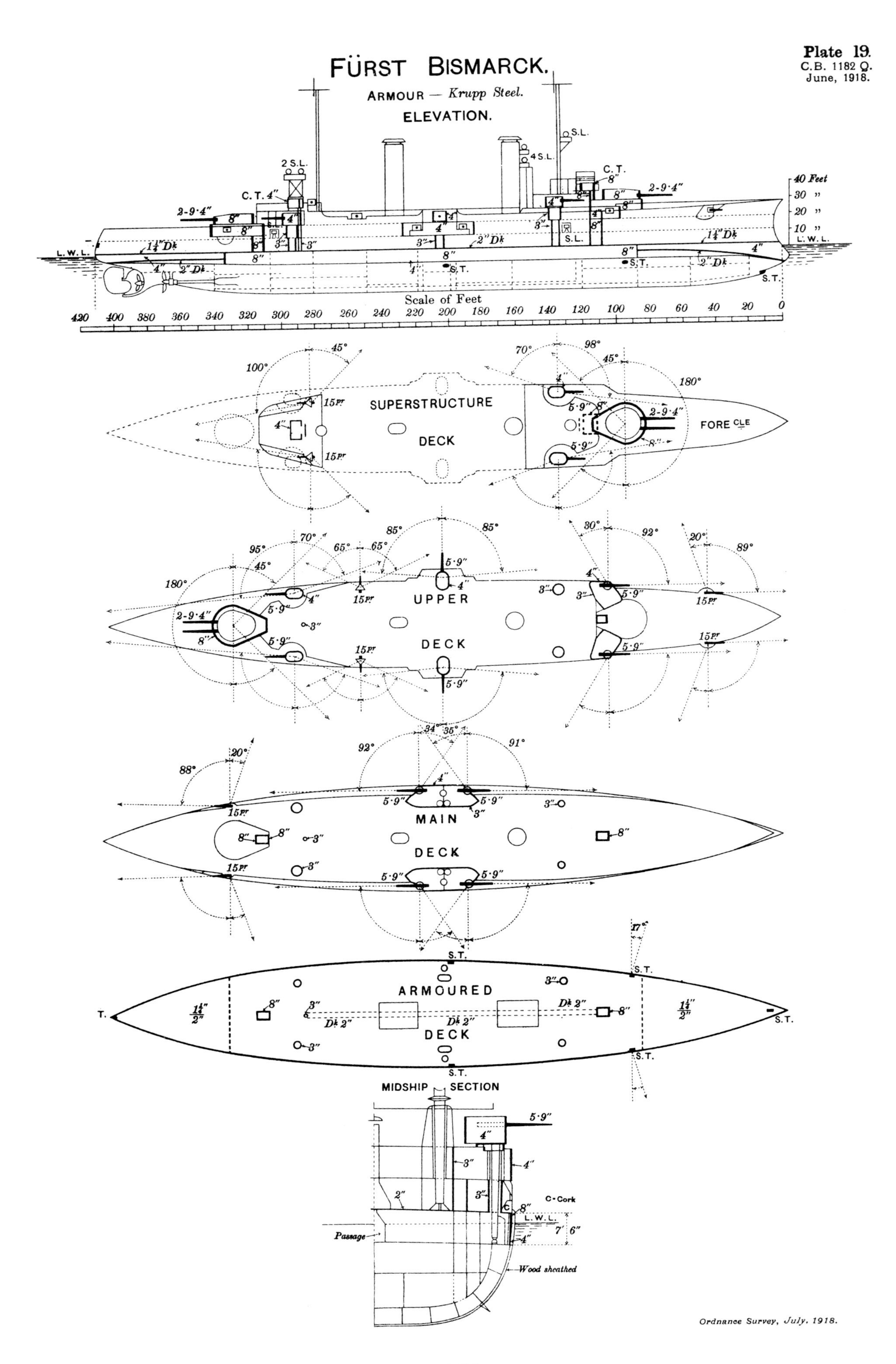
FÜRST BISMARCK.
ARMOUR — Krupp Steel.
ELEVATION.
Plate 19.
C.B. 1182 Q.
June, 1918.
Scale of Feet
SUPERSTRUCTURE DECK
FORE CLE
UPPER DECK
MAIN DECK
ARMOURED DECK
MIDSHIP SECTION
C - Cork
Passage
Wood sheathed
Ordnance Survey, July, 1918.

Plate 20.
C.B. 1182 Q.
June, 1918.

VICTORIA LOUISE
HERTHA
FREYA
VINETA
HANSA

ARMOUR - *Krupp Steel*

ELEVATION

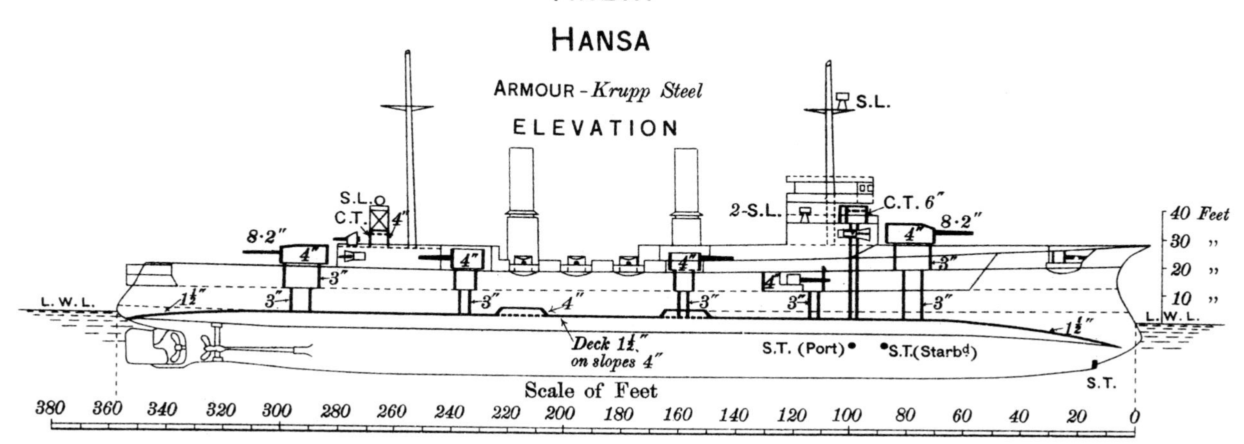

Note :- *"Vineta" has only* 2-S.T.

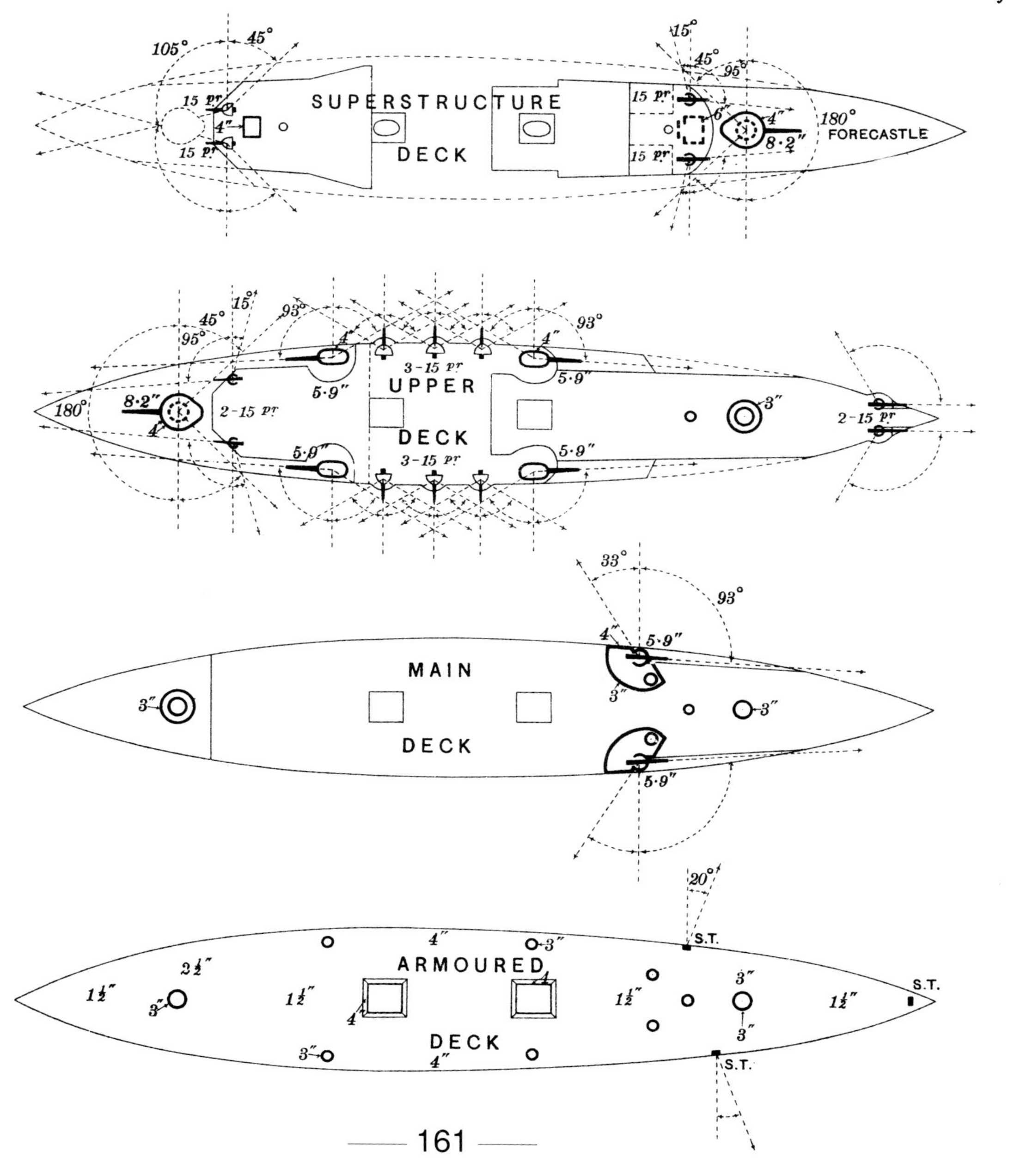

Ordnance Survey, July, 1918.

KAISERIN AUGUSTA

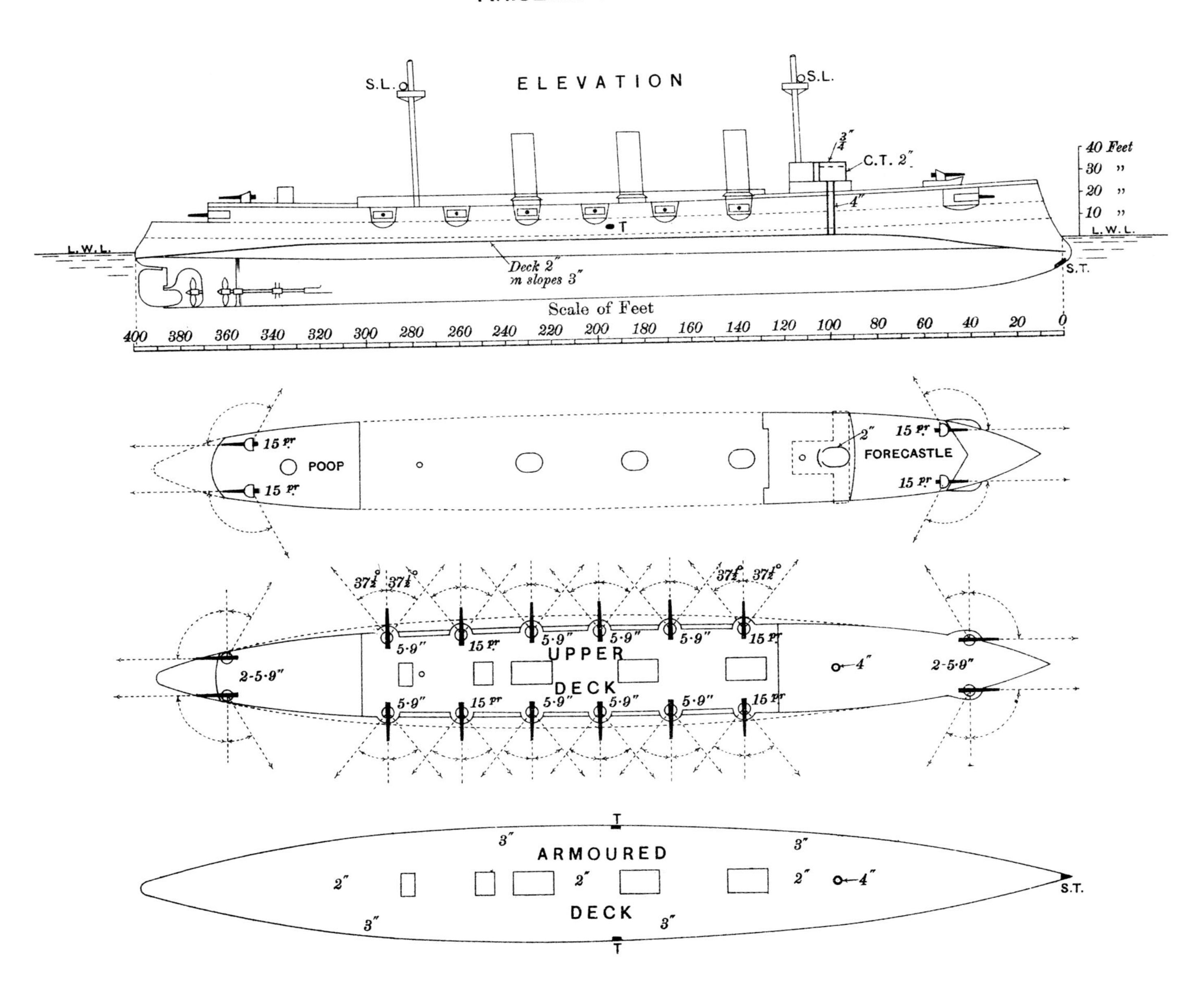

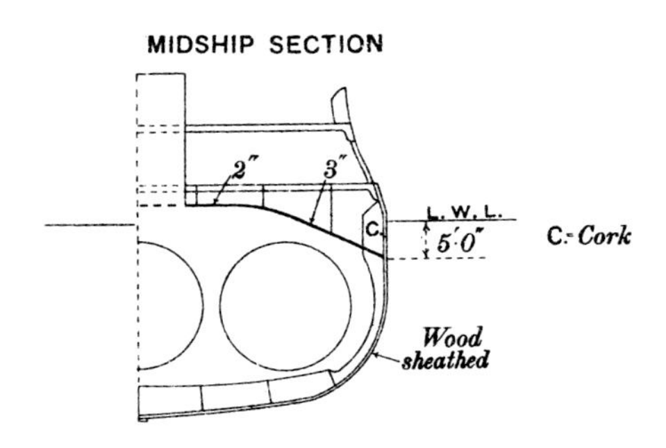

Ordnance Survey, July, 1918.

SECTION IV
DESTROYERS AND TORPEDO BOATS

NOTE:- The Vessels are arranged generally in order of date, recent boats first.

B 109-112. B 97-98. G 101-104.

S 49-52. V 47.

S 13-24. G 7-12.

T 173-G 175. T 169-172.

T 137. T 132-136

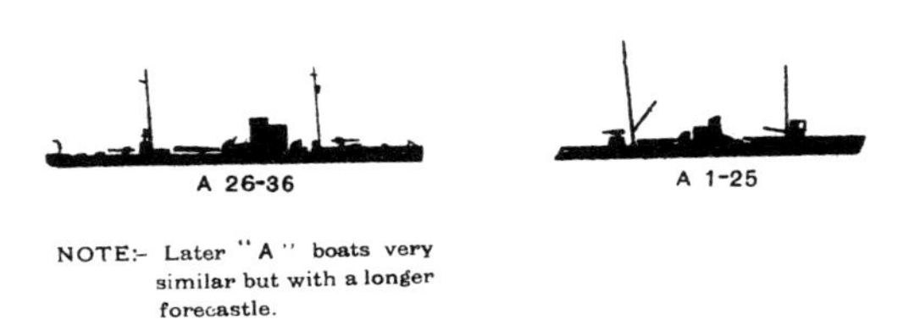

A 26-36 A 1-25

NOTE:- Later "A" boats very similar but with a longer forecastle.

T 88-89.

DESTROYERS & TORPEDO BOATS.

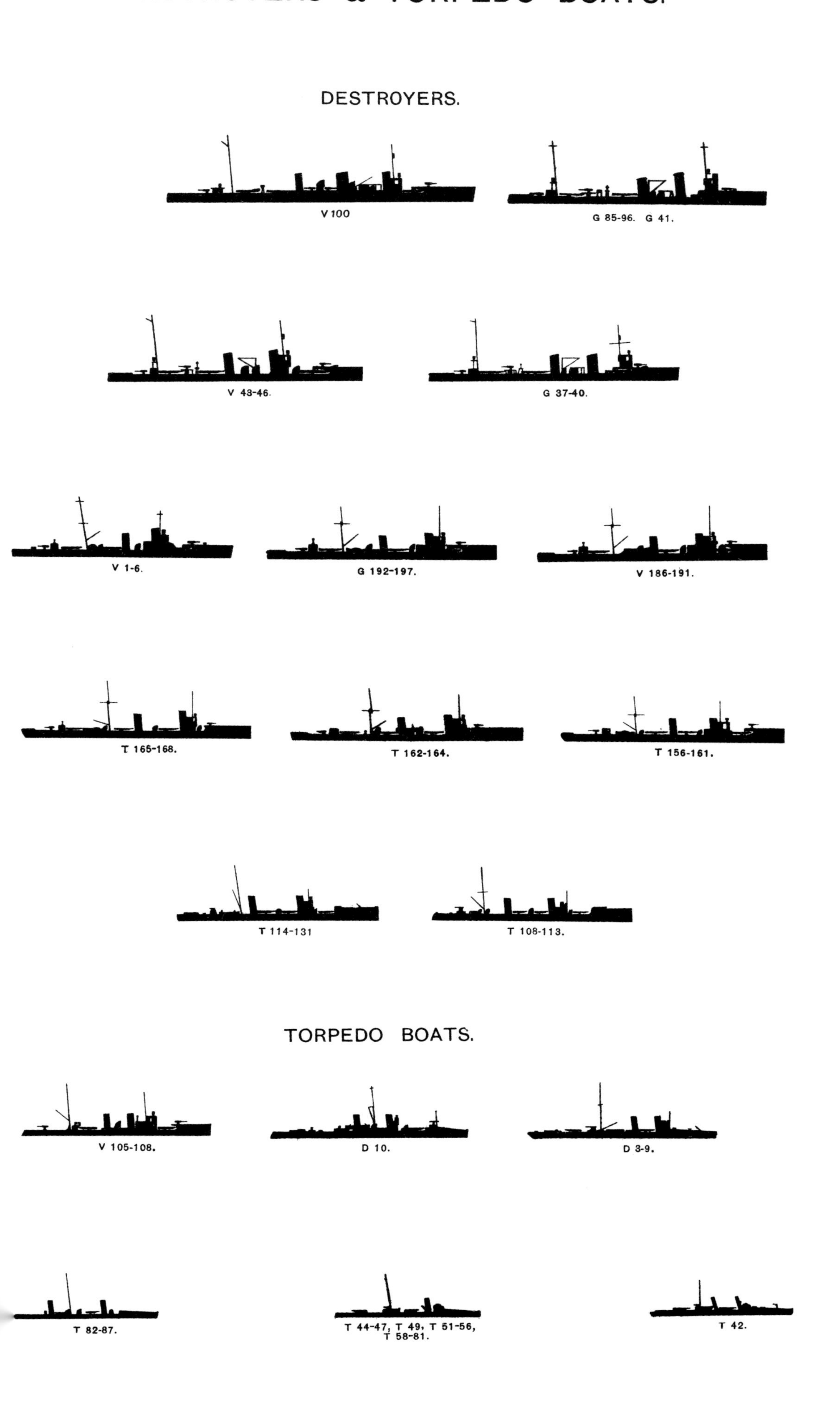

C.B. 1182R.
I.D. 1159.
August 1918.

NOTE:- A number of the later two funnel German Destroyers including "G" as well as "V" and "S" boats have had their foremost Torpedo Tubes removed, *i.e.*, those immediately abaft the forecastle.

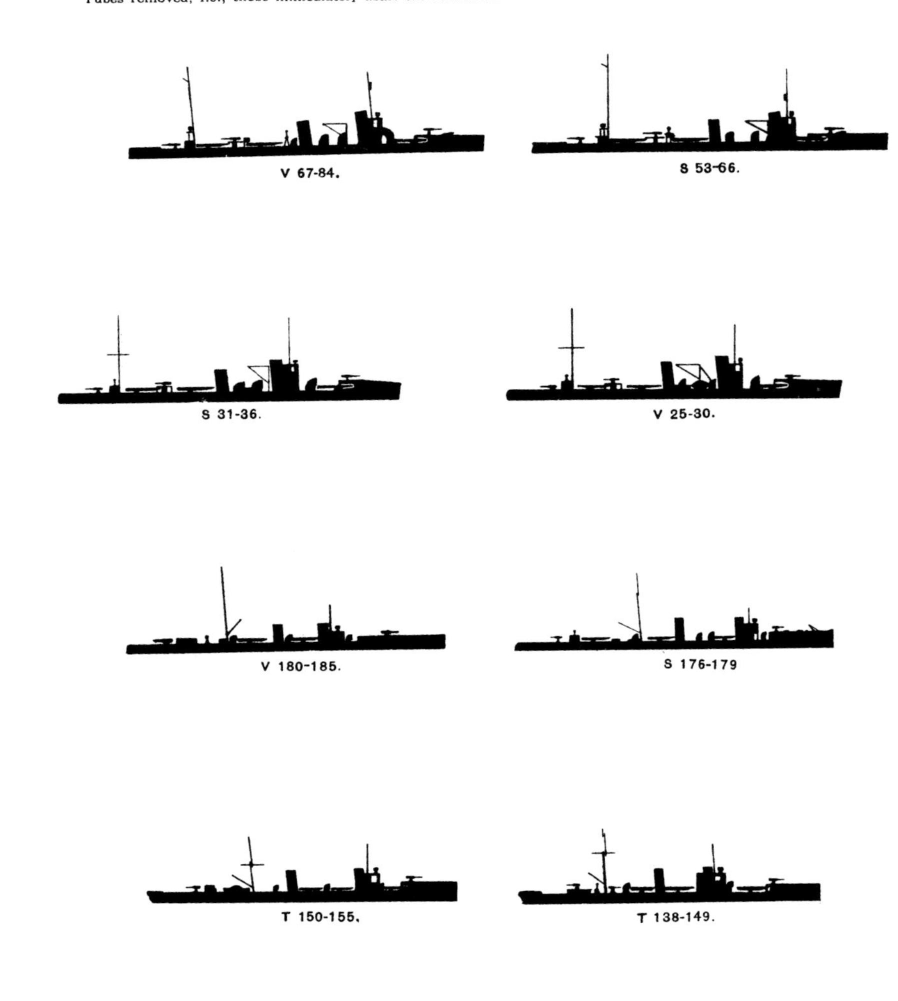

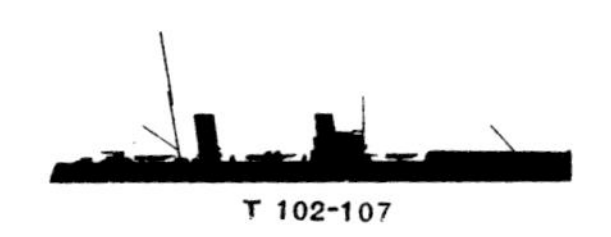

T 102-107

T 91-101.

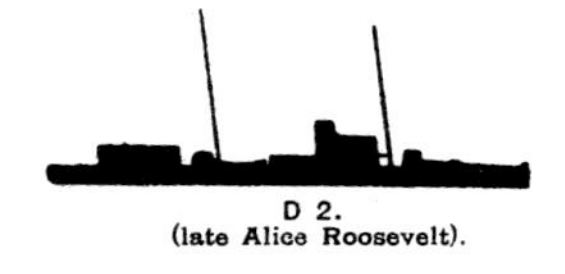

D 2.
(late Alice Roosevelt).

D 1.

T 11, T 13-16, T 20-22,
T 24, T 27-31, T 33-40.

Ordnance Survey, Sept, 1918.

GERMAN NAVY.

PART III.

SECTION 4.

DESTROYERS AND TORPEDO BOATS.

SUBJECT.

O (33) AS 4924—22 Pk 2972 2000 9/18 E | & S A

DESTROYERS AND TORPEDO BOATS.

General Remarks.

Classification.

German torpedo craft are classified officially as small torpedo boats and large torpedo boats.

"*T. 11*" to "*T. 89*" (the oldest boats) are classified as small torpedo boats, whilst the so-called divisional boats, "*D. 1–10*," and all boats subsequent to "*T. 89*," are classified as large torpedo boats.

In the present pamphlet (as in all other I.D. reports) the large torpedo boats (divisional boats excepted) are styled destroyers; but this title is not applied to them by the Germans, who prefer the term torpedo boats as being more correctly descriptive of their purpose.

The Germans sometimes use the name of destroyer for the big 3-funnel boats, "*G. 101–104*" (originally building for the Argentine), "*B. 97–98*," and "*109–112*," &c., which correspond roughly (in dimensions, speed, and armament) to our earlier flotilla leaders.

Nomenclature.

When Germany commenced seriously building torpedo craft in 1883, a system of consecutive numbering in a series from 1 upwards was adopted.

In order to encourage a spirit of emulation among the builders, however, each boat carried the initial letter of her building yard before her serial number. These initial letters were "G," for Germania Yard, Kiel; "S" for Schichau Yard, Elbing; and "V" for Vulcan Yard, Stettin. Thus No. 1 became "*S. 1*," No. 88 became "*G. 88*," and so on. Carefully managed press notices of such vessels as had distinguished themselves in the matter of speed or seaworthiness soon rendered the plan a complete success.

When the original series had reached "*G. 197*," and the earlier vessels were no longer likely to break any speed records, the authorities apparently considered that the numbering had gone high enough, and they commenced altering the initial letters of the older boats to "T" (standing for torpedo boat) in order to avoid confusion with the new series, which was commenced with "*V. 1*" (1911–12 programme). Up to date (August 1918) this re-lettering process has reached the destroyer "*G. 173*," which recently became "*T. 173*."

During the war additional initial letters have appeared in the shape of "B" for Blohm and Voss Yard, Hamburg; "H" for Howaldt's Yard, Kiel; and "WW" for Weser Yard, Bremen; whilst the letter "V" has been used for destroyers built at Vulcan Yard, Hamburg, as well as for those built at Stettin.

All German torpedo craft are built in private yards.

Besides the two main series, the old and the new, two entirely separate series have been built. The first carries the initial letter "D," is numbered 1–10, and consists of old divisional boats originally intended to lead divisions of small torpedo boats. The other series carries the initial letter "A," is numbered 1–100 and beyond, and includes three or more classes of torpedo boats built since the outbreak of war and used mainly for minesweeping and patrol work.

Total Numbers.

In August 1918 Germany possessed, completed, about 190 destroyers and 150 torpedo boats of all types.

RECOGNITION.

(*See* Plates 1–7 and Silhouettes.)

Destroyers.

All German destroyers have either two or three funnels. The former present certain broad characteristics and form a type very distinct from British boats. The latter, of which a round dozen have been completed, can be fairly easily identified by certain salient details from those of our own boats which may resemble them.

Nearly all German destroyers carry on their masts prominent flotilla marks, viz., circles, triangles, &c., and also two horizontal hoops for night recognition signals.

Colour.

All destroyers are now painted light grey, with the exception of certain boats employed in the Sound, which have recently been reported as dazzle-painted.

Three-Funnel Destroyers.

There are three existing classes, viz., "B," "V," and "G" boats, all of the same general type but differing in detail. The "B" and "V" boats have a characteristic well between the forecastle and bridge, three funnels of equal height but varying thicknesses, and two masts, the fore being the shorter. The "G" boats, on the other hand, have :—

(*a*) A shorter forecastle on which the bridge is built.

(*b*) Funnels of equal thickness, the fore funnel being raised.

The position of the boat derrick is different in each class (*see* Plate 1 and Silhouettes).

Two-Funnel Destroyers.

Pre-War Types.

These are distinguishable by :—

(*a*) The low freeboard.

(*b*) The well deck between the forecastle and navigating bridge, which in the majority of boats is of sufficient length to permit of torpedo tubes being mounted there.

(*c*) The position of the bridge, close back against the foremost funnel.

(*d*) The rig—a short signal pole on the bridge and a tall mainmast, usually well abaft the after funnel.

War Types.

These are somewhat enlarged editions of the older types, having the same general features, excepting that :—

(*a*) The signal pole is replaced by a regular foremast, usually fitted with a crow's nest. The mainmast is taller and stepped well aft by the wheel house.

(*b*) A samson post with boat derrick is carried between the funnels.

(*c*) The fore funnel is higher than the after one.

(*d*) The armament has been increased from two guns and four tubes to three guns and six tubes.

(*e*) "*V. 67–84*" have the foremost ventilating cowl replaced by a conspicuous trunk curving upwards and aft into the bridge structure.

In one class only is there no well between the forecastle and bridge, viz., in "*G. 37–42*" and "*G. 85–96.*" In these boats the fore funnel is about midway between the bridge and the after funnel, and the samson post is fitted nearer to the after than to the fore funnel, which is the reverse of what obtains in "S" and "V" boats.

Torpedo Boats.

With the exception of "*D. 3–9*" and "*V. 105–108,*" which look like small editions of destroyers, German torpedo boats have :—

(*a*) Only one funnel with either one or two masts ; or

(*b*) Two funnels widely separated (in a few cases with a mast between them).

The forecastles are long and low, and there is no gun platform such as is fitted in British boats.

Colour.

Most, if not all, torpedo boats are now painted, like destroyers, light grey.

Series "A" Torpedo Boats.

These boats have all been built since the outbreak of war ; about a hundred are known to have been completed. They all have one funnel and two masts.

The earliest class are numbered "*A. 1–25,*" the second class "*A. 26–36*" (possibly higher). The rest form a third and probably a fourth class, representing slight advances on "*A. 26–36.*"

A large proportion of these boats have had their torpedo tubes removed and replaced by a powerful winch for minesweeping purposes.

A 2

General Details.

Accommodation.

Generally good. The "G" boats are considered the most comfortable. In war time the crews of German torpedo boats always live on board, unless the boat is in dockyard hands. All their uniform is carried with them, and they have to be in possession of their No. 1 rig. Clothes are kept in bags stowed in iron racks fixed close to the ship's side abreast the men's messes.

In the modern "G" boats, *G. 37–42* and *G. 85–96* (complement, 98 officers and men), the accommodation is arranged as follows, from aft forward :—

Petty Officers' Mess.—Right aft, extending five scuttles on each side. All the petty officers of the ship mess together, about 24 in all. About 19 or 20 collapsible bunks are fitted; the remaining P.O.'s sleep in hammocks.

Warrant Officers' Mess.—Forward of the petty officers' mess. Each warrant officer has a cabin to himself.

Officers' Mess.—Amidships, immediately before the warrant officers' mess. Four or five officers are carried, and each has a cabin to himself.

Seamen's Mess.—The seamen's mess deck is under the forecastle. Ten collapsible bunks are fitted, about 20 men sleeping in hammocks.

Stoker's Mess.—The stokers' mess deck is immediately below the seamen's mess deck.

See also under Steam Heating, &c. (p. 10).

Armament.

Up to the outbreak of war all German torpedo craft were designed as torpedo boats proper, and not as destroyers, and hence carried a strong torpedo armament with a relatively weak gun armament. The lessons of the war, however, and particularly the Battle of Jutland, have led to the abandonment of this system and the introduction of a considerably heavier gun armament. Prior to the war, two 22-pr. (8·8-cm.) L/30 guns formed the standard armament for destroyers. In the boats completed towards the end of 1914, the number of 22-prs. was increased to three. In the course of 1915, a new 4·1″ (10·5-cm.) destroyer gun was introduced, and during 1916 boats appeared armed with three of these guns. Though most, if not all, of the first war series of destroyers, up to *G. 96*, were originally designed for three 22-prs. only, the majority of them have since been armed or re-armed with three 4·1″ guns, and this has now become the standard armament. So far, only the large 3-funnel boats carry *four* 4·1″ guns. No gun shields are fitted.

All modern boats also carry two machine guns, for which tripod mountings are provided in several different positions.

Ammunition Supply.

Fixed ammunition is used for both 4·1″ and 22-pr. guns. 22-pr. ammunition is supplied to the gun in cylindrical leather cases, each containing five rounds. Metal racks for these cases are fitted abreast the guns, providing a ready supply. The 4·1″ ammunition is supplied to the gun in boxes, each containing three rounds. Six such boxes are kept stowed in the vicinity of the gun. The foremost gun is supplied from a magazine immediately below it; the after gun or guns from a magazine aft, the ammunition being passed up by hand through the after wheel-house.

Gun Control.

Fire is controlled by means of electric transmitter, telephone, or flexible voice-piping. For rangefinders, *see* p. 9.

An electric receiving apparatus (*Geber*) is fixed above each gun in front of the breech. It shows (*a*) range, to the nearest 100 metres; (*b*) deflection, in red figures for port, green figures for starboard; (*c*) bearing in degrees, so that the gun may be laid on a particular bearing (a graduated bearing ring being fitted on the pedestal of the gun); (*d*) orders for firing. The last-named are as follows :—Rapid Fire (*Schnellfeuer*), Independent (*Selbständig*), Local Control (*Geschützweise*), and Cease Firing (*Batterie Halt*).

Fire gongs are also fitted.

Salvo firing is always used as long as the control remains intact, and salvo firing in half flotillas appears to be practised considerably.

Torpedo Tubes and Torpedoes.

All destroyers prior to *G. 174* (1909–10 programme) carry three 17·7″ revolving tubes mounted in the centre line, so that all three will bear on either broadside. Five torpedoes are carried, three in the tubes and two spare.

All destroyers from *G. 174* (old series) to *S. 24* (new series) inclusive carry four 19·7″ tubes.* The two foremost tubes are mounted one on either side, directly abaft the forecastle, with an arc of training of about 45° from nearly right ahead to about 40° before the beam. The other two are mounted in the centre line for all-round training. This disposition permits of the use of three tubes on either beam. Five torpedoes are carried—four in the tubes and one in a watertight holder on deck.

All destroyers from *V. 25* onwards originally carried six 19·7″ torpedo tubes, viz., one on either side, directly abaft the forecastle (as in *G. 174–S. 24*), the other four mounted in pairs on the centre line; the two tubes of each pair diverge at an angle of about 12°. Seven torpedoes were carried, six in the tubes and one spare. In a number of these boats, however, from *G. 37* onwards, the two foremost tubes have now been entirely removed, this reduction of armament presumably having been found necessary in order to balance the considerable increase in weight due to substituting three 4·1″ guns for the three 22-prs. for which the boats were all (or nearly all) designed.

In all modern destroyers, along one side of the upper deck, rails are fitted for a torpedo trolley. They are usually on the port side in "S." and "V." boats, and on the starboard side in "G." boats.

Torpedo Control.

Torpedoes are usually fired electrically from the bridge, a director being fitted on either side, probably on the lower bridge. As a rule, the two foremost tubes of modern boats are kept locked on the extreme forward bearing, owing to their being inaccessible in anything but fine weather; the two pairs of tubes aft are trained to the most suitable bearing shortly before firing.

Mines.

See under Minelaying, p. 9.

Boats.

Most destroyers carry two boats—a dinghy and a whaler—which, in destroyers up to *S 24*, are stowed just before the mainmast, but in later vessels are stowed either abreast one of the funnels or between the funnels. They are hoisted out by derrick. Some leaders of flotillas carry a motor boat instead of a whaler.

Colour.

See under Recognition, p. 3.

Compasses.

In the majority of recent destroyers, five liquid compasses are carried, mounted as follows:—

(1) On upper bridge.
(2) On lower bridge.
(3) Under break of forecastle.
(4) In after wheel-house.
(5) The standard compass, on a metal platform abaft after funnel.

In *G. 37–42*, *G. 85–96*, and *G. 101–104*, there is no compass under the break of the forecastle.

The bridge compasses are stated to be perfectly reliable, in spite of the proximity of the funnel in all German destroyers except recent "G." boats.

Gyro compasses are not fitted.

Complement.

Recent destroyers carry four or five officers; viz., three executive officers, one engineer, and sometimes a surgeon†; about 4 warrant officers, about 45 deck ratings and 45 engine-room and other ratings; total, 98–100 officers and men.

* *S. 13–24* are believed to have had the two foremost tubes removed on being rearmed with 4·1″ guns.

† Leaders of flotillas carry a *Stabs-arzt* or senior *Oberassistenzarzt*, leaders of half-flotillas a junior *Assistenzarzt*. Other destroyers do not carry medical officers.

A 3

Constructive Details.

General.

In the endeavour to ensure the continuance of the essential characteristic of German torpedo craft, viz., high speed, without unduly increasing their size and visibility, the construction of destroyers has been kept admittedly light. It is claimed, nevertheless, that by a careful choice of material and a scientific system of construction, the requisite strength has been secured; and this appears to be borne out on the whole, so far as the hull is concerned, by the tests of active service.

Frames.

The inset plan shows a section through Engine and Boiler Rooms of "*V. 1–6*" class:—

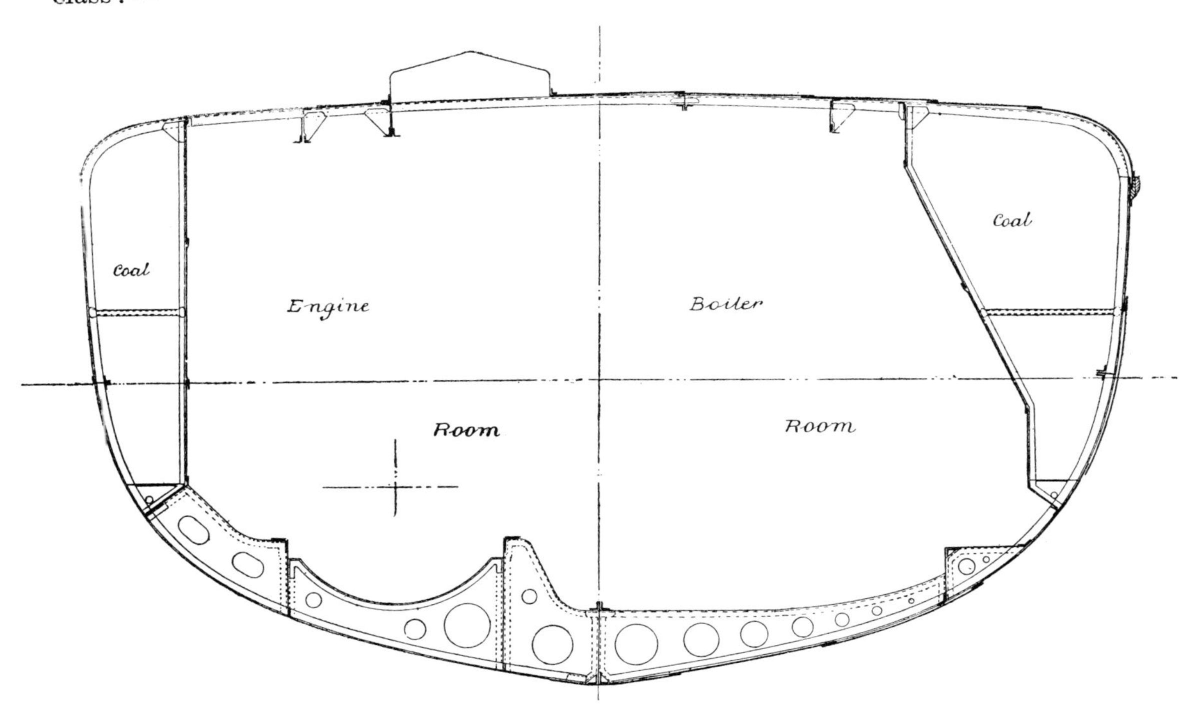

The frames are 19·7 ins. apart amidships.

Plating.

The plating, which is of Siemens-Martin steel, has apparently increased in thickness. It is reported as being, for—

T. 114–131 (1902 to 1904–5 programmes) - - - - -	bottom, plating - -	·16″
	side plating - -	·14″
T. 138–149 (1906–7 programme) -	plating amidships -	·29″
V. 1–6 (1911–12 programme) - -	side and bottom plating	·37″ to ·14″

The figures for *V. 1–6* are authentic.

Weights.

Great economy in weights, both above and below deck, is observed, and to this end aluminium has been introduced to an increasing extent. In *T. 138–149* all panels and lockers are made of it; in *T. 165–168* the boiler casings in part also; and in *V. 1–6* and *G. 7–12* many fittings for which steel was formerly used are made of aluminium.

Subdivision.

Internally the hull is divided into 12 or more water-tight compartments. In all modern destroyers each boiler is in a separate stokehold.

Metacentric Height.

In *V. 1–6* the metacentric height is about 32½ inches.

Cost.

The *average* cost of one destroyer of each year's programme was :—

Series.	Programme Year.	Hull, Machinery, &c.	Gun Armament.	Torpedo Armament.	Total Cost.	Remarks.
		£	£	£	£	
T. 90—T. 95	1898–99	39,742	3,474	4,892	48,108	Reciprocating engines. (T. 90—T. 95 to T. 120—125)
T. 96—101	1899–1900	39,742	3,474	4,892	48,108	
T. 102—107	1900–01	47,619	3,474	4,892	55,985	
T. 108—113	1901–02	47,619	3,474	4,892	55,985	
T. 114—119	1902–03	45,173	3,474	4,892	53,539	
T. 120—125	1903–04	45,173	3,474	4,892	53,539	T. 125 turbines.
T. 126—131	1904–05	45,173	3,474	7,338	55,985	New type of torpedo introduced, and number of torpedoes increased.
T. 132—137	1905–06	45,173	6,279	7,338	58,790	T. 137 alone received turbines. Gun armament of all vessels increased.
T. 138—149	1906–07	58,709	6,930	7,338	72,977	
T. 150—161	1907–08	61,155	6,930	7,338	75,423	T. 161 alone received turbines. Price of boilers increased.
T. 162—164 T. 165—168 T. 169—173	1908–09	77,870	7,746	9,377	94,993	Displacement increased. All destroyers of this and succeeding programmes fitted with turbines. An improvement was effected at this date in torpedoes.
G. 174—175 S. 176—179 V. 180—185	1909–10	77,870	7,746	13,698	99,314	The number of torpedo tubes was increased from three to four, and the calibre to 19·7 in.
V. 186—191 G. 192—197	1910–11	77,870	7,746	13,698	99,314	
V. 1— 6 G. 7— 12	1911–12	67,271	7,746	13,698	88,715	The displacement was reduced from about 640 to about 560 tons.
S. 13— 24	1912–13	67,271	7,746	13,698	88,715	
V. 25— 30 S. 31— 36	1913–14	80,724	7,746	17,539	106,009	Displacement again increased. "All oil" fuel introduced.

Depth Charges.

Destroyers and some of the torpedo boats operating in the North Sea usually carry from two to six cylindrical depth charges, for which a launching trough is fitted on either quarter. These depth charges are about 3 ft. in length and 16 ins. in diameter, and are made in two parts, comprising a buoy and a charge case, which are dropped together. The charge case, which contains about 56 lbs. of explosive, is released from the buoy after a certain number of seconds, according to the adjustment of an indicator on the buoy; the *depth* at which the charge explodes is adjusted by a separate indicator on the charge case. These indicators are set by orders from the bridge, and the depth charges are dropped by hand, two sets of hand-release gear being fitted at the stern of the boat.

Draught Marks.

The draught is marked in decimetres (1 dcm. = 3·93 ins.).

Explosive Kite.

German torpedo craft are now supplied with an anti-submarine explosive kite, about 4 ft. in length and 2 ft. across, which can be towed at any speed between 6 and 21 knots. It has some form of inertia pistol, and is towed by means of an armoured electric cable which is carried aft in the boat on a drum of special construction; the drum contains a dry battery and is fitted with switches, pilot lamp, &c. The kite is used when searching for a submarine which has been reported or sighted in the neighbourhood and has dived. It is towed from the minesweeping fairlead aft (*see* p. 9), where fitted, otherwise from a hinged derrick at the stern of the boat. Three or four of these kites are usually carried.

Fuel.

Coal only is used in all destroyers up to and including *T. 161* (1907–08 programme). In *T. 162–173* a small supply of oil fuel was first carried in addition

A 4

to coal, and a separate boiler for oil-firing only was fitted. Its introduction was contemporaneous with the final adoption of turbine propulsion in destroyers. The practice of carrying coal as well as oil was followed until 1913, when oil-firing was adopted in *V. 25–S. 36* (1913–14 programme). Great importance had previously been attached to the protection against gunfire afforded by the coal bunkers, but the advantages of oil were evidently at length recognised, and no subsequent boats carry coal. During the war, the coal-burning boats have fallen lower and lower in estimation, and in a number of cases their boilers have now been converted to burn coal and oil (mixed firing).

Coal.

In Home waters, Westphalian coal is used exclusively.

It is generally supplied ready, in bags of thick jute (having two rope loops at the top), which are estimated to hold 220 lbs.

Destroyers up to *T. 114* are allowed a deck cargo of 10 tons; *T. 120–149*, 15 tons; and later types, 20 tons.

In peace time, whenever flotillas were attached to a fleet, all battleships (except the Fleet Flagship) and all battle-cruisers and cruisers were each obliged to keep 20 tons of coal in bags ready for immediate delivery to the destroyers.

Oil Fuel.

In the 2-funnel boats burning oil only the fuel stowage is 300–325 tons; in the 3-funnel boats it is somewhat higher.

Funnels.

See under Recognition, pp. 2–3.

Gas Masks.

Sufficient gas masks for each man in the complement are supplied to German destroyers.

Hydrophones.

There is no evidence of German destroyers or torpedo boats being fitted with hydrophones.

Machinery and Boilers.

Main Engines.

The destroyers up to *T. 160* inclusive were fitted with two vertical, triple-expansion engines, except *T. 125* (lost 1916) and *T. 137*. These two, and also *T. 161*, were experimental boats fitted with turbine machinery.

From *T. 162* (1908–09 programme) onwards, turbine machinery only was fitted; and in *T. 173* and *G. 174–5*, Parsons turbines, the only system previously fitted on three shafts, were brought into line with the other systems and fitted to work on two shafts.

A further development was the fitting of three boats of the 1913–14 programme (*V. 25–S. 36*) with Föttinger transformers; these transmit only about 88 per cent. of the power developed by the turbines, but, being reversible, do away with the necessity of installing separate turbines for going astern, besides rendering a very high power available for that purpose, and allow the turbines to be driven at a much greater number of revolutions than the propeller shafts.

This experiment has apparently been considered a success, as all the information available up to the present points to the conclusion that the majority of the destroyers of the 1914–15 and later programmes have also been fitted with this arrangement in preference to gearing: in fact, there is no evidence of geared turbines being utilised except, perhaps, experimentally in two or three cases.

Boilers.

In all destroyers up to and including *T. 136* (1905–06 programme), three water-tube boilers are fitted. From *T. 137* to *G. 197* four are fitted, of which, from *T. 162* inclusive onwards, one is fitted to burn oil and three to burn coal only. This arrangement has now been extended to some boats prior to *T. 162*. In the first of the new series of boats, from *V. 1* to *S. 24*, there are three boilers, the foremost of which is double-ended and fitted to burn oil only, whilst the other two were originally fitted to burn coal, but in some, if not all, cases have now been converted to burn coal and oil. All destroyers from *V. 25* to *G. 96* have three boilers fitted to burn oil only, the two after ones being double-ended. The 3-funnel boats, *B. 97–98* and *B. 109–112*, *V. 100* and *G. 101–104*, have four double-ended boilers, burning oil only.

The life of boiler tubes in German destroyers is estimated at three to four years.

Dynamos.

The voltage is 110.

Masts.

See under Recognition, p. 3.

Distinctive Marks.

The destroyers of a flotilla generally all carry a shaped frame or frames with the centre cut out (triangle, circle, diamond or half-moon) on the foremast as flotilla sign, and a second shape of varying description on the mainmast as individual sign. In some cases the flotilla is distinguished by the absence of the foremost sign ; in June 1917 the 1st and 2nd Flotillas were thus distinguished. Leaders of flotillas generally carry the flotilla sign only (on the foremast), and no individual sign.

All destroyers usually carry in addition, two conspicuous horizontal hoops, as described below under Recognition Signals.

Minelaying.

Destroyers carry mines only when they are detailed for that particular purpose, and whilst carrying them are considered to be precluded from undertaking other offensive operations, and are usually provided with an escort.

When detailed for mining, each destroyer carries 12 mines, six on either side amidships, lashed as convenient. There are no special fittings for them, except in a few of the older boats. Rollers are fitted on the under side of the sinkers to facilitate moving the mines about the deck. The mines are laid at a speed of about 15 knots. They are simply pushed overboard, alternately from the starboard and the port side.

Minesweeping.

A considerable number of destroyers, including some of the most recent classes, are fitted for minesweeping. A large double fairlead is fitted aft, as shown in Plates 2 and 11 (*V. 67–84*). Minesweeping is carried out by half-flotillas in line abreast, the two ends of each sweep being towed by adjacent destroyers. The destroyers steam 164 yards apart.

All torpedo boats, except those few which are used as despatch boats and tenders, are fitted for minesweeping and in war time are very largely employed on minesweeping duties. In a number of these boats the torpedo tubes have been replaced by winches.

Provisions and Stores.

Under war conditions, provisions are taken in about once a week and stores about every two months. Bread is not baked on board.

Rangefinders.

In the newer destroyers a 1-metre (3¼ feet) rangefinder is mounted on the searchlight platform above the bridge ; some recent destroyers are reported to have a second instrument, mounted aft. A specially trained able seaman rangetaker is carried in each destroyer. It is said to be very seldom possible to take ranges in action, owing to the excessive vibration.

Torpedo boats of the "A" class are also reported to carry a 1-metre (3¼-feet) rangefinder.

Recognition Signals.

For recognition signals by day, German torpedo craft hoist a combination of two cones or balls, or flash a letter on the searchlight. At night they generally use systems of coloured lanterns, three of which are fitted on the foremast as follows :—

(1) Night signalling apparatus, which consists of three (double) lanterns, arranged vertically ; the upper half of each lantern is fitted with white glass, the lower half with red glass, and only one half is switched on at a time.

(2) Two conspicuous horizontal hoops, each carrying a number of electric lamps, which may be red, green, or white, but are all of one colour in one hoop.

(3) Four small lanterns on a fixed yard on the fore side of the foremast, which are also reported to be red, green, or white. These are probably used for flashing a recognition signal.

Refits and Repairs,

Repairs and refits are usually carried out in the Imperial dockyards at Kiel or Wilhelmshaven, but several extensive repairs have been effected at Ostend, Bruges, and Antwerp in the case of destroyers and torpedo boats of the Flanders flotillas.

Searchlights.

In all destroyers up to and including *S. 36* (1913–14 programme), one 23″ searchlight is carried on a platform above the fore-bridge. More recent destroyers carry two 35″ searchlights, one above the fore-bridge, the other on a platform over the after wheel-house. They are controlled by telephone from the fore-bridge.

Signals.

In peace time, under normal conditions, destroyers communicate visually by flag signals, searchlights, star signals, and the night signalling apparatus, consisting of three vertical lights, red and white. But every endeavour is made to reduce signalling to a minimum and to confine it to the use of hand flags, hand-flashing lamps, and megaphones.

In war time, communication *between* destroyers is almost entirely by megaphone, supplemented by whistle for helm signals, and shaded stern light for speed signals.

Smoke-producing Apparatus.

All modern destroyers carry, aft on either quarter, a metal cylinder, about 2 feet in length, mounted on a sloping platform. These two cylinders, which contain some chemical composition, are in connection with the compressed air service, and are used for developing a smoke-screen. The apparatus is controlled from the fore-bridge or after wheel-house. The smoke produced is greyish-white in colour.

The boilers are also utilised to produce a smoke-screen when required.

Speed (Cruising).

Before the war, the cruising speed of destroyers was fixed as follows :—

(*a*) For the boats from *T. 90* to *T. 136* inclusive - 14 knots.
(*b*) For the boats from *T. 137* onwards - - 17 knots.

In turbine boats with two shafts, for speeds of 12 knots and under, one shaft only was to be used, if possible, the other being allowed to run idle.

Steam Heating, &c.

Most, if not all, German torpedo craft are fitted with steam heating installations for the mess decks and cabins, together with arrangements whereby one vessel can supply steam for heating one or more other boats lying alongside. All German torpedo-craft are also lined with either wood or linoleum to prevent sweating.

Steering Gear and Steering Positions.

The great majority of destroyers and torpedo boats are fitted with a bow as well as a stern rudder. In destroyers the bow rudder is situated below the keel about 12 feet abaft the stem. It can be raised into a recess in the hull by means of a fine-pitched screw, and has to be so raised before the boat can be taken in tow. It is always used when manœuvring, as the steering capabilities are insufficient without it.

In some of the very latest destroyers, however, no bow rudder is fitted.

Destroyers are fitted with two steering engines, the main steering engine being forward, under the break of the forecastle, and the second, or reserve steering engine aft. The steering positions, usually three in number, are situated—

(1) on the bridge ;
(2) under the break of the forecastle, with the main steering engine ;
(3) in the after wheel-house, with the reserve steering engine.

In *G. 37–42*, *G. 85–96* and *G. 101–104*, there are only two steering positions, viz., Nos. 1 and 3 above.

Nos. 1 and 2 are connected with the main steering gear, forming a system of rods and chains, fitted with bottle screws at intervals, which is carried along on both sides of the upper deck, on rollers, to the tiller.

No. 3 works a duplicate set of gear, also carried along the upper deck.

Each of these positions can be connected up with the bow rudder, where fitted, by special chain-and-rod gearing. When Nos. 1 and 2 are in use, the gear of No. 3 leading to both bow and stern rudders is disconnected, and *vice versâ*.

Submarine Attack.

See under Depth Charges and Explosive Kite, p. 7.

Trials.

The following are some of the trials carried out by newly-built destroyers in peace time :—

(1) Acceptance trial.
(2) Three hours' full power trial.
(3) Measured mile trials.
(4) Twelve hours' trial (? fuel consumption trial).
(5) Storm trials.

With the exception of (5) all the above are carried out in the Baltic ; the storm trials are carried out in the North Sea. The measured mile trials are carried out chiefly on the two-mile course near Neukrug in the Gulf of Danzig.

The measured mile trials, it is stated, are carried out entirely with naval ratings, and with the full equipment of stores, ammunition, and water, and the normal supply of fuel, the vessels being structurally complete otherwise. Any deficiency in numbers of crew or in stores is made up with ballast.

The 12 hours' trial is carried out with the destroyer fully loaded at the start, *i.e.*, with all fuel, &c., and provisions for one month.

Storm trials are carried out by newly-built destroyers in the North Sea, between Cuxhaven and Heligoland, to test their seaworthiness. At least one destroyer of every batch delivered is so tested. It is reported that the test has to be carried out at 17 knots in a gale for a few hours, the behaviour of the vessel being tested with the sea on the bow, beam, and quarter. The force of the wind has been variously stated as 8–10 and 7 ; the latter is apparently the approximate average.

Wireless Telegraphy.

In destroyers from *V. 186* to *S. 24*, a wireless house, about 6 ft. by 4 ft., is fitted just abaft the fore funnel. In destroyers from *V. 25* onwards the wireless room is usually situated under the bridge. (In the 3-funnel boats *B. 97–98*, *V. 100*, and *B. 109–112* it is apparently between the first and second funnels.)

The usual destroyer wireless set has a normal effective night range of about 200 miles. As a rule, destroyers use a wave-length of 2,625 ft. (800 m.).

One petty officer and four other telegraphist ratings are usually carried.

The following detailed description of the destroyer series "*V. 1–6*" is authentic, and this series may be regarded as typical of modern *pre-war* German destroyers.

Destroyers "V. 1—6."*

(See Plates 3 and 12.)

General Remarks.

These six destroyers formed half of the 1911–12 programme, and were built at the Vulcan Works at Stettin.

Cost.

For average cost, *see* page 7.

Complement.

In peace time, 73 officers and men. This is no doubt augmented under war conditions.

General Dimensions, &c.

Length	between perpendiculars	223 ft. 1 ins.
	L.W.L.	230 ,, 3 ,,
	over all	233 ,, 1 ,,
Breadth, extreme, at upper deck		24 ,, 11 ,,
Height of	forecastle above W.L.	13 ,, 1 ,,
	freeboard amidships at centre line	6 ,, 10 ,,
	,, ,, at side	6 ,, 3 ,,
	deck above keel amidships	14 ,, 5 ,,

* *V. 4* was lost on 31 May or 1 June 1916.

Designed Load Draught*	to bottom of keel - - -	7 ft. 7 ins.
	to bottom of rudder - - -	10 „ 0 „
Designed Displacement	at load draught* - - -	561 tons.
	at extreme draught - - -	689 „
Tons per inch immersion - - - - -		about 9·2
Speed, designed - - - - - -		32·5 knots.

Fuel Stowage.

Coal	(*V. 1–4*) - -	141 tons.	Oil	(*V. 1–4*) -	54 tons.
	(*V. 5–6*) - -	139 „		(*V. 5–6*) -	68 „
	Deck cargo -	20 „			

Hull.

Material.

Siemens-Martin steel, tensile strength 35 tons per sq. in.

Constructive Details.

Vertical keel—·25″ plating, height under engine room 23″.
Flat keel—·31″ plating, decreasing to ·21″ aft and ·18″ forward, width 3′ 8″.
Vertical keel plates connected with double butt straps are double-riveted.
Angle-bar for vertical keel—upper 2·1″ × 2·1″ × ·23″.
lower 2·9″ × 2·5″ × ·27″.
Bulkheads—lower plates ·16″, upper plates ·12″.
Frames—consist of angle-bar 2·4″ × 1·6″ × ·2″ generally placed 19·7″ apart, but closer at after end.
Side and bottom plating—·37″ to ·14″.
Deck plating—·16″, strengthened in way of heavy weights.

Armament.

2—22-pr. (8·8-cm.) Q.F., L/30 (1 on forecastle, 1 abaft deck-house aft).

2—machine guns.

4—19·7″ torpedo tubes (1 on either side of deck directly abaft forecastle—arc of training about 45°, *i.e.*, from almost right ahead to about 40° before the beam; 1 tube directly abaft the second funnel and 1 directly before deck-house aft, both on centre line, with all-round training).

At least five torpedoes are carried, four in the tubes and one in a watertight holder the starboard side abreast second funnel.

Searchlights.

1—23″, on chart-house forward; 60 ampères and 48 volts.

Communications.

Voice-pipes from bridge to engine-room, chart-house, wireless room, and steering positions forward and aft.

Wireless Telegraphy.

The wireless room, which is situated directly abaft the fore funnel, is insulated against heat and sound.

Rudders.

A bow rudder is fitted which can be raised into a recess in the hull when boat is to be towed, &c.

Steering Gear.

2 steam steering engines and 3 steering positions, one of which is on the bridge, and one at each steering engine.

Machinery and Boilers.

Machinery.

2 Curtis-A.E.G.-Vulcan turbines, in two compartments, one abaft the other.
Each shaft is fitted with an astern turbine.
Designed S.H.P. 15,000. Pressure 272 lbs. per sq. in.

---

* With 70 tons of coal and 20 of oil on board.

Boilers.

3 Schulz-Vulcan, in separate compartments. The foremost boiler is double-ended and fitted to burn oil only—heating surface 7,320 sq. ft. Centre and after boiler fitted to burn coal*—total heating surface of these two boilers 9,688 sq. ft., total grate surface 182 sq. ft.

Propellers.

2 in number, bronze, 3-bladed.

Auxiliary Machinery.

2 air-pumps, vertical rotary turbine-driven.
2 circulating pumps.
2 evaporators, capacity 18 tons in 24 hrs. - (In engine-room.)
1 distilling apparatus, capacity 2 tons in 24 hrs. (In engine-room.)
2 main duplex feed-pumps - - - (In engine-room.)
1 auxiliary duplex feed-pump - - - (In engine-room.)
2 duplex suction pumps - - - - (In engine-room.)
2 electric fans.
3 duplex auxiliary feed-pumps.
1 ash-ejector fitted with bilge suction.
2 ash-ejectors.
1 oil-heating apparatus.
2 turbo-dynamos, 110 volts, for all purposes, each capable of supplying current for all lighting purposes, except searchlight. When the searchlight is in use, both are required. One turbo-dynamo is fitted in the fore and the other in the after engine-room.

Miscellaneous.

3 feed tanks in boiler-room, and 1 overflow tank in engine-room, total capacity 19·5 tons.
Fresh water tanks—3 tons. Small hand-pump in galley.
Living spaces are steam-heated and electrically lighted. Five portable coal stoves are fitted in various parts of the boat.

* It is believed that in some, if not all, cases these two boilers have now been converted to burn coal and oil.

B 3

TABULAR

DESTROYERS.

Official Number.	Where built.	Date of Launch.	Complement.	Hull. Length. (a) P.P. (b) L.W.L. (c) extreme.	Breadth (extreme).	Designed Load. Draught.	Displacement.	Armament.
				Ft. Ins.	Ft. Ins.	Ft. Ins.	Tons (Eng.).	
H 169* - - H 168* - - H 167* - - H 166* - -	Howaldt, Kiel.	—	—	—	—	—	—	—
V 165* - - V 164* - - V 163* - - V 162* - - V 161* - - V 160* - - V 159* - - V 158* - -	Vulcan, Hamburg.	—	—	—	—	—	—	—
S 157* - - S 156* - - S 155* - - S 154* - - S 153* - - S 152* - -	Schichau, Elbing.	—	—	—	—	—	—	—
WW 151* -	Weser, Bremen.	—	—	—	—	—	—	—
G 150* - - G 149* - - G 148* - -	Germania, Kiel.	—	—	—	—	—	—	—
H 147 - - H 146 - - H 145 - -	Howaldt, Kiel.	—	—	—	—	—	—	—

* Completing in August 1918.

DETAILS.

DESTROYERS.

Machinery and Boilers.				Fuel Supply.		Appearance, &c.	OFFICIAL NUMBER.
No. of Propellers.	(a) Engines. (b) Boilers.	Horse Power. (a) Designed. (b) Mean on trial.	Speed. (a) Designed. (b) Mean on trial.	(a) Coal (max.). (b) Oil.	Endurance at 15 Knots.		
			Knots.	Tons.	Miles.		
—	—	—	—	—	—	These vessels are probably two-funnelled boats of an improved *G. 85–96*, *V. 67-84*, or *S. 53-66* class (*see* pp. 21 and 23).	H 169. H 168. H 167. H 166.
—	—	—	—	—	—		V 165. V 164. V 163. V 162. V 161. V 160. V 159. V 158.
—	—	—	—	—	—		S 157. S 156. S 155. S 154. S 153. S 152.
—	—	—	—	—	—		WW 151.
—	—	—	—	—	—		G 150. G 149. G 148.
—	—	—	—	—	—		H 147. H 146. H 145.

Destroyers—

Official Number.	Where Built.	Date of Launch.	Complement.	Hull. Length. (a) P.P. (b) L.W.L. (c) extreme.	Hull. Breadth (extreme)	Hull. Designed Load. Draught	Hull. Designed Load. Displacement.	Armament.
				Ft. ins.	Ft. ins.	Ft. ins.	Tons (Eng.).	
V 144, V 143, V 142, V 141, V 140	Vulcan, Hamburg.	—	—	—	—	—	—	—
S 139, S 138, S 137, S 136, S 135, S 134, S 133, S 132, S 131	Schichau, Elbing.	—	—	—	—	—	—	—
V 130, V 129, V 128, V 127, V 126, V 125	Vulcan, Hamburg.	—	—	—	—	—	—	—
B 124*, B 123*, B 122*	Blohm and Voss, Hamburg.	—	—	—	—	—	—	
G 121*, G 120*, G 119*	Germania, Kiel.	—	—	—	—	—	—	—
V 118*, V 117*, V 116*	Vulcan, Hamburg.	—	—	—	—	—	—	—

* Completing in August 1918.

continued.

Machinery and Boilers.				Fuel Supply.		Appearance, &c.	OFFICIAL NUMBER.
No. of Propellers.	(a) Engines. (b) Boilers.	Horse Power. (a) Designed. (b) Mean on trial.	Speed. (a) Designed. (b) Mean on trial.	(a) Coal (max.). (b) Oil.	Endurance at 15 Knots.		
			Knots.	Tons.	Miles.		
—	(a) Schichau turbines. (b) 3 Schulz.	—	—	—	—	These are probably two-funnelled boats of an improved *V. 67–84* or *S. 53–66* class (*see* pp. 21 and 23).	V 144. V 143. V 142. V 141. V 140.
—	—	—	—	—	—		S 139. S 138. S 137. S 136. S 135. S 134. S 133. S 132. S 131.
—	—	—	—	—	—		V 130. V 129. V 128. V 127. V 126. V 125.
—	—	—	—	—	—	These are probably large three-funnelled boats of an improved *V. 100*, *G. 101-104*, or *B. 109–112* class. None are believed (Aug. 1918) to be yet in commission.	B 124. B 123. B 122.
—	—	—	—	—	—		G 121. G 120. G 119.
—	—	—	—	—	—		V 118. V 117. V 116.

Destroyers—

Official Number.	Where Built.	Date of Launch.	Complement.	Hull. Length. (a) P.P. (b) L.W.L. (c) extreme.		Breadth (extreme).		Designed Load. Draught.		Displacement.	Armament.
				Ft.	Ins.	Ft.	Ins.	Ft.	Ins.	Tons (Eng.)	
S 115*	Schichau, Elbing.	—	—	—		—		—		—	—
S 114*											
S 113*											
B 112	Blohm & Voss, Hamburg.	1915	—	(c) 325	0	32	0†	11	0†	1,300†	4—4·1" - - 2—m. 6—T. (19·7")
B 111											
B 110											
B 109											
G 104‡	Germania, Kiel.	1914	100†	(c) 312	8	30	1	9	4	1,250	4—4·1" - - 2—m. 6—T. (19·7")
G 103‡											
G 102‡											
G 101‡											
V 100	Vulcan, Hamburg.	1915	—	(c) 328	0	32	0†	11	0†	1,300†	4—4·1" - - 2—m. 6—T. (19·7")

* Believed to be completing in August 1918. † Approximate only. ‡ Originally building for Argentina.

continued.

Machinery and Boilers.				Fuel Supply.		Appearance, &c.	Official Number.
No. of Propellers.	(*a*) Engines. (*b*) Boilers.	Horse Power. (*a*) Designed. (*b*) Mean on trial.	Speed. (*a*) Designed. (*b*) Mean on trial.	(*a*) Coal (max.). (*b*) Oil.	Endurance at 15 Knots		
			Knots.	Tons.	Miles.		
—	—	—	—	—	—	These are probably large three-funnelled boats of an improved *V. 100* or *B. 109–112* class. None are believed (Aug. 1918) to be yet in commission.	S 115. S 114. S 113.
2	(*a*) Turbines. (*b*) 4 Schulz double ended.	—	(*a*) 34·0†	Oil only	—	*See* Plates 1 & 8 & Silhouette. High forecastle cut away on both sides for torpedo tubes; well-deck between forecastle and bridge which is against fore funnel. 3 funnels, the centre one being much broader than the other two. 4·1″ guns all on centre line; 1 on forecastle, 1 between 1st and 2nd funnels, 1 between 2nd and 3rd funnels, 1 abaft after wheel-house. 2 tubes, one on either side, abaft forecastle; 4 tubes mounted in pairs on the centre line abaft 3rd funnel.	B 112 B 111. B 110. B 109.
2	(*a*) Germania turbines. (*b*) 4 Schulz double ended.	(*a*) 24,000 (T.).	(*a*) 32·0 (*b*) 34·0†	(*a*) Nil (*b*) 345	3,200	*See* Plates 1 & 9 & Silhouette. High forecastle extending to after edge of fore funnel. 3 funnels of equal size, the foremost being higher than the others. 4·1″ guns all on centre line; 1 on forecastle, 1 between 2nd and 3rd funnels, 1 before and abaft the after wheel-house. 2 single tubes between 1st and 2nd funnels, one on either side of boat skids. 4 tubes in pairs on the centre line abaft 3rd funnel. 2 searchlights, one above bridge, one on after wheel-house. Standard compass on platform abaft after funnel.	G 104 G 103 G 102 G 101.
2	(*a*) A.E.G.—Vulcan turbines.* (*b*) 4 Schulz double-ended.	—	(*a*) 34·0†	(*b*) 400†	—	*See* Plate 1 and Silhouette. High forecastle, cut away on both sides for torpedo tubes; well deck between forecastle and bridge, which is against fore funnel. 3 funnels, the centre one being much broader than the other two. 4·1″ guns all on centre line; 1 on forecastle, 2 between 2nd and 3rd funnels, 1 abaft after wheel-house. 2 single tubes, one on either side forward of bridge. 4 tubes in pairs on centre line between 3rd funnel and mainmast.	V 100.

* Turbines originally built for Russia. † Approximate only.

Destroyers—

Official Number.	Where Built.	Date of Launch.	Complement.	Hull. Length. (a) P.P (b) L.W.L. (c) extreme.	Breadth (extreme).	Designed Load. Draught.	Designed Load. Displacement.	Armament.
				Ft. Ins.	Ft. Ins.	Ft. Ins.	Tons (Eng.).	
B 98, B 97	Blohm & Voss, Hamburg.	1915	—	(c) 325 0	32 0†	11 0†	1,300†	4—4·1" - - 2—m. 6—T. (19·7").
G 96, G 95, G 92, G 91, G 90, G 89, G 86	Germania, Kiel.	1915–16	98	(c) 272 0	—	—	750†	3—4·1" - - 2—m. *4 or 6—T. (19·7"). 5 or 7 torpedoes.
V 83, V 82, V 81, V 80, V 79, V 78, V 77, V 74, V 73	Vulcan, Stettin.	1915–16	98	(c) 270 0	—	—	750†	3—4·1" - - 2—m. *4 or 6—T. (19·7"). 5 or 7 torpedoes.

* These boats all originally carried 6 torpedo tubes, but in a good many cases the two foremost tubes have been removed. (*See* p. 5.)
† Approximate only.

continued.

Machinery and Boilers.				Fuel Supply.		General Remarks.	Official Number.
No. of Propellers.	(*a*) Engines. (*b*) Boilers.	Horse Power. (*a*) Designed. (*b*) Mean on trial.	Speed. (*a*) Designed. (*b*) Mean on trial.	(*a*) Coal (max.) (*b*) Oil	Endurance at 15 Knots.		
			Knots.	Tons.	Miles.		
2	(*a*) Turbines. (*b*) 4 Schulz double-ended.	—	(*a*) 34·0†	Oil only	—	Same type as B. 109–112 (*see* p. 19).	B 98. B 97.
2	(*a*) Germania turbines. (*b*) 3 Schulz (2 being double-ended.	(*a*) 23,000†	(*a*) 34·0†	(*b*) 325†	3,400	*See* Plates 2 & 10 & Silhouette. Forecastle cut away on both sides for torpedo tubes. 2 funnels, the fore funnel higher than the after one. 2 masts (foremast immediately abaft bridge, mainmast just before after wheel-house). Bridge and forecastle in one; interval between bridge and fore funnel. 1—4·1″ on forecastle. 1—4·1″ on platform between after funnel and mainmast. 1—4·1″ abaft after wheel-house. 2 tubes, one on either side, abreast bridge. 4 tubes, mounted in pairs on the centre line before and abaft midship gun. 2 searchlights, one above bridge, one on platform over after wheel-house. Standard compass on platform abaft after funnel.	G 96. G 95. G 92. G 91. G 90. G 89. G 86.
2	(*a*) Turbines. (*b*) 3 Schulz (2 being double-ended).	(*a*) 23,000†	(*a*) 34·0†	(*b*) 325†	3,400	*See* Plates 2 & 11 & Silhouette. Differ from G. 85–96 (*see* above) in having a well-deck between forecastle and bridge, with a conspicuous ventilating trunk curving upwards and aft into the bridge structure, also in the position of the samson post and derrick.	V 83. V 82. V 81. V 80. V 79. V 78. V 77. V 74. V 73.

C 3

Destroyers—

Official Number.	Where Built.	Date of Launch.	Complement.	Hull. Length. (a) P.P. (b) L.W.L. (c) extreme.	Breadth (extreme).	Designed Load. Draught.	Displacement.	Armament.
				Ft. Ins.	Ft. Ins.	Ft. Ins.	Tons (Eng.).	
V 72 - - V 71 - - V 70 - - V 69 - - V 68 - - V 67 - -	Vulcan, Hamburg.	1915–16	98	(c) 270 0	—	—	750†	3—4·1″. 2—m. *4 or 6—T. (19·7″). 5 or 7 torpedoes.
S 65 - - S 64 - - S 63 - - S 61 - - S 60 - - S 59 - - S 55 - - S 54 - - S 53 - -	Schichau, Elbing.	1916	98	(c) 272 0	—	—	750†	3—4·1″ - - 2—m. *4 or 6—T. (19·7″). 5 or 7 torpedoes.
S 52 - - S 51 - - S 50 - - S 49 - -	Schichau, Elbing.	1913–15	98	(c) 262 0	—	—	750†	3—4·1″ (*or* 22-pr.) 2—m. 4 or 6—T. (19·7″). 5 or 7 torpedoes.

* These boats all originally carried 6 torpedo tubes, but in a good many cases the two foremost tubes have been removed (*see* p. 5).
† Approximate only.

continued.

Machinery and Boilers.				Fuel Supply.		Appearance, &c.	Official Number.
No. of Propellers.	(*a*) Engines. (*b*) Boilers.	Horse Power. (*a*) Designed. (*b*) Mean on trial.	Speed. (*a*) Designed. (*b*) Mean on trial.	(*a*) Coal (max.). (*b*) Oil.	Endurance at 15 Knots.		
			Knots.	Tons.	Miles.		
2	(*a*) Turbines. (*b*) 3 Schulz (2 being double-ended).	(*a*)23,000†	(*a*) 34·0†	(*b*) 325†	3,400	Practically identical with *V. 73–84*. *See* Plates 2 & 11 & Silhouette.	V 72. V 71. V 70. V 69. V 68. V 67.
2	(*a*) Turbines. (*b*) 3 Schulz (2 being double-ended).	(*a*)23,000†	(*a*) 34·0†	(*b*) 325†	3,400	Same general type as *V. 73–84*, but without the conspicuous ventilating trunk forward. *See* Plate 2 and Silhouette.	S 65. S 64. S 63. S 61. S 60. S 59. S 55. S 54. S 53.
2	(*a*) Turbines. (*b*) 3 Schulz (2 being double-ended).	(*a*)23,000†	(*a*) 34·0†	(*b*) 300†	3,250	Same general type as *V. 73–84*, but without the conspicuous ventilating trunk forward. *See* Silhouette.	S 52. S 51. S 50. S 49.

† Approximate only.

C 4

Destroyers—

Official Number.	Where Built.	Date of Launch.	Complement.	Hull. Length. (a) P.P. (b) L.W.L. (c) extreme.	Hull. Breadth (extreme).	Hull. Designed Load. Draught.	Hull. Designed Load. Displacement.	Armament.
				Ft. Ins.	Ft. Ins.	Ft. Ins.	Tons (Eng.).	
1914–15 PROGRAMME.								
V 47	Vulcan, Stettin.	1915	98	(*c*) 274 0	—	—	750†	**3**—4·1″ *or* 22-pr. **2**—m. ***4** or **6**—T (19·7″). **5** or **7** torpedoes.
V 46, **V 45**, **V 44**, **V 43**	Vulcan, Stettin.	1915	98	(*c*) 270 0	—	—	750†	**3**—4·1″ *or* 22-pr. **2**—m. ***4** or **6**—T. (19·7″). **5** or **7** torpedoes.
G 41	Germania, Kiel.	1915	98	(*c*) 272 0	—	—	750†	**3**—4·1″ *or* 22-pr. **2**—m. ***4** or **6**—T. (19·7″). **5** or **7** torpedoes.
G 40, **G 39**, **G 38**	Germania, Kiel.	1915	98	(*c*) 266 0	—	—	700†	**3**—4·1″ *or* 22-pr. **2**—m. ***4** or **6**—T. (19·7″). **5** or **7** torpedoes.
1913–14 PROGRAMME.								
S 36, **S 34**, **S 33**, **S 32**	Schichau, Elbing.	1913–14	90†	(*c*) 262 0	—	—	700†	**3**—22-pr. **2**—m. **6**—T. (19·7″). **7** torpedoes.

* These boats all originally carried 6 torpedo tubes, but in a good many cases the two foremost tubes have been removed (*see* p. 5).
† Approximate only.

—*continued.*

Machinery and Boilers.				Fuel Supply.		Appearance, &c.	Official Number.
No of Propellers.	(*a*) Engines. (*b*) Boilers.	Horse Power. (*a*) Designed. (*b*) Mean on trial.	Speed. (*a*) Designed. (*b*) Mean on trial.	(*a*) Coal. (max.). (*b*) Oil.	Endurance at 15 Knots.		
			Knots.	Tons.	Miles.		**1914–15 PROGRAMME.**
2	(*a*) **A.E.G.—Vulcan** turbines. (*b*) 3 Schulz (2 being double-ended).	(*a*) 23,000†	(*a*) 34·0†	(*a*) Nil (*b*) 300†	3,250	Closely resemble *V. 67–84* (*see* p. 21), but without the conspicuous ventilating trunk forward. *See* Silhouettes.	**V 47.**
2	(*a*) A.E.G.—Vulcan turbines. (*b*) 3 Schulz (2 being double-ended).	(*a*) 23,000†	(*a*) 34·0†	(*a*) Nil (*b*) 300†	3,250		**V 46.** **V 45.** **V 44.** **V 43.**
2	—	(*a*) 23,000†	(*a*) 34·0†	(*a*) Nil (*b*) 325†	3,400	Same general type as *G. 85–96* (*see* p. 21).	**G 41.**
2	(*a*) Germania turbines. (*b*) 3 Schulz (2 being double-ended).	(*a*) 23,000†	(*a*) 34·0†	(*a*) Nil (*b*) 300	3,250		**G 40.** **G 39.** **G 38.**
2	(*a*) Schichau turbines. (*b*) 3 Schulz (2 being double-ended).	(*a*) 20,000	(*a*) 32·5	(*a*) Nil (*b*) 300	3,250	*See* Plate 3 and Silhouette. Forecastle cut away on both sides for torpedo tubes. 2 funnels well forward, fore funnel higher than after funnel. 2 masts (foremast immediately abaft bridge, mainmast just before after wheel-house). Bridge against fore funnel; well-deck between bridge and forecastle. 1—22-pr. on forecastle. 1—22-pr. on platform between after funnel and mainmast. 1—22-pr. abaft after wheel-house. 2 tubes, one on either side, under break of forecastle. 4 tubes, mounted in pairs, on the centre line, before and abaft midship gun. 1 searchlight above bridge. Standard compass on after wheel house.	**1913–14 PROGRAMME.** **S 36.** **S 34.** **S 33.** **S 32.**

† Approximate only.

O AS 4924—22 D

Destroyers—

Official Number.	Where Built.	Date of Launch.	Complement.	Hull: Length. (a) P.P. (b) L.W.L. (c) extreme.	Hull: Breadth (extreme).	Hull: Designed Load. Draught.	Hull: Designed Load. Displacement.	Armament.
				Ft. Ins.	Ft. Ins.	Ft. Ins.	Tons (Eng.)	
1913–14 PROGRAMME —*cont.*								
V 30, V 28, V 26	Vulcan, Stettin.	1914	90†	(c) 254 0†	—	—	640	3—22-pr. 2—m. 6—T. (19·7″). 7 torpedoes.
1912–13 PROGRAMME.								
S 24, S 23, S 19, S 18, S 15	Schichau, Elbing.	1912–13	78	(c) 234 7	24 4	10 0	555	2—4·1″ or 22-pr. 2—m. *2 or 4—T. (19·7″). 3 or 5 torpedoes.
1911–12 PROGRAMME.								
G 11, G 10, G 9, G 8, G 7	Germania, Kiel.	1911–12	73	(c) 233 1	24 11	10 0	555	2—22-pr. 2—m. 4—T. (19·7″). 5 torpedoes.

* These boats originally carried two 22-pr. guns and four torpedo tubes, but it is believed that they have all now been rearmed with 4·1″ guns and have had their two foremost tubes removed.

† Approximate only.

continued.

Machinery and Boilers.				Fuel Supply.		Appearance, &c.	OFFICIAL NUMBER.
No. of Propellers.	(*a*) Engines. (*b*) Boilers.	Horse Power. (*a*) Designed. (*b*) Mean on trial.	Speed. (*a*) Designed. (*b*) Mean on trial.	(*a*) Coal (max.). (*b*) Oil.	Endurance at 15 Knots.		
			Knots.	Tons.	Miles.		
							1913–14 PROGRAMME —*cont.*
2	(*a*) A.E.G.—Vulcan turbines. (*b*) 3 Schulz (2 being double-ended).	(*a*) 20,000	(*a*) 32·5	(*a*) Nil (*b*) 240†	3,250	*See* Silhouette. Same general type as *S. 31–36* (*see* p. 25).	V 30. V 28. V 26.
							1912–13 PROGRAMME.
2	(*a*) Schichau turbines. (*b*) 3 Schulz.	(*a*)15,000	(*a*) 32·5 (*a*) 32·5 (*b*) 35·6 (*a*) 32·5 (*a*) 32·5 (*a*) 32·5 (*a*) 32·5 (*b*) 34·0	(*a*) 132 + 20‡ (*b*) 55	1,700	*See* Plate 3 and Silhouettes. Forecastle cut away on both sides for torpedo tubes. 2 funnels, well forward. 2 masts (foremast by bridge, mainmast abaft 2nd funnel). Bridge against 1st funnel; well-deck between bridge and forecastle. Wireless house abaft 1st funnel. After wheel-house well abaft mainmast. 1—22-pr. on forecastle. 1—22-pr. abaft deck-house aft. 2 tubes, one on either side, abaft forecastle. 1 tube on centre line, abaft 2nd funnel. 1 tube in centre line before after wheel-house. 1 searchlight above bridge. Standard compass on after wheel-house.	S 24. S 23. S 19. S 18. S 15.
							1911–12 PROGRAMME.
2	(*a*) Germania turbines. (*b*) 3 Schulz.	(*a*)15,000	(*a*) 32·5 (*a*) 32·5 (*a*) 32·5 (*a*) 32·5 (*a*) 32·5	(*a*) 136 + 20‡ (*b*) 53	1,620		G 11. G 10. G 9. G 8. G 7.

‡ Deck cargo.

Destroyers—

Official Number.	Where Built.	Date of Launch.	Complement.	Hull. Length. (a) P.P. (b) L.W.L. (c) extreme.	Hull. Breadth (extreme).	Hull. Designed Load. Draught.	Hull. Designed Load. Displacement.	Armament.
				Ft. Ins.	Ft. Ins.	Ft. Ins.	Tons (Eng.).	
1911–12 PROGRAMME *—cont.*								
V 6*		1913						
V 5*		1913						
V 3	Vulcan, Stettin.	1911	73	(a) 223 1 (b) 230 3 (c) 233 1	24 11	10 0	561	2—22-pr. 2—m. 4—T. (19·7″). 5 torpedoes.
V 2		1911						
V 1		1911						
1910–11 PROGRAMME.								
G 197								
G 196								
G 195	Germania, Kiel.	1910–11	83	(a) 232 11 (c) 242 10	25 11	10 6	638	2—22-pr. 2—m. 4—T. (19·7′ 5 torpedoes.
G 193								
G 192								
V 190		1911						
V 189	Vulcan, Stettin.	1911	83	(a) 232 11 (c) 242 6	25 11	10 6	646	2—22-pr. 2—m. 4—T. (19·7″). 5 torpedoes.
V 186		1910						

* Built to replace two similarly numbered which were building for the German Navy, and which the Contractors were permitted to sell to Greece in 1912—delivery to the German naval authorities not having taken place.

continued.

Machinery and Boilers.				Fuel Supply.		Appearance, &c.	OFFICIAL NUMBER.
No. of Propellers.	(a) Engines (b) Boilers.	Horse Power. (a) Designed. (b) Mean on trial.	Speed. (a) Designed. (b) Mean on trial.	(a) Coal (max.). (b) Oil.	Endurance at 15 Knots.		
			Knots.	Tons.	Miles.		1911–12 PROGRAMME —*cont.*
2	(a) A.E.G.—Vulcan turbines. (b) 3 Schulz-Vulcan.	(a) 15,000	(a) 32·5 (b) 33·4 (a) 32·5	(a) 139 +20‡ (b) 68	1,830	*See* Plates 3 and 12 and Silhouette. Forecastle cut away on both sides for torpedo tubes. 2 funnels well forward. 2 masts (foremast by bridge, mainmast abaft second funnel). Bridge against 1st funnel; well-deck between bridge and forecastle. Wireless house abaft 1st funnel. After wheel-house well abaft mainmast. 1—22-pr. on forecastle. 1—22-pr. abaft deck-house aft. 2 tubes, one on either side under break of forecastle. 1 tube on centre line abaft 2nd funnel. 1 tube on centre line before after wheel-house. 1 searchlight above bridge. Standard compass on after wheel-house.	V 6*. V 5.*
			(a) 32·5 (b) 33·6 (a) 32·5 (a) 32·5 (b) 34·2	(a) 141 +20‡ (b) 54	1,730		V 3. V 2. V 1.
							1910–11 PROGRAMME.
2	(a) Germania turbines. (b) 4 Schulz Navy type.	(a) 16,000	(a) 32·5	(a) 168 +20‡ (b) 54	1,260	*See* Plate 4 and Silhouettes. Forecastle cut away on both sides for torpedo tubes. 2 funnels, well forward. 2 masts (foremast by bridge, mainmast abaft 2nd funnel). Bridge against 1st funnel; well-deck between bridge and forecastle. Wireless house abaft 1st funnel. After wheel-house well abaft mainmast. 1—22-pr. on forecastle. 1—22-pr. abaft after wheel-house. 2 tubes, one on either side, under break of forecastle. 1 tube on centre line before 2nd funnel. 1 tube on centre line before after wheel-house. 1 searchlight above bridge. Standard compass on after wheel-house.	G 197. G 196. G 195. G 193. G 192.
2	(a) A.E.G.—Vulcan turbines. (b) 4 Schulz Navy type.	(a) 16,000	(a) 32·5	(a) 160 +20‡ (b) 53	1,230		V 190. V 189. V 186.

‡ Deck cargo

D 3

Destroyers—

Official Number.	Where Built.	Date of Launch.	Complement.	Hull. Length. (a) P.P. (b) L.W.L. (c) extreme.	Breadth (extreme).	Designed Load. Draught.	Displacement.	Armament.
				Ft. Ins.	Ft. Ins.	Ft. Ins.	Tons (Eng.).	
1909–10 PROGRAMME.								
V 185, V 184, V 183, V 182, V 181, V 180	Vulcan, Stettin.	1909–10	83	(c) 242 6	25 11	10 2	626	2—22-pr. 2—m. 4—T. (19·7"). 5 torpedoes.
S 179, S 178, S 176	Schichau, Elbing.	1910	83	(c) 243 5	25 11	10 2	625	2—22-pr. 2—m. 4—T. (19·7"). 5 torpedoes.
G 175, G 174	Germania, Kiel.	1910	83	(b) 241 2 (c) 242 10	25 11	10 2	643	2—22-pr. 2—m. 4—T. (19·7"). 5 torpedoes.
1908–09 PROGRAMME.								
T 173, T 172, T 170, T 169	Germania, Kiel.	1908–09	83	(a) 234 7 (c) 242 10	25 11	10 2	625	2—22-pr. 2—m. 3—T. (17·7"). 5 torpedoes.
T 168*, T 167*, T 166*, T 165*	Schichau, Elbing.	1910–11	83	(c) 243 5	25 11	9 10	605	2—22-pr. 2—m. 3—T. (17·7"). 5 torpedoes.
T 164, T 163, T 162	Vulcan, Stettin.	1909	83	(c) 242 6	25 11	9 10	603	2—22-pr. 2—m. 3—T. (17·7"). 5 torpedoes.

* Built to replace four (numbered S. 165–168), which were building for the German Navy and which the contractors were permitted to sell to Turkey in 1910, delivery to the German naval authorities not having taken place.

continued.

Machinery and Boilers.				Fuel Supply.		Appearance, &c.	Official Number.
No. of Propellers.	(*a*) Engines. (*b*) Boilers.	Horse-Power. (*a*) Designed. (*b*) Mean on trial.	Speed. (*a*) Designed. (*b*) Mean on trial.	(*a*) Coal (max.). (*b*) Oil.	Endurance at 15 Knots.		
			Knots.	Tons.	Miles.		
							1909–10 PROGRAMME.
2	(*a*) A.E.G.—Vulcan turbines. (*b*) 4 Schulz Navy type.	(*a*) 16,000	(*a*) 32·5 (*a*) 32·5 (*a*) 32·5 (*b*) 32·15 (*a*) 32·5 (*b*) 32·2 (*a*) 32·5 (*b*) 32·6 (*a*) 32·5 (*b*) 32·7	(*a*) 148 +20‡ (*b*) 40	1,300	*See* Plates 4 & 5 & Silhouettes. 2 funnels, well forward. 2 masts (stump foremast by bridge, mainmast abaft 2nd funnel). Bridge against 1st funnel; very short well-deck between bridge and forecastle. After wheel-house abaft mainmast. 1—22-pr. on forecastle. 1—22-pr. abaft deck house aft. 2 broadside tubes, one on either side, abaft 1st funnel. 1 tube on centre line abaft 2nd funnel. 1 tube on centre line before after wheel-house. 1 searchlight above bridge. Standard compass abaft mainmast, in some cases on after wheel-house.	V 185. V 184. V 183. V 182. V 181. V 180.
2	(*a*) Melms and Pfenniger turbines. (*b*) 4 Schulz Navy type.	(*a*) 16,000	(*a*) 32·0 (*a*) 32·0 (*a*) 32·0 (*b*) 33·0	(*a*) 148 +20‡ (*b*) 50	1,460		S 179. S 178. S 176.
2	(*a*) Parsons turbines (modified), (*b*) 4 Schulz, Navy type.	(*a*) 16,000	(*a*) 31·5	(*a*) 189 +20‡ (*b*) 47	1,240		G 175. G 174.
							1908–09 PROGRAMME
3 (G. 173 has 2.)	† (*a*) Parsons turbines. *Note.*—T. 173 has Zoelly turbines. (*b*) 4 Schulz.	(*a*) 14,000 (*b*) 16,400 (*a*) 14,000 (*a*) 14,000 (*a*) 14,000	(*a*) 30·0 (*b*) 32·9 (*a*) 30·0 (*a*) 30·0 (*b*) 33·7 (*a*) 30·0 (*b*) 33·8	(*a*) 159 +20‡ (*b*) 40	1,260	*See* Plate 5 and Silhouettes. 2 funnels, well forward. 2 masts (stump foremast by bridge, mainmast abaft 2nd funnel). Bridge against 1st funnel; very short well-deck between bridge and forecastle. After wheel-house abaft mainmast. 1—22-pr. on forecastle. 1—22-pr. abaft after wheel-house. 1 tube between funnels. 1 tube abaft 2nd funnel. 1 tube before after wheel-house. 1 searchlight above bridge. Standard compass on after wheel-house.	T 173. T 172. T 170. T 169.
2	(*a*) Schichau turbines. (*b*) 4 Schulz.	(*a*) 14,000	(*a*) 32·0	(*a*) 150 +20‡ (*b*) 40	1,200		T 168*. T 167*. T 166*. T 165*.
2	(*a*) A.E.G. turbines. *Note.*—T. 164 has A.E.G. Vulcan turbines. (*b*) 4 Schulz.	(*a*) 14,000	(*a*) 30·0 (*b*) 34·4 (*a*) 30·0 (*b*) 33·0 (*a*) 30·0 (*b*) 32·2	(*a*) 150 +20‡ (*b*) 40	1,200		T 164. T 163. T 162.

† In two engine rooms; three shafts, except G. 173, which has two shafts.

‡ Deck cargo.

D 4

Destroyers—

Official Number.	Where Built.	Date of Launch.	Complement.	Hull. Length. (a) P.P. (b) L.W.L. (c) Extreme.	Breadth (extreme).	Designed Load. Draught.	Designed Load. Displacement.	Armament.
				Ft. Ins.	Ft. Ins.	Ft. Ins.	Tons (Eng.).	
1907–08 PROGRAMME.								
T 161, T 160, T 159, T 158, T 157, T 156, T 155, T 154, T 153, T 152, T 151	Vulcan, Stettin.	1907–08	83	(a) 228 0 (c) 236 7	25 7	10 2	545	2—15-pr. 2—m. 3—T. (17·7″). 5 torpedoes.
1906–07 PROGRAMME.								
T 148, T 147, T 146, T 145, T 144, T 143, T 142, T 141, T 140, T 139, T 138	Schichau, Elbing.	1906–07	80	(a) 229 8 (c) 231 0	25 7	8 10	520	1—15-pr. 3—4-pr. 2—m. 3—T. (17·7″). 5 torpedoes.

continued.

Machinery and Boilers.				Fuel Supply.		Appearance, &c.	Official Number.
No. of Propellers.	(a) Engines. (b) Boilers.	Horse Power. (a) Designed. (b) Mean on trial.	Speed. (a) Designed. (b) Mean on trial.	(a) Coal (max.). (b) Oil.	Endurance at 15 Knots.		
			Knots.	Tons.	Miles.		
							1907–08 PROGRAMME.
2	(a) Vertical triple expansion. *Note.*—T. 161 has A.E.G. turbines. (b) 4 Schulz.	(a) 10,250 (b) 13,000	(a) 30·0 (b) 32·1	(a) 165 + 20‡. (b) Nil	1,575	*See* Plate 5 and Silhouettes. 2 funnels, well forward. 2 masts (stump foremast by bridge, mainmast abaft 2nd funnel). Bridge against 1st funnel; well-deck between bridge and forecastle. After wheel-house abaft main-mast. 1—15-pr. on forecastle. 1—15-pr. abaft after wheel-house. 1 tube abaft forecastle. 2 tubes, on centre line, between funnels. 1 searchlight above bridge. Standard compass fitted just before after wheel-house.	T 161.
		(a) 10,250	(a) 30·0	(a) 158 + 20‡. (b) Nil	1,450		T 160.
		(a) 10,250	(a) 30·0				T 159.
		(a) 10,250	(a) 30·0				T 158.
		(a) 10,250	(a) 30·0				T 157.
		(a) 10,250	(a) 30·0				T 156.
		(a) 10,250	(a) 30·0				T 155.
		(a) 10,250	(a) 30·0				T 154.
		(a) 10,250	(a) 30·0				T 153.
		(a) 10,250	(a) 30·0 (b) 31·55				T 152.
		(a) 10,250	(a) 30·0 (b) 31·12				T 151.
							1906–07 PROGRAMME.
2	(a) Vertical triple expansion. (b) 4 Schulz.	(a) 10,000	(a) 30·0	(a) 190 + 15‡. (b) Nil	2,500	*See* Plate 6 and Silhouettes. 2 funnels, well forward. 2 masts (stump foremast by bridge, mainmast abaft 2nd funnel). Bridge against fore funnel; well-deck between bridge and forecastle. 1—4-pr. at either end of bridge. 1—4-pr. before mainmast (on platform in some cases). 1—15-pr. on platform abaft mainmast. 1 tube abaft forecastle. 1 tube between funnels. 1 tube abaft 2nd funnel. 1 searchlight above bridge. Standard compass on platform abaft mainmast.	T 148. T 147. T 146. T 145. T 144. T 143. T 142 T 141 T 140. T 139. T 138.

‡ Deck cargo.

Destroyers—

Official Number.	Where Built.	Date of Launch.	Complement.	Hull. Length. (a) P.P. (b) L.W.L. (c) extreme.	Hull. Breadth (extreme).	Hull. Designed Load. Draught.	Hull. Designed Load. Displacement.	Armament.
				Ft. Ins.	Ft. Ins.	Ft. Ins.	Tons (Eng.).	
1905–1906 PROGRAMME.								
T 137	Germania, Kiel.	1907	80	(a) 224 9	25 1	9 10	560	1—15-pr. 3—4-pr. 2—m. 3—T. (17·7″). 5 torpedoes.
T 136	Germania, Kiel.	1906	68	(a) 207 0	23 0	7 7	480	4—4-pr. 2—m. 3—T. (17·7″). 5 torpedoes.
T 135	Germania, Kiel.	1906	68	(a) 207 0	23 0	7 7	480	1—15-pr. 2—4-pr. 2—m. 3—T. (17·7″). 5 torpedoes.
T 134	Germania, Kiel.	1906	68	(a) 207 0	23 0	7 7	480	4—4-pr. 2—m. 3—T. (17·7″). 5 torpedoes.
T 133	Germania, Kiel.	1906	68	(a) 207 0	23 0	7 7	480	
T 132	Germania, Kiel.	1906	68	(a) 207 0	23 0	7 7	480	
1904–1905 PROGRAMME.								
T 131 T 130 T 128 T 127 T 126*	Schichau, Elbing.	1905	60	(a) 207 8	23 0	7 7	480	3—4-pr. 2—m. 3—T. (17·7″). 5 torpedoes.
1903–1904 PROGRAMME.								
T 123 T 122 T 121 T 120	Schichau, Elbing.	1904	60	(a) 205 1	23 0	7 7	465	3—4-pr. 2—m. 3—T. (17·7″). 5 torpedoes.

* Sunk November 1905; raised May 1906 in two portions.

continued.

Machinery and Boilers.				Fuel Supply.			
No. of Propellers.	(*a*) Engines. (*b*) Boilers.	Horse-Power. (*a*) Designed. (*b*) Mean on trial.	Speed. (*a*) Designed. (*b*) Mean on trial.	(*a*) Coal (max.). (*b*) Oil.	Endurance at 15 Knots.	Appearance, &c.	OFFICIAL NUMBER.
			Knots.	Tons.	Miles.		
3	(*a*) Parsons turbines. (*b*) 4 Schulz-Thornycroft.	(*a*) 10,000 (*b*) 13,800 (est.)	(*a*) 30·0 (*b*) 33·0	(*a*) 165 + 15‡ (*b*) Nil	1,500	*See* Plate 6 and Silhouette. Same as *T. 132–134* (*see* below), but has— **1**—4 pr. at either end of bridge. **1**—4-pr. on platform between 2nd funnel and mainmast. **1**—15-pr. on platform abaft mainmast. 1 tube directly abaft forecastle. 1 tube before 2nd funnel. 1 tube abaft mainmast.	**1905–1906 PROGRAMME.** **T 137.**
2	(*a*) Vertical triple expansion. (*b*) 3 Schulz.	(*a*) 6,000	(*a*) 27·0	(*a*) 129 + 15‡ (*b*) Nil.	1,280	*See* Plate 6 and Silhouette. 2 slightly raking funnels. 2 masts (stump foremast by bridge, mainmast abaft 2nd funnel). Bridge against 1st funnel; well-deck between bridge and forecastle. **1**—4-pr. at either end of bridge. **2**—4-prs. on platforms on centre line abaft mainmast. 1 tube abaft forecastle. 2 tubes, on centre line, between funnels. 1 searchlight above bridge. (*T. 135* has **1**—15-pr. in place of **2**—4-prs. on platform abaft mainmast.	**T 136.**
		(*a*) 6,000	(*a*) 27·0				**T 135.**
		(*a*) 6,000	(*a*) 27·0				**T 134.**
		(*a*) 6,000	(*a*) 27·0				**T 133.**
		(*a*) 6,000 (*b*) 6,500	(*a*) 27·0 (*b*) 29·0				**T 132.**
2	(*a*) Vertical triple expansion. (*b*) 3 Schulz-Thornycroft.	(*a*) 6,500	(*a*) 27·0	(*a*) 113 + 15‡ (*b*) Nil.	1,640	*See* Silhouette. 2 slightly raking funnels. 2 masts (stump foremast by bridge; mainmast abaft 2nd funnel). Bridge against 1st funnel; well-deck between bridge and forecastle. **1**—4-pr. on either side abreast the bridge. **1**—4-pr. abaft mainmast. 1 tube abaft forecastle. 2 tubes, on centre line, between funnels. 1 searchlight above bridge.	**1904–1905 PROGRAMME.** **T 131.**
		(*a*) 6,500 (*b*) 6,600					**T 130.**
		(*a*) 6,500					**T 128.**
		(*a*) 6,500					**T 127.**
		(*a*) 6,500					**T 126.***
2	(*a*) Vertical triple expansion. (*b*) 3 Schulz-Thornycroft.	(*a*) 6,500 (*b*) 6,800	(*a*) 27·0 (*b*) 28·3†	(*a*) 113 + 15‡	1,890	Same type as *T. 126–131* (*see* above).	**1903–1904 PROGRAMME.** **T 123.**
		(*a*) 6,500 (*b*) 6,600	(*a*) 27·0 (*b*) 27·0†	(*a*) 113 + 15‡	1,890		**T 122.**
		(*a*) 6,500 (*b*) 6,800	(*a*) 27·0 (*b*) 27·1†	(*a*) 113 + 15‡	1,890		**T 121.**
		(*a*) 6,500 (*b*) 6,750	(*a*) 27·0 (*b*) 27·01†	(*a*) 113 + 15‡	1,890		**T 120.**

† Mean of three hours' trial. ‡ Deck cargo.

E 2

Destroyers—

Official Number.	Where Built.	Date of Launch.	Complement.	Hull: Length. (a) P.P. (b) L.W.L. (c) extreme.	Hull: Breadth (extreme).	Hull: Designed Load. Draught.	Hull: Designed Load. Displacement.	Armament.
				Ft. Ins.	Ft. Ins.	Ft. Ins.	Tons (Eng.).	
1902–1903 PROGRAMME.								
T 114 - -	Schichau, Elbing.	1902	56	(a) 201 2	23 0	7 7	413	3—4-pr. - - 2—m. 3—T. (17·7″). 5 torpedoes.
1901–1902 PROGRAMME.								
T 113 - - T 112 - - T 111 - - T 110 - - T 109 - - T 108 - -	Germania, Kiel.	1901–02	56	(a) 207 8	22 0	7 7	394	3—4-pr. - - 2—m. 3—T. (17·7″). 5 torpedoes.
1900–1901 PROGRAMME.								
T 107 - - T 106 - - T 105 - - T 104 - - T 103 - - T 102 - -	Schichau, Elbing.	1901	56	(a) 200 2 (c) 210 0	23 0	7 7	394	3—4-pr. - - 2—m. 3—T. (17·7″) 5 torpedoes.
1899–1900 PROGRAMME.								
T 101 - - T 100 - - T 99 - - T 98 T 97 (Sleipner) T 96 - -	Schichau, Elbing.	1900–01	56	(a) 200 2 (c) 210 0	23 0	7 7	394	3—4-pr. - - 2—m. 3—T. (17·7″) 5 torpedoes.

continued.

Machinery and Boilers.				Fuel Supply.		Appearance, &c.	Official Number.
No. of Pro-pellers.	(*a*) Engines. (*b*) Boilers.	Horse Power. (*a*) Designed. (*b*) Mean on trial.	Speed. (*a*) Designed. (*b*) Mean on trial.	(*a*) Coal (max.). (*b*) Oil.	Endurance at 15 Knots.		
			Knots.	Tons.	Miles.		
							1902–1903 PROGRAMME.
2	(*a*) Vertical, triple expansion. (*b*) 3 Schulz-Thornycroft.	(*a*) 6,200	(*a*) 26·0	(*a*) 98 + 10‡ (*b*) Nil	1,350	Same type as *T. 126–131* (*see* above).	T 114.
							1901–1902 PROGRAMME.
2	(*a*) Vertical triple expansion. (*b*) 3 Thornycroft.	(*a*) 6,000	(*a*) 26·0 (*a*) 26·0 (*a*) 26·0 (*a*) 26·0 (*a*) 26·0 (*a*) 26·0 (*b*) 29·1	(*a*) 110 + 10‡ (*b*) Nil	1,840	*See* Silhouette. 2 slightly raking funnels, rather far apart. 2 masts (stump foremast by bridge, mainmast abaft 2nd funnel). Bridge against 1st funnel; well-deck between bridge and forecastle. 1—4-pr. on either side nearly abreast bridge. 1—4-pr. abaft mainmast. 1 tube abaft forecastle. 2 tubes, on centre line, between funnels.	T 113. T 112. T 111. T 110. T 109. T 108.
							1900–1901 PROGRAMME.
2	(*a*) Vertical, triple expansion. (*b*) 3 Thornycroft.	(*a*) 5,400	(*a*) 26·0 (*b*) 28·3 (*a*) 26·0 (*a*) 26·0 (*a*) 26·0 (*b*) 28·9 (*a*) 26·0 (*a*) 26·0	(*a*) 92 + 10‡ (*b*) Nil	1,300	*See* Silhouette. 2 raking funnels, rather far apart. 2 masts (stump foremast by bridge, mainmast abaft 2nd funnel). Bridge against 1st funnel; well-deck between bridge and forecastle. 1—4-pr. on either side, abreast bridge. 1—4-pr. abaft mainmast. 1 tube abaft forecastle. 1 tube before and 1 tube abaft 2nd funnel.	T 107. T 106. T 105. T 104. T 103. T 102
							1899–1900 PROGRAMME.
2	(*a*) Vertical, triple expansion. (*b*) 3 Thornycroft.	(*a*) 5,400	(*a*) 26·0 (*a*) 26·0 (*a*) 26·0 (*a*) 26·0 (*a*) 26·0 (*b*) 28·0 (*a*) 26·0	(*a*) 92 + 10‡ (*b*) Nil	1,300	*See* Silhouette. Same general type as *T. 102–107* (*see* above); but *T 97*, which before the war was used as the Emperor's despatch boat, may differ slightly, and probably still has one mast only, abaft forecastle.	T 101. T 100 T 99. T 98. T 97 (Sleipner). T 96.

‡ Deck cargo.

E 3

Destroyers—

Official Number.	Where Built.	Date of Launch.	Complement.	Hull. Length. (a) P.P. (b) L.W.L. (c) extreme.	Breadth (extreme).	Designed Load. Draught.	Designed Load. Displacement.	Armament.
				Ft. Ins.	Ft. Ins.	Ft. Ins.	Tons. (Eng.)	
1898–1899 PROGRAMME.								
T 95 - - T 94 - - T 93 - - T 92 - - T 91 - -	Schichau, Elbing.	1899–1900.	56	(a) 200 2 (c) 210 0	23 0	7 7	394	3—4-pr. - - 2—m. 3—T. (17·7"). 5—torpedoes.

continued.

Machinery and Boilers.				Fuel Supply.		Appearance, &c.	OFFICIAL NUMBER.
No. of Pro-pellers.	(a) Engines. (b) Boilers.	Horse Power. (d) Designed. (b) Mean on trial.	Speed. (a) Designed. (b) Mean on trial.	(a) Coal (max.). (b) Oil.	Endurance at 15 Knots.		
			Knots.	Tons.	Miles.		
							1898–1899 PROGRAMME.
2	(a) Vertical, triple expansion. (b) 3 Thornycroft.	(a) 5,400	(a) 26·0	(a) 92 + 10‡ (b) Nil.	1,300	*See* Silhouette. Same general type as *T. 102–107* (*see* p. 37).	**T 95.** **T 94.** **T 93.** **T 92.** **T 91.**

‡ Deck cargo.

E 4

TORPEDO

Official Number.	Where Built.	Date of Launch.	Complement.	Hull. Length. (a) P.P. (b) L.W.L. (c) extreme.		Breadth (extreme).		Designed Load. Draught.		Displacement.	Armament.
				Ft.	Ins.	Ft.	Ins.	Ft.	Ins.	Tons (Eng.).	
SPECIAL CLASS (DUTCH).*											
V 108, V 106, V 105	Vulcan, Stettin.	1914	—	(c) 201	1	20	4	6	1	318	2—13-pr. 4—T. (17·7″).
DIVISIONAL (OLD).											
D 10	Thornycroft, Chiswick.	1898	60	(c) 210	0	19	4	7	7	349	5—4-pr. 2—m. 2—T. (17·7″). 5 torpedoes.
D 9	Schichau, Elbing.	1894	59	(c) 196	10	25	3	12	2	374	3—4-pr. 2—m. 1—S.T. (17·7″). 2—T. (17·7″). 5 torpedoes.
D 8, D 7	"	1891	49	(c) 195	10	23	0	11	6	344	3—4-pr. 2—m. 1—S.T. (17·7″). 2—T. (17·7″). 5 torpedoes.
D 6, D 5	"	1888–89	49	(c) 190	0	23	0	11	6	315	3—4-pr. 2—m. 1—S.T. (17·7″). 2—T. (17·7″). 5 torpedoes.
D 4, D 3	"	1887	49	(c) 181	5	21	8	10	6	295	3—4-pr. 2—m. 1—S.T. (17·7″). 2—T. (17·7″). 5 torpedoes.
D 2 (*late* Alice Roosevelt), D 1	"	1886	46	(c) 183	0	21	8	9	9	226	3—4-pr. 2—m. 1—S.T. (17·7″). 2—T. (17·7″). 5 torpedoes.

* Originally building for Holland, but commandeered by Germany in August 1914. There were originally four of these boats, but *V. 107* has been lost.

BOATS.

Machinery and Boilers.				Fuel Supply.		
No. of Propellers.	(*a*) Engines. (*b*) Boilers.	Horse Power. (*a*) Designed. (*b*) Mean on trial.	Speed. (*a*) Designed. (*b*) Mean on trial.	(*a*) Coal (max.). (*b*) Oil.	Appearance, &c.	OFFICIAL NUMBER.
			Knots.	Tons.		
						SPECIAL CLASS (DUTCH).
2	(*a*) A.E.G. Vulcan turbines. (*b*) 3 (2 coal-burning, 1 oil).	(*a*) 5,300	27·0	—	*See* Plate 13 and Silhouette. Short high forecastle. Well-deck between forecastle and bridge. 2 raking masts; foremast close abaft bridge, mainmast well aft. 2 raking funnels.	V 108. V 106. V 105.
						DIVISIONAL (OLD).
2	(*a*) Triple expansion. (*b*) 3 Thornycroft.	(*a*) 5,500	(*a*) 26·0	(*a*) 79	*See* Plate 7 and Silhouette. Turtle-back forecastle, with raised platform on which are mounted 2—4-pr. guns. 2 raking funnels, with mast between; stump foremast on forecastle.	D 10.
2	(*a*) Triple expansion. (*b*) 3 locomotive.	(*b*) 4,040	(*b*) 24·5	(*a*) 105	*See* Silhouette. Turtle-back forecastle. 2 raking funnels. 1 mast abaft 2nd funnel. In some of these vessels there is a stump foremast also, and in one at least an additional foremast has been filled at break of forecastle. High bridge before fore funnel. 2—4-prs. on forebridge. 1—4-pr. on conning tower aft. Searchlight above forebridge.	D 9.
2	(*a*) Triple expansion. (*b*) 2 locomotive.	(*a*) 4,000	(*a*) 26·0	(*a*) 75		D 8. D 7.
2	(*a*) Triple expansion. (*b*) Schulz. (*a*) Triple expansion. (*b*) Locomotive.	(*a*) 3,600	(*a*) 23·0	(*a*) 90		D 6. D 5.
2	(*a*) Triple expansion. (*b*) Locomotive.	(*a*) 2,500	(*a*) 21·0	(*a*) 65		D 4. D 3.
2	(*a*) Triple expansion. (*b*) Locomotive.	(*a*) 1,800	(*a*) 21·0	(*a*) 55	*See* Silhouette. 2 pole masts, 1 funnel; large deck house abaft the mainmast. *See* Silhouette. 1 mast and 1 funnel; 2 conning towers. Bridge against funnel, with searchlight on port side. Turtle-back forecastle.	D 2 (*late* Alice Roosevelt). D 1.

TORPEDO

Official Number.	Where Built.	Date of Launch.	Complement.	Hull. Length. (a) P.P. (b) L.W.L. (c) extreme	Hull. Breadth (extreme).	Hull. Designed Load. Draught.	Hull. Designed Load. Displacement.	Armament.
				Ft. Ins.	Ft. Ins.	Ft. Ins.	Tons (Eng.).	
1st CLASS (NEW).								
A 150	Pola	1917	—	—	—	—	—	**2**—22-pr.
A 100								
A 99								
A 98								
A 97								
A 96								
A 95								
A 94								
A 93								
A 92								
A 91								
A 90								
A 89								
A 88	—	1917	—	—	—	—	200†	**2**—22-pr. **1**—m. **1**—T.
A 87								
A 86								
A 85								
A 84								
A 83								
A 82								
A 81								
A 80								
A 78								
A 76								
A 75								
A 74								

(A 100 to A 74 are bracketed together; the particulars in the A 88 row apply to the whole group.)

† Approximate only.

BOATS.

Machinery and Boilers.				Fuel Supply.	Appearances, &c.	OFFICIAL NUMBER.
No. of Propellers.	(*a*) Engines. (*b*) Boilers.	Horse Power. (*a*) Designed. (*b*) Mean on trial.	Speed. (*a*) Designed. (*b*) Mean on Trial.	(*a*) Coal (max.). (*b*) Oil.		
			Knots.	Tons.		1st CLASS (NEW).
—	(*a*) Turbines (*b*) Two.	—	23†	Oil only	Turtle-back deck forward 1 mast. 1 funnel. No tubes.	A 150.
—	(*a*) Turbines (*b*) 2 Schulz.	—	24†	Oil only	1 funnel amidships 2 masts. Long forecastle, breaking about abreast of funnel. 1—22-pr. on forecastle. 1—22-pr. aft. 1 tube abaft funnel.	A 100. A 99. A 98. A 97. A 96. A 95. A 94. A 93. A 92. A 91. A 90. A 89. A 88. A 87. A 86. A 85. A 84. A 83. A 82. A 81. A 80. A 78. A 76. A 75. A 74.

F 2

Torpedo

Official Number.	Where Built.	Date of Launch.	Complement.	Hull. Length. (a) P.P. (b) L.W.L. (c) Extreme.	Hull. Breadth (extreme).	Hull. Designed Load. Draught.	Hull. Designed Load. Displacement.	Armament.
				Ft. Ins.	Ft. Ins.	Ft. Ins.	Tons (Eng.).	
1st CLASS (NEW)—*cont.*								
A 70								
A 69								
A 68								
A 67								
A 66								
A 65								
A 64								
A 63								
A 62								
A 61								
A 59								
A 58								
A 55								
A 54								
A 53	—	1917	35	—	—	—	200†	2—22-pr. 1—m. 1—T.
A 52								
A 51								
A 50								
A 49								
A 48								
A 47								
A 46								
A 45								
A 44								
A 42								
A 41								
A 40								
A 39								
A 38								
A 37								

† Approximate only.

Boats—*continued.*

Machinery and Boilers.				Fuel Supply.	Appearance, &c.	OFFICIAL NUMBER.
No. of Propellers.	(a) Engines. (b) Boilers.	Horse Power. (a) Designed. (b) Mean on trial.	Speed. (a) Designed. (b) Mean on trial.	(a) Coal (max.). (b) Oil.		
			Knots.	Tons.		**1st CLASS (NEW)**—*cont.*
—	(a) Turbines. (b) 2 Schultz.	—	24†	Oil only	One funnel amidships. Two masts. Long forecastle, breaking about abreast of funnel. 1—22-pr. on forecastle. 1—22-pr. aft. 1 tube abaft funnel.	A 70. A 69. A 68. A 67. A 66. A 65. A 64. A 63. A 62. A 61. A 59. A 58. A 55. A 54. A 53. A 52. A 51. A 50. A 49. A 48. A 47. A 46. A 45. A 44. A 42. A 41. A 40. A 39. A 38. A 37.

† Approximate only.

F 3

Torpedo

OFFICIAL NUMBER.	Where Built.	Date of Launch.	Complement.	Hull. Length. (a) P.P. (b) L.W.L. (c) extreme.	Hull. Breadth (extreme).	Hull. Designed Load. Draught.	Hull. Designed Load. Displacement.	Armament.
1st CLASS. (NEW)—*cont.*				Ft. Ins.	Ft. Ins.	Ft. Ins.	Tons (Eng.).	
A 36, A 35, A 34, A 33, A 32, A 31, A 30, A 29, A 28, A 27, A 26	Vulcan, Hamburg.	1915–16	35	(*c*) 170 0†	—	—	200†	**2**—22-pr. **1**—m. **1**—T. (17·7″). **2** torpedoes.
A 25, A 24, A 23, A 22, A 21, A 20, A 19, A 18, A 17, A 16, A 15, A 14, A 13, A 12, A 11, A 9, A 8, A 7, A 5, A 4, A 3, A 1	Vulcan, Hamburg.*	1915	29	(*c*) 140 0	19 6	8 2†	80†	**1**—4-pr. **1**—m. ‡**2**—T. (17·7″). **2** torpedoes.

* Some of these boats were sent in sections by rail from Hamburg to Antwerp and there put together. † Approximate only.
‡ In most cases only one of these tubes remains, the other having been removed to make room for a minesweeping winch.

Boats—*continued.*

Machinery and Boilers.				Fuel Supply.	Appearance, &c.	Official Number.
No. of Propellers.	(*a*) Engines. (*b*) Boilers.	Horse Power. (*a*) Designed. (*b*) Mean on trial.	Speed. (*a*) Designed. (*b*) Mean on trial.	(*a*) Coal (max.). (*b*) Oil.		
			Knots.	Tons.		1st CLASS (NEW)—*cont.*
—	(*a*) Turbines (*b*) 2 Schulz.	—	24†	Oil only.	*See* Plate 7 and Silhouette. One funnel amidships with bridge built close back against it. Two masts. Well deck between forecastle and bridge. Straight stem. 1—22-pr. on forecastle. 1—22-pr- aft. 1 tube abaft funnel. Searchlight on deckhouse at foot of mainmast.	A 36. A 35. A 34. A 33. A 32. A 31. A 30. A 29. A 28. A 27. A 26.
1	(*a*) Vertical (*b*) 1 Schulz.	—	15†	(*a*) 27	*See* Plate 7 and Silhouette. Turtle - backed forecastle with bridge close abaft it. 1 raking funnel; 2 masts. A small conning tower is fitted below bridge. 1—4-pr. on platform abaft mainmast. 1 tube between funnel and foremast. 1 tube between funnel and mainmast. (*One of these tubes has been removed from a number of these boats, perhaps all.*) Standard compass on pedestal close before funnel.	A 25. A 24. A 23. A 22. A 21. A 20. A 19. A 18. A 17. A 16. A 15. A 14. A 13. A 12. A 11. A 9. A 8. A 7. A 5. A 4. A 3. A 1.

† Approximate only.

F 4

Torpedo Boats

Official Number.	Where Built.	Date of Launch.	Complement.	Hull. Length. (a) P.P. (b) L.W.L. (c) extreme.		Breadth (extreme).		Designed Load. Draught.		Designed Load. Displacement.	Armament.
1st CLASS (OLD)—*cont.*				Ft.	Ins.	Ft.	Ins.	Ft.	Ins.	Tons (Eng.).	
T 89, T 88	Germania, Kiel.	1897–98	24	(*a*) 154	2	16	5	9	6	152	1—4-pr. on after conning tower. *2—T. (17·7″). 4 torpedoes.
T 87, T 86, T 85, T 84, T 83, T 82	Schichau, Elbing.	1897	24	(*a*) 152	7	16	6	7	3	152	1—4-pr., on after conning tower. 1—S.T. 17·7″ in bow. *2—T. 17·7″. 4 torpedoes.
T 81, T 80, T 79, T 77, T 76, T 75	Schichau, Elbing.	1894–95	24	(*a*) 152	7	16	6	7	3	150	1—4-pr., on after conning tower. 1—S.T. 17·7″ in bow. *2—T. 17·7″. 4 torpedoes.
T 74	Schichau, Elbing.	1895	24	(*a*) 152	7	16	6	7	3	143	1—4-pr. on conning tower aft. 1—S.T. 17·7″ in bow. *2—T. 17·7″. 4 torpedoes.
T 73, T 72, T 71, T 70, T 69, T 68, T 67, T 66	Schichau, Elbing.	1893	21	(*a*) 152	7	16	6	7	3	167	1—4-pr. on conning tower aft. 1—S.T. (17·7″) in bow. *2—T. (17·7″). 4 torpedoes.

* In most cases only one of these tubes remains, the other having been removed to make room for a minesweeping winch.

—continued.

Machinery and Boilers.				Fuel Supply.		
No. of Propellers.	(a) Engines. (b) Boilers.	Horse-Power. (a) Designed. (b) Mean on trial.	Speed. (a) Designed. (b) Mean on trial.	(a) Coal (max.). (b) Oil.	Appearance, &c.	OFFICIAL NUMBER.
			Knots.	Tons.		**1st CLASS (OLD)**—*cont.*
1	(a) Triple expansion. (b) Thorny-croft-Schulz.	(a) 1,800	(a) 25	(a) 30 (b) 7	*See* Silhouette - - - 1 mast between 2 funnels, which are wide apart. Turtle-back to fore conning tower which is just before fore funnel. After conning tower abaft after funnel.	T 89. T 88.
1	(a) Triple expansion. (b) Thornycroft.	(a) 1,800	(a) 25	(a) 30*	*See* Silhouette - - - 1 mast between 2 funnels which are wide apart. Rounded stem, turtle back to fore conning tower which is just before fore funnel. After conning tower abaft after funnel.	T 87. T 86. T 85. T 84. T 83. T 82.
1	(a) Triple expansion. (b) 2 Locomotive.	(a) 1,800	(a) 25	(a) 30*	*See* Silhouette - - - 1 mast aft, 1 funnel, 2 conning towers.	T 81. T 80. T 79. T 77. T 76. T 75.
1	(a) Triple expansion. (b) Thornycroft.	(a) 1,800	(a) 26	(a) 30*	*See* Silhouette - - - 1 mast aft, 1 funnel, 2 conning towers.	T 74.
1	(a) Triple expansion. (b) 2 Locomotive.	(a) 1,800	(a) 22	(a) 30*	*See* Silhouette - - - 1 mast aft, 1 funnel. Rounded stem, partial turtle back to fore conning tower. After conning tower abaft mast.	T 73. T 72. T 71. T 70. T 69. T 68. T 67. T 66.

* N.B.—Some of these boats have been fitted to burn oil fuel.

Torpedo Boats

Official Number.	Where Built.	Date of Launch.	Complement.	Hull. Length. (a) P.P. (b) L.W.L. (c) extreme.	Breadth (extreme).	Designed Load. Draught.	Displacement.	Armament.
				Ft. Ins.	Ft. Ins.	Ft. Ins.	Tons (Eng.).	
1st CLASS (OLD)—*cont.*								
T 65 - - T 64 - - T 63 - - T 62 - - T 61 - - T 60 - - T 59 - - T 58 - -	Schichau, Elbing.	1891–92	16	(*a*) 152 7	16 6	7 3	151	1—4-pr. on conning tower aft. 1—S.T. (13·8″) in bow. *2—T. (13·8″). 4 torpedoes.
T 56 - - T 55 - - T 53 - - T 52 - - T 51 - - T 49 - -	Schichau, Elbing.	1889–90	16	(*a*) 152 7	16 6	7 3	128	1—4-pr. on conning tower aft. 1—S.T. (13·8″) in bow. *2—T. (13·8″). 4 torpedoes.
T 47 - - T 46 - - T 45 - - T 44 - - T 42 - -	Schichau, Elbing.	1889	16	(*a*) 144 4	16 5	7 3	143	1—4-pr. on conning tower aft. 1—S.T. (13·8″) in bow. *2—T. (13·8″). 4 torpedoes.

* In most cases only one of these tubes remains, the other having been removed to make room for a minesweeping winch.

—continued.

Machinery and Boilers.				Fuel Supply.	Appearance, &c.	OFFICIAL NUMBER.
No. of Pro-pellers.	(a) Engines. (b) Boilers.	Horse Power. (a) Designed. (b) Mean on Trial.	Speed. (a) Designed. (b) Mean on Trial.	(a) Coal (max.). (b) Oil.		
			Knots.	Tons.		**1st CLASS (OLD)**—*cont.*
1	(a) Triple expansion. (b) Locomotive.	(a) 1,800	(a) 23	(a) 30*	*See* Silhouette - - 1 mast aft, 1 funnel. Rounded stem, partial turtle back to fore conning tower. After conning tower abaft mast. Bunkers are abreast the engines.	T 65. T 64. T 63. T 62. T 61. T 60. T 59. T 58.
1	(a) Triple expansion. (b) Locomotive.	(a) 1,500	(a) 22	(a) 42*	*See* Silhouette - - - 1 mast aft, 1 funnel. Rounded stem, turtle-back forecastle. After conning tower abaft mast.	T 56. T 55. T 53. T 52. T 51. T 49.
1	(a) Triple expansion. (b) Locomotive.	(a) 1,350	(a) 22	(a) 30*	*See* Silhouette - - - 1 mast aft, 1 funnel. Rounded stem, turtle-back forecastle. After conning tower abaft mast. *See* Silhouette - - - 2 thin raking funnels, 1 upright mast abaft them: otherwise resembles T. 44. (Alterations made when boat was re-boilered.)	T 47. T 46. T 45. T 44. T 42.

* N.B.—Some of these boats have been fitted to burn oil fuel.

O AS 4924—22 H

Torpedo Boats—

Official Number.	Where Built.	Date of Launch.	Complement.	Hull. Length. (a) P.P. (b) L.W.L. (c) extreme.	Hull. Breadth (extreme).	Hull. Designed Load. Draught.	Hull. Designed Load. Displacement.	Armament.
				Ft. Ins.	Ft. Ins.	Ft. Ins.	Tons (Eng.).	
2nd and 3rd CLASSES (OLD).								
T 40, T 39, T 38, T 37, T 36, T 35, T 34, T 33, T 31, T 30, T 29, T 28, T 27	Schichau, Elbing.	1887	16	(a) 131 3	14 9	5 11	84	1—4-pr. on conning tower aft. 1—S.T. (13·8″) in bow. *2—T. (13·8″).
T 24, T 22, T 21, T 20, T 16, T 15, T 14, T 13, T 11	Schichau, Elbing.	1885–86	16	(a) 123 10	15 9	6 6	84	1—4-pr. on conning tower aft. 1—S.T. (13·8″) in bow. *2—T. (13·8″).

* In most cases only one of these tubes remains, the other having been removed to make room for a minesweeping winch.

continued.

Machinery and Boilers.				Fuel Supply.	Appearance, &c.	OFFICIAL NUMBER.
No. of Propellers.	(*a*) Engines. (*b*) Boilers.	Horse Power. (*a*) Designed. (*b*) Mean on trial.	Speed. (*a*) Designed. (*b*) Mean on trial.	(*a*) Coal (max.). (*b*) Oil.		
			Knots.	Tons.		2nd and 3rd CLASSES (OLD).
1	(*a*) Triple expansion. (*b*) 1 Locomotive.	(*a*) 1,000	(*a*) 20	(*a*) 34*	*See* Silhouette - - - †1 mast aft; 1 funnel, large, raking. Deck flat with rounded edges, forward partially turtle-backed to conning tower. After conning tower abaft mainmast.	T 40. T 39. T 38. T 37. T 36. T 35. T 34. T 33. T 31. T 30. T 29. T 28. T 27.
1	(*a*) Triple expansion. (*b*) 1 Locomotive.	1,000	20	(*a*) 26	See Silhouette - - - †1 mast aft; 1 funnel, large, raking. Deck flat with rounded edges, forward partially turtle-backed to conning tower. After conning tower abaft mainmast.	T 24. T 22. T 21. T 20. T 16. T 15. T 14. T 13. T 11.

* N.B.—Some of these boats have been fitted to burn oil fuel. † Some of these boats are reported now to have 2 masts.

DESTROYERS.

B. 109-112 and B. 97-98.

G. 101-104.

V. 100.

Ordnance Survey, September, 1918.

DESTROYERS.

G. 85-96 and G. 41.
(G. 37-40 very similar).

V. 67-84.

S. 53-66.

Ordnance Survey, September, 1918.

V. 25-30.
(S. 31-36 very similar).

G. 7-12.
(S. 13-24 very similar).

V. 16.

Ordnance Survey, September, 1918.

DESTROYERS.

G. 196 (G. 192-197).

V. 186 (V. 186-191).

V. 181 (V. 180-185).

Ordnance Survey, September, 1918

DESTROYERS.

S. 177 (S. 176-179).

T. 172 (T. 169-172).

T. 151 (T. 150-155).

Ordnance Survey, September, 1918.

DESTROYERS.

T. 149 (T. 138-149).

T. 137.

T. 134 (T. 132-136).

Ordnance Survey, September, 1918

D. 10.

A. 26-36.

A. 1-25.

Ordnance Survey, September, 1918.

Plate 8.
C.B. 1182 R.
August, 1918.

TORPEDO BOAT DESTROYERS B.109-112, B.97-98.

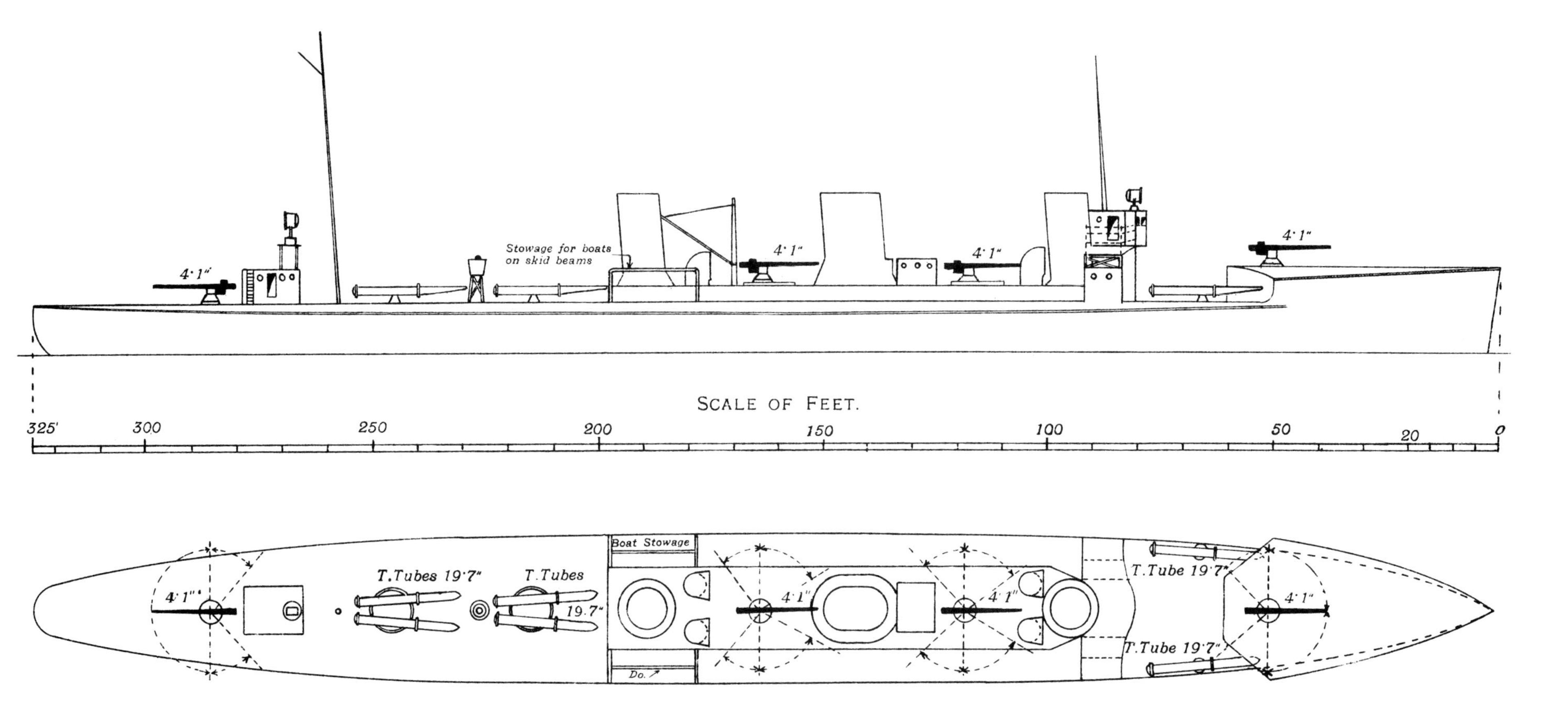

Ordnance Survey, Sept 1918.

TORPEDO BOAT DESTROYERS G.101-104.

Originally building for ARGENTINA.

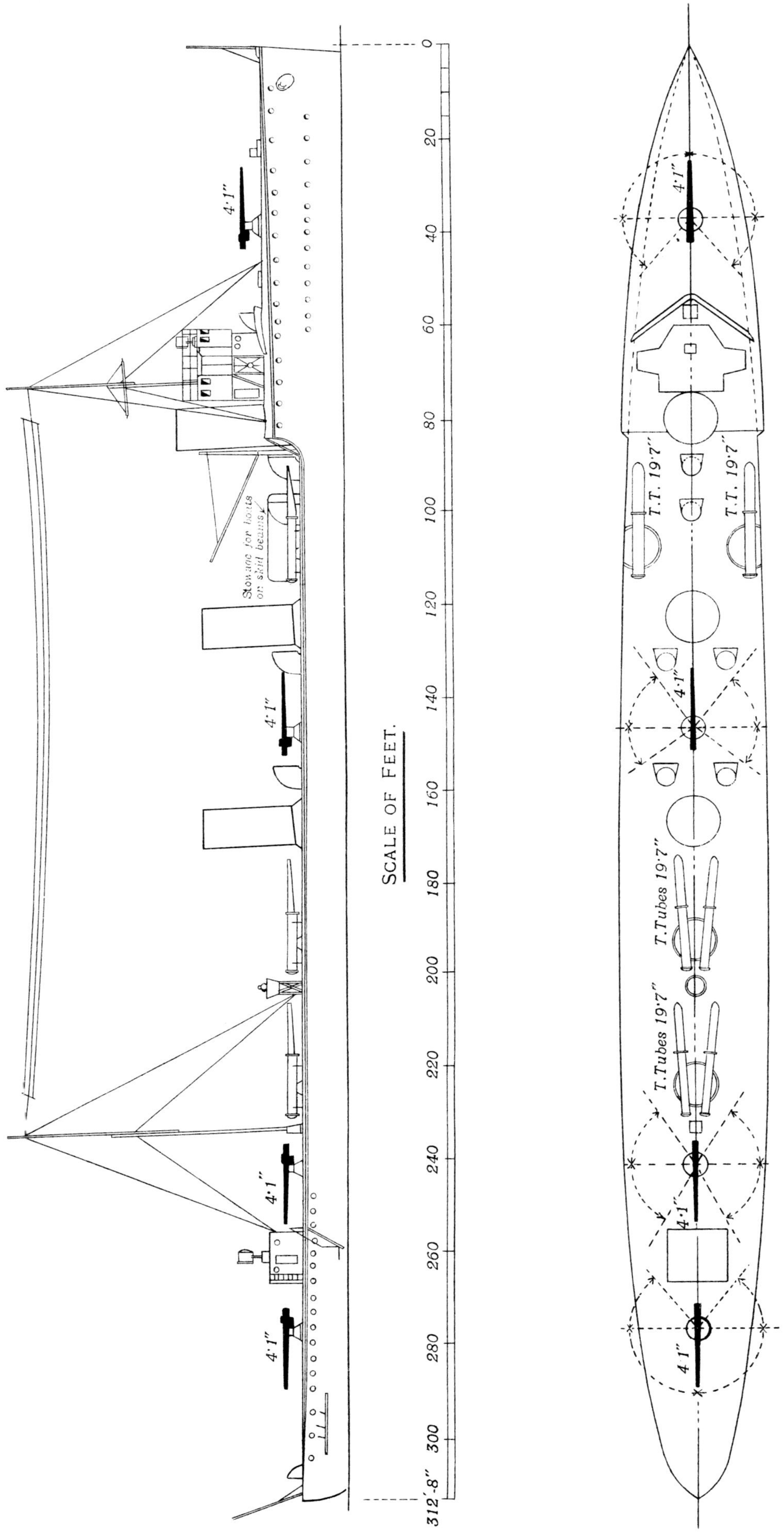

4·1"
4·1"
4·1"
4·1"
Stowage for boats on skid beams

SCALE OF FEET.

312'-8" 300 280 260 240 220 200 180 160 140 120 100 80 60 40 20 0

4·1"
4·1"
T.Tubes 19·7"
T.Tubes 19·7"
4·1"
T.T. 19·7"
T.T. 19·7"
4·1"

Ordnance Survey, Sept 1918.

Plate 10.
C.B. 1182 R.
August, 1918.

TORPEDO BOAT DESTROYERS G.85-96, G.41.

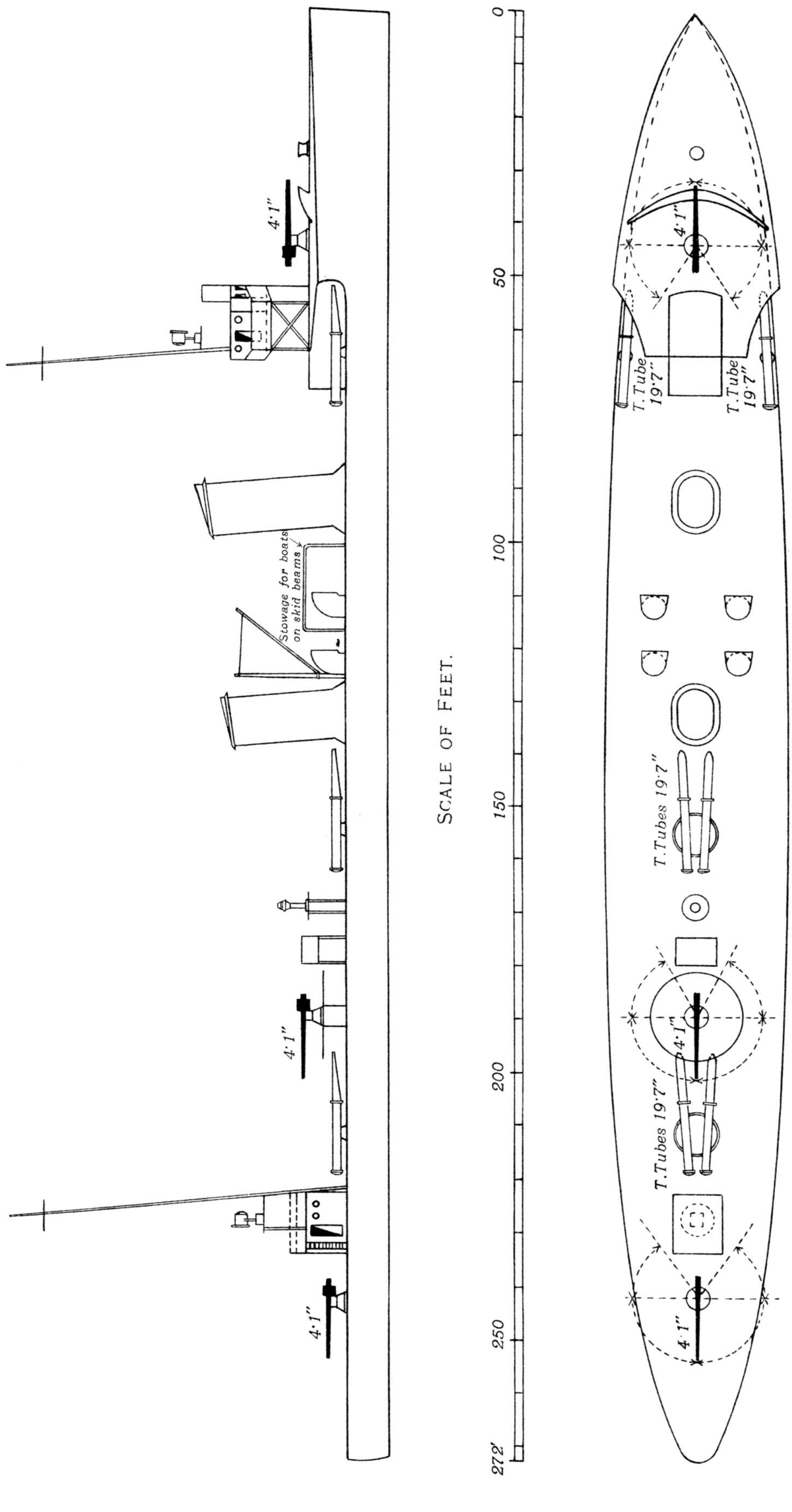
4·1"
Stowage for boats on skid beams
4·1"
4·1"
SCALE OF FEET.
0
50
100
150
200
250
272'
T.Tube 19·7"
T.Tube 19·7"
T.Tubes 19·7"
T.Tubes 19·7"
4·1"
4·1"
4·1"

Ordnance Survey, Sept 1918.

Plate 11.
C.B.1182.R.
August, 1918.

TORPEDO BOAT DESTROYERS V.67-84.

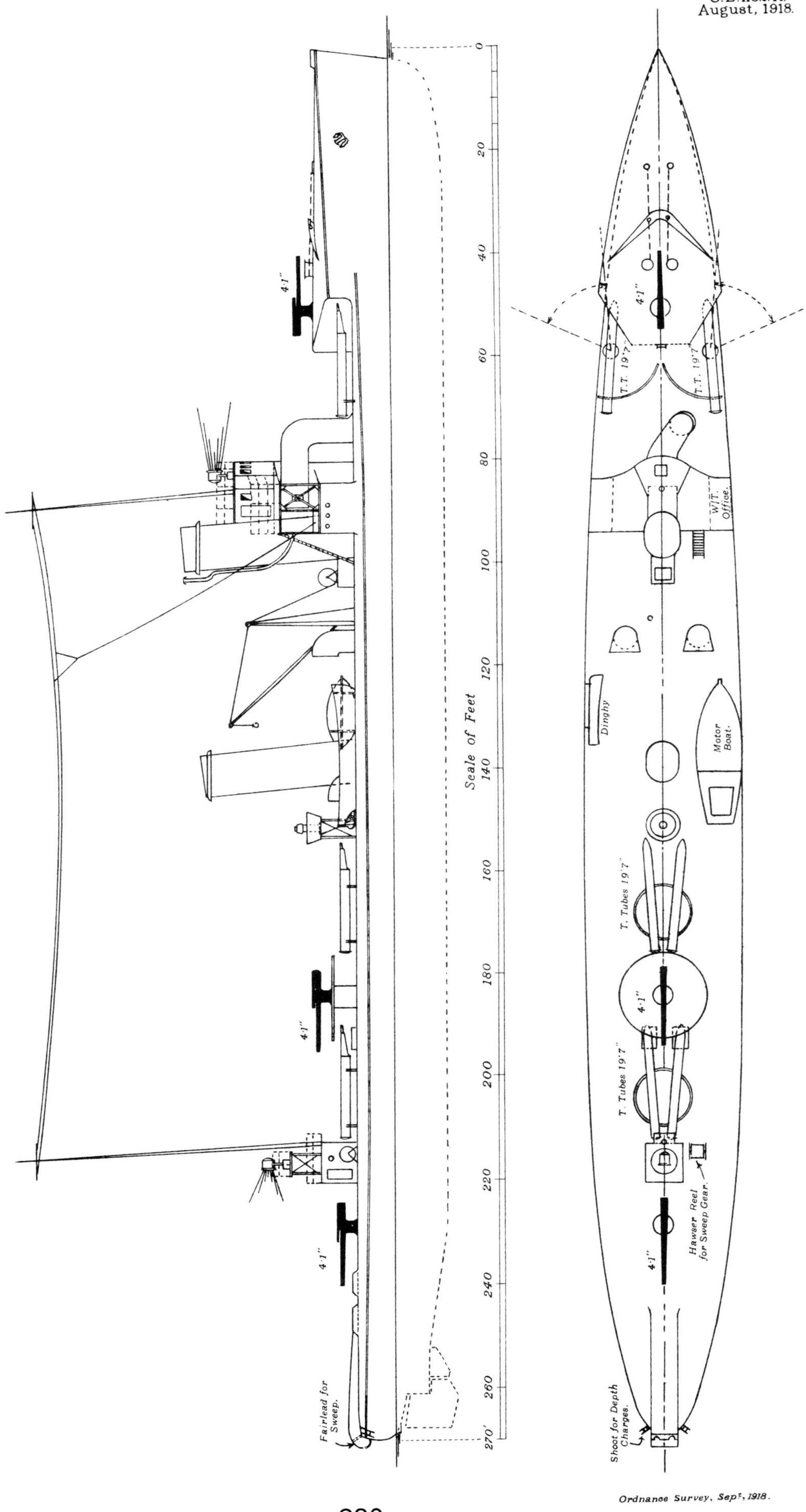

Fairlead for Sweep.
4·1"
4·1"
4·1"

Scale of Feet
270' 260 240 220 200 180 160 140 120 100 80 60 40 20 0

Shoot for Depth Charges.
4·1"
Hawser Reel for Sweep Gear.
T. Tubes 19'7"
4·1"
T. Tubes 19'7"
Dinghy
Motor Boat.
W/T. Office.
T.T. 19'7"
4·1"
T.T. 19'7"

Ordnance Survey, Sept. 1918.

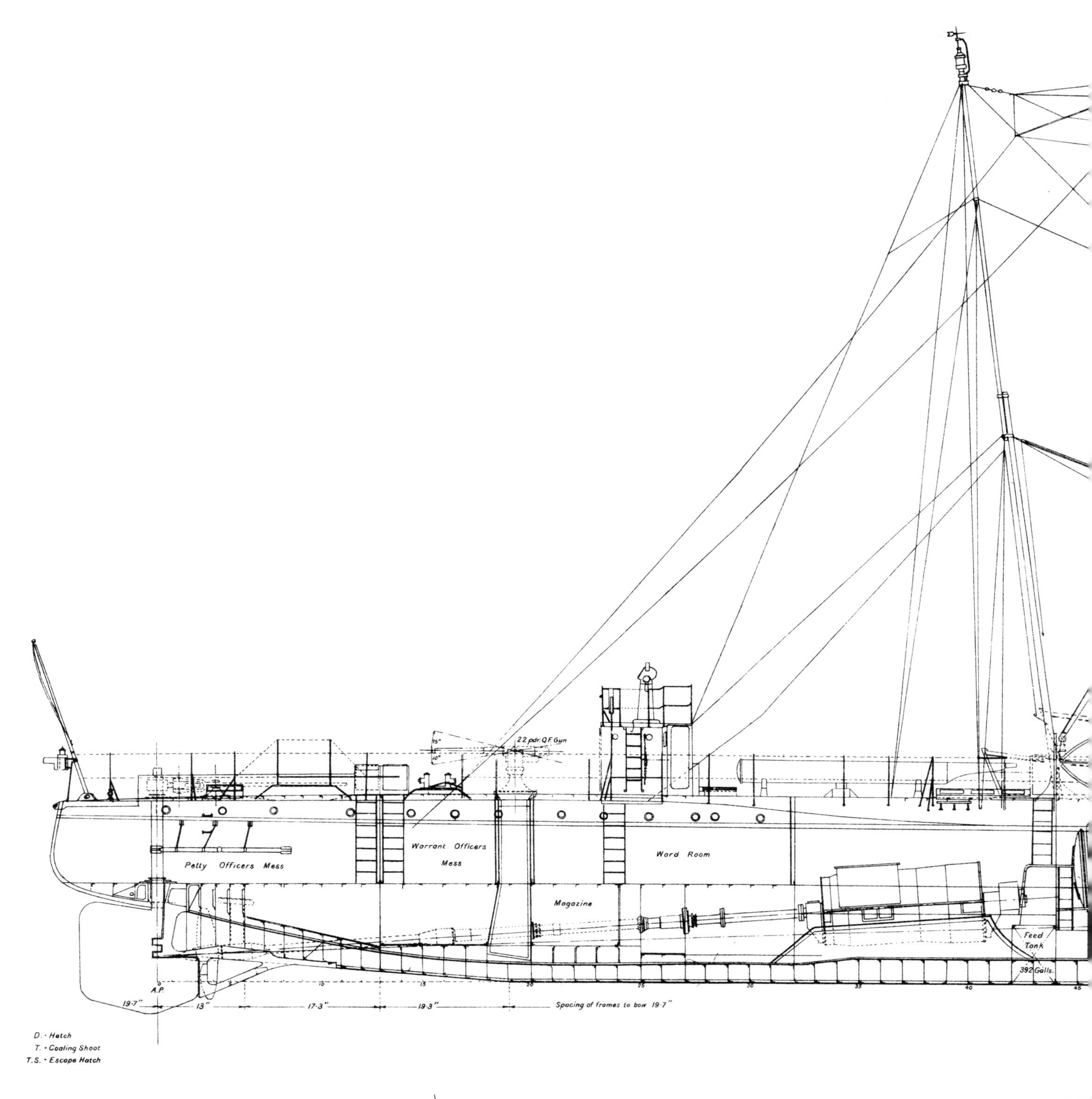

22 pdr. Q.F. Gun
Petty Officers Mess
Warrant Officers
Mess
Ward Room
Magazine
Feed
Tank
392 Galls.
A.P.
19·7"
13"
17·3"
19·3"
Spacing of frames to bow 19·7"
D. - Hatch
T. - Coaling Shoot
T.S. - Escape Hatch

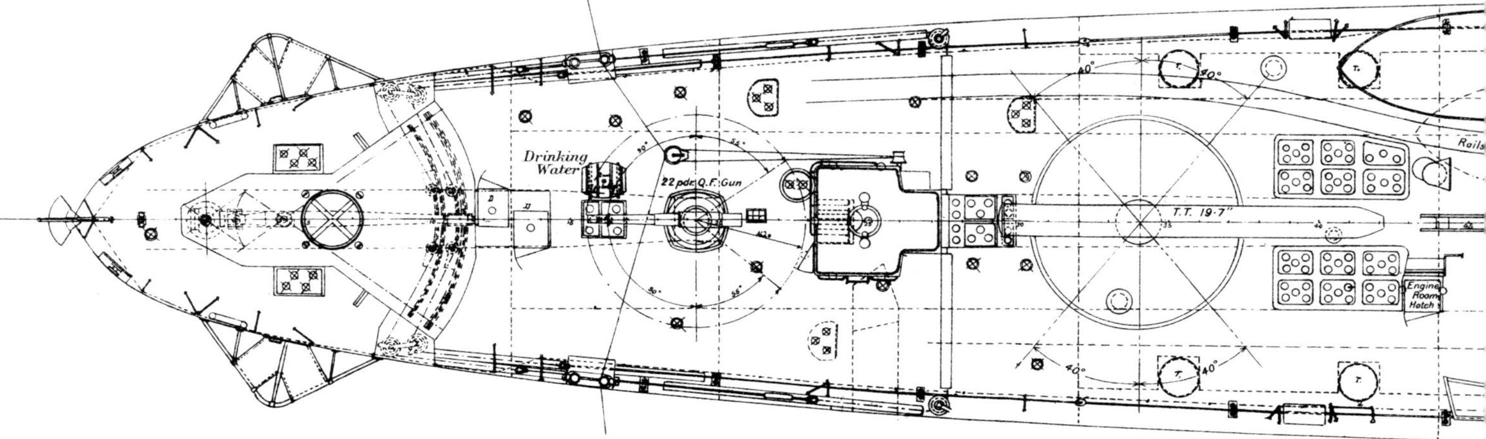

Drinking
Water
22 pdr. Q.F. Gun
T.T. 19·7"
Rails
Engine
Room
Hatch

TORPEDO BOAT DESTROYERS V.1-6.

Scale 1:100.

Longitudinal Section

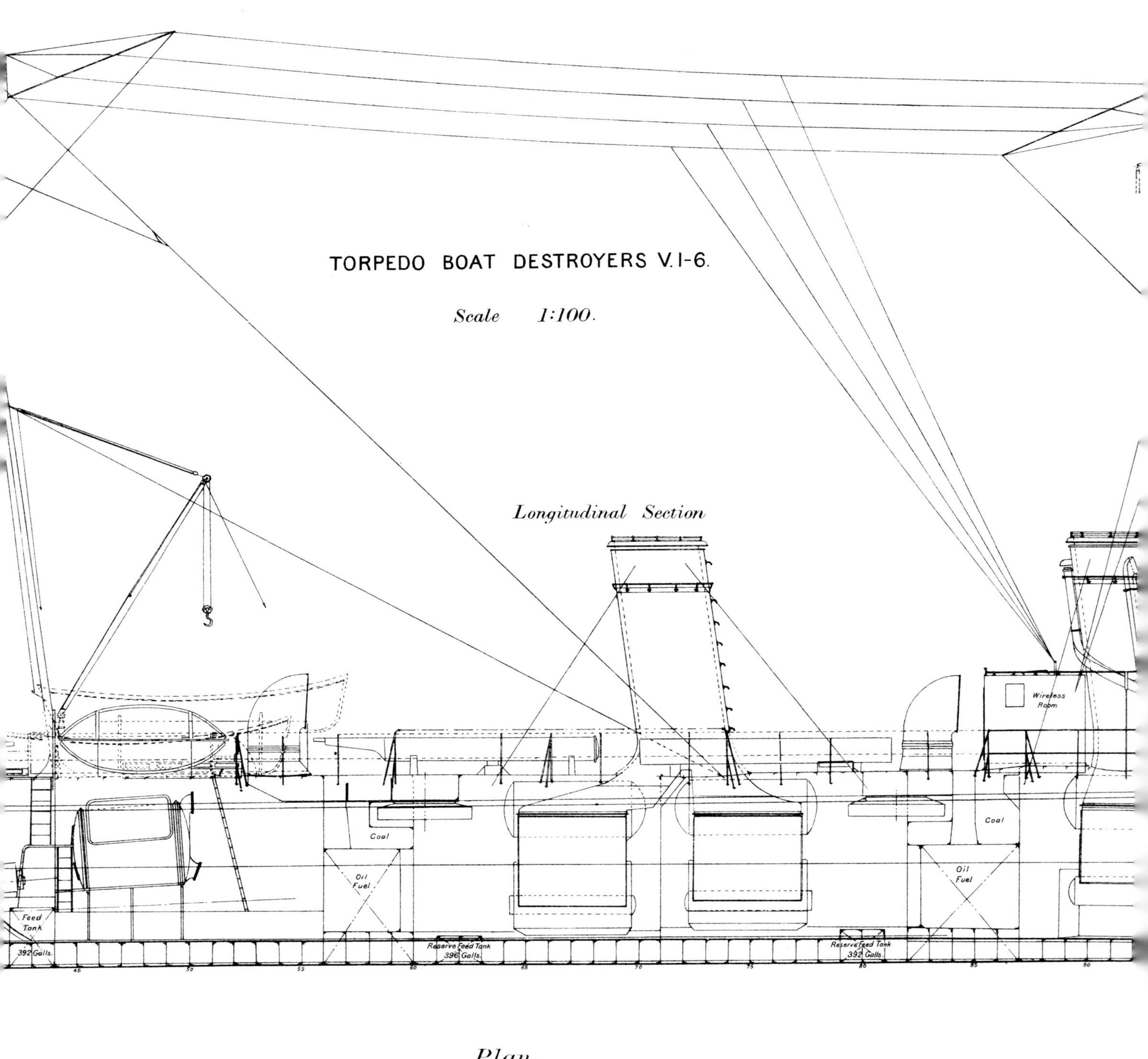

Plan

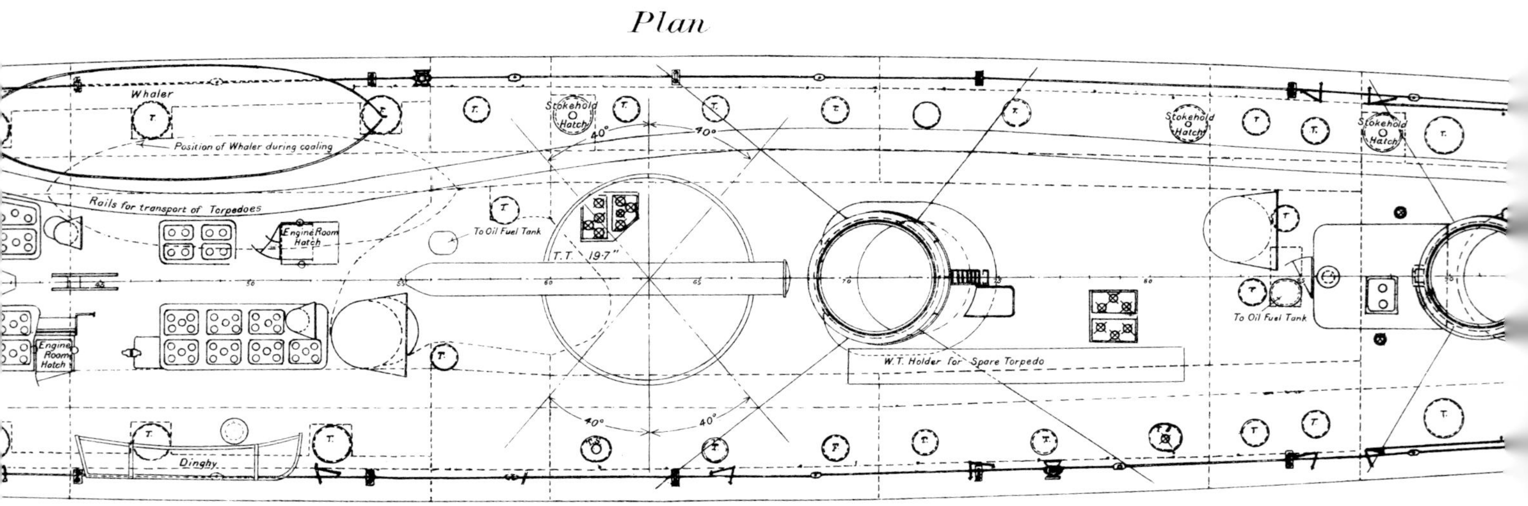

Plate 12.
C.B. 1182 R.
August, 1918.

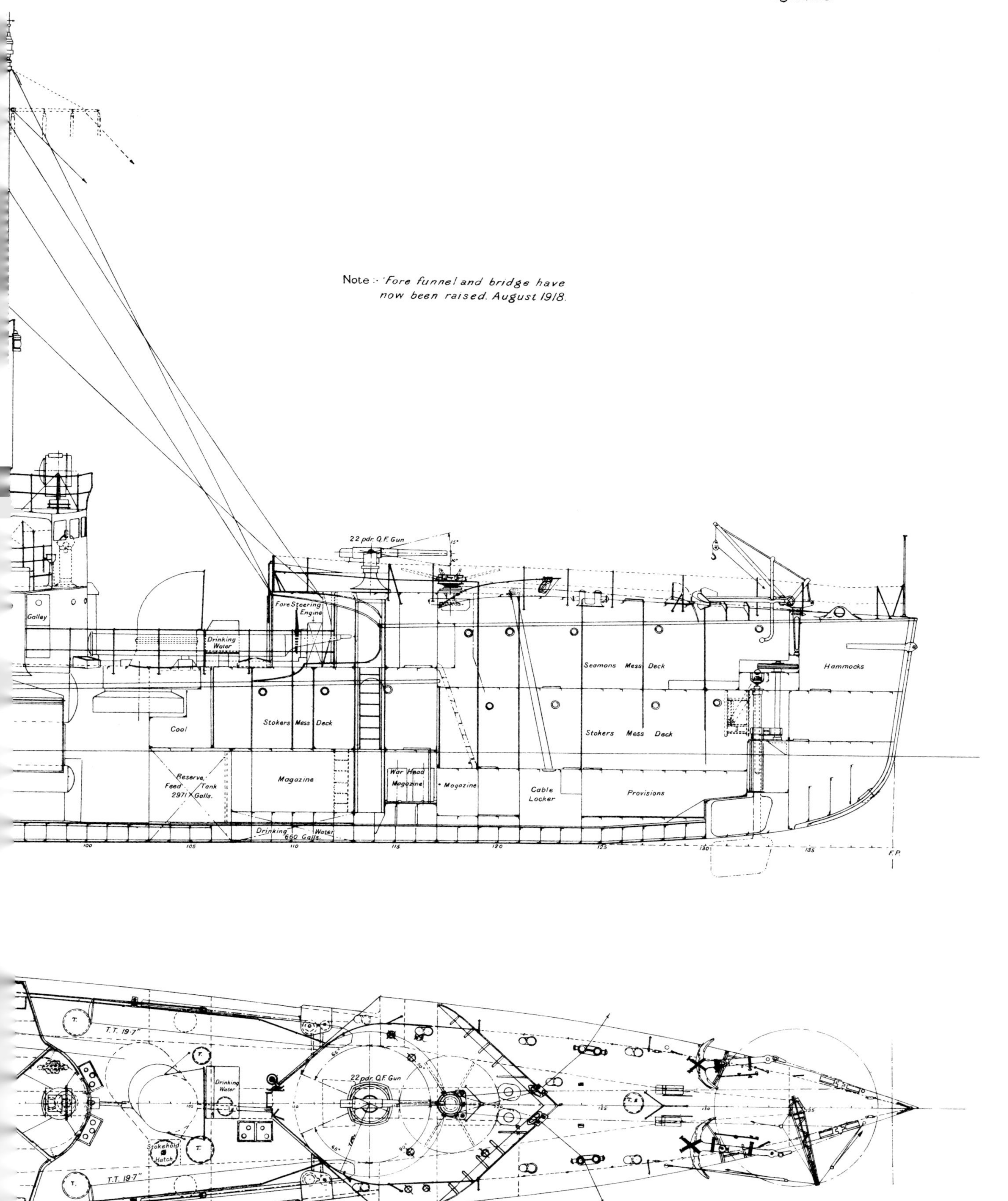

Ordnance Survey, Sept. 1918.

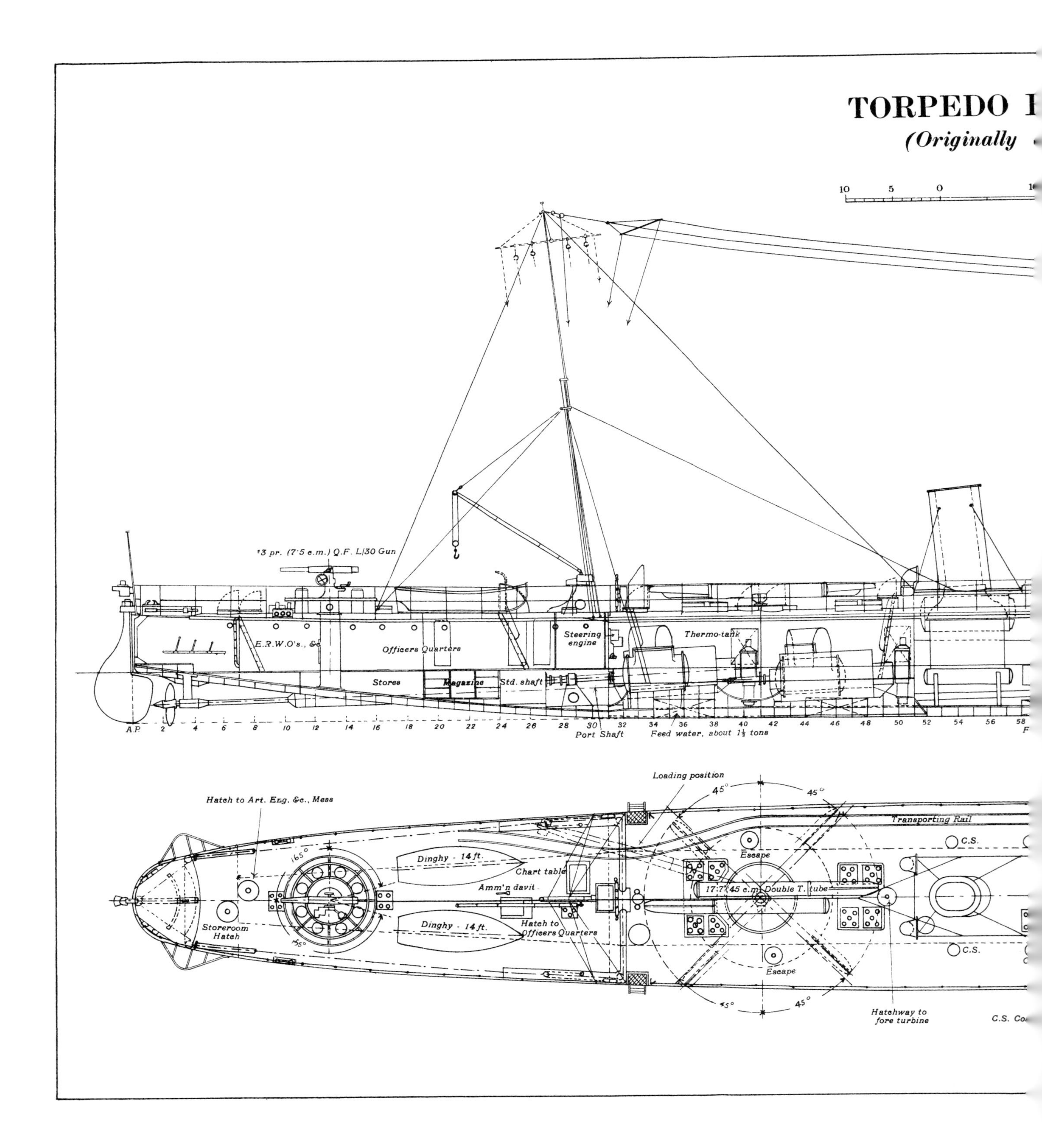

TORPEDO I
(Originally
10 5 0 10
3 pr. (7·5 c.m.) Q.F. L/30 Gun
E.R.W.O's., &c
Officers Quarters
Steering engine
Thermo-tank
Stores
Magazine
Std. shaft
A.P.
Port Shaft
Feed water, about 1½ tons
Loading position
Hatch to Art. Eng. &c., Mess
Transporting Rail
C.S.
Escape
Dinghy - 14 ft.
Chart table
Amm'n davit
17.7/45 c.m. Double T. tube
Dinghy - 14 ft.
Hatch to Officers Quarters
Storeroom Hatch
Hatchway to fore turbine
C.S. Co

Plate 13.
C.B. 1182R.
August, 1918.

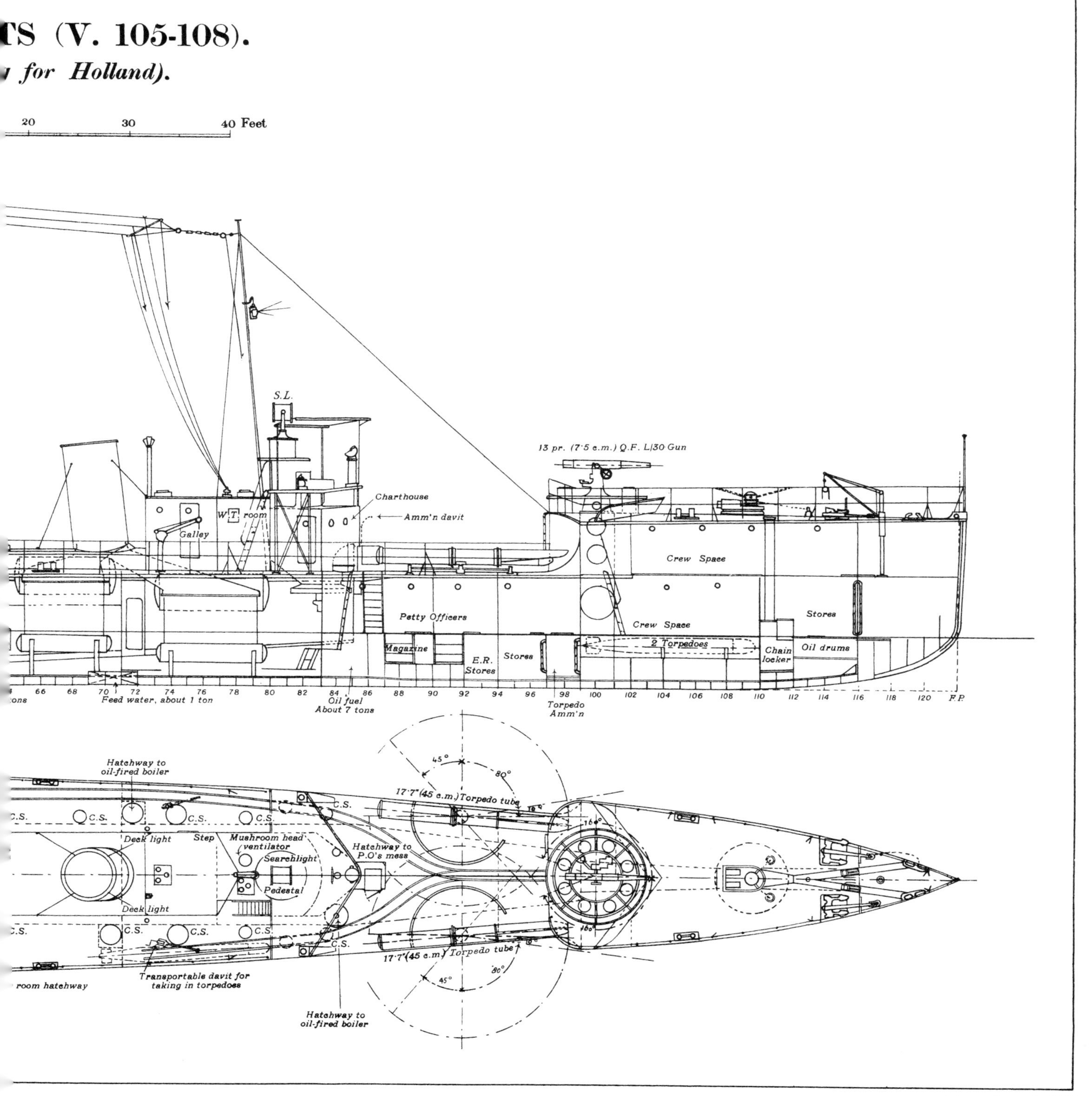

Ordnance Survey, Sept. 1918.

SECTION V
SUBMARINES

GERMAN NAVY.

PART III.
SECTION 5.

SUBMARINES.

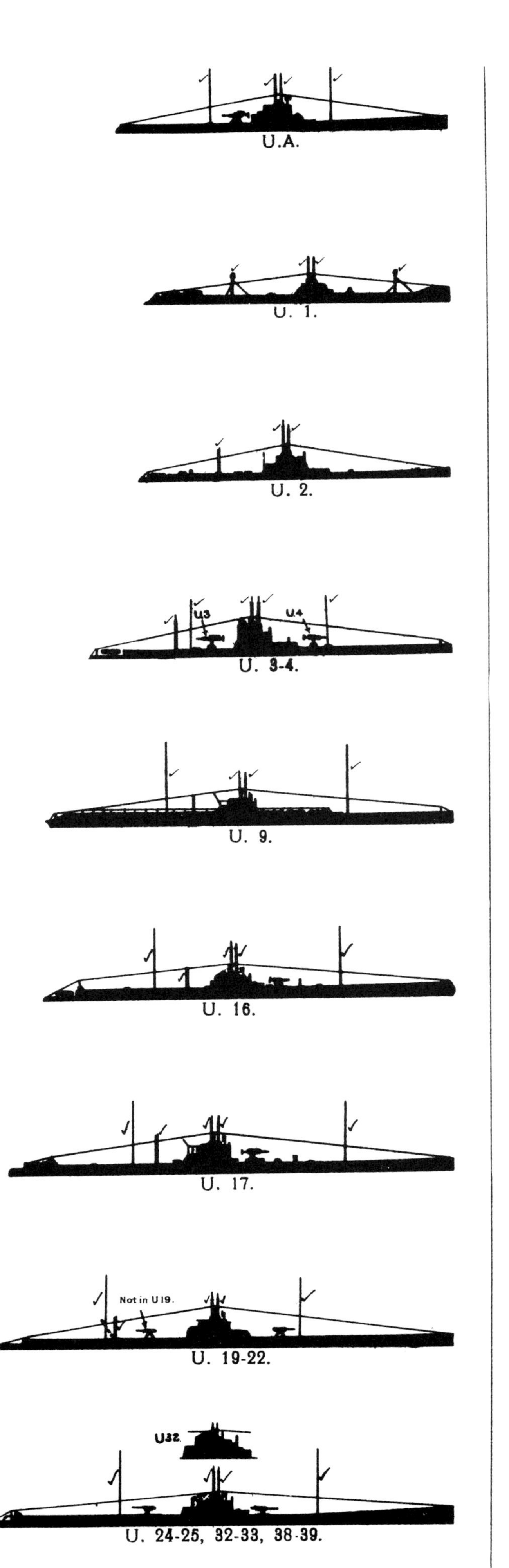

U.A.
U. 1.
U. 2.
U3
U4
U. 3-4.
U. 9.
U. 16.
U. 17.
Not in U19.
U. 19-22.
U32.
U. 24-25, 32-33, 38-39.

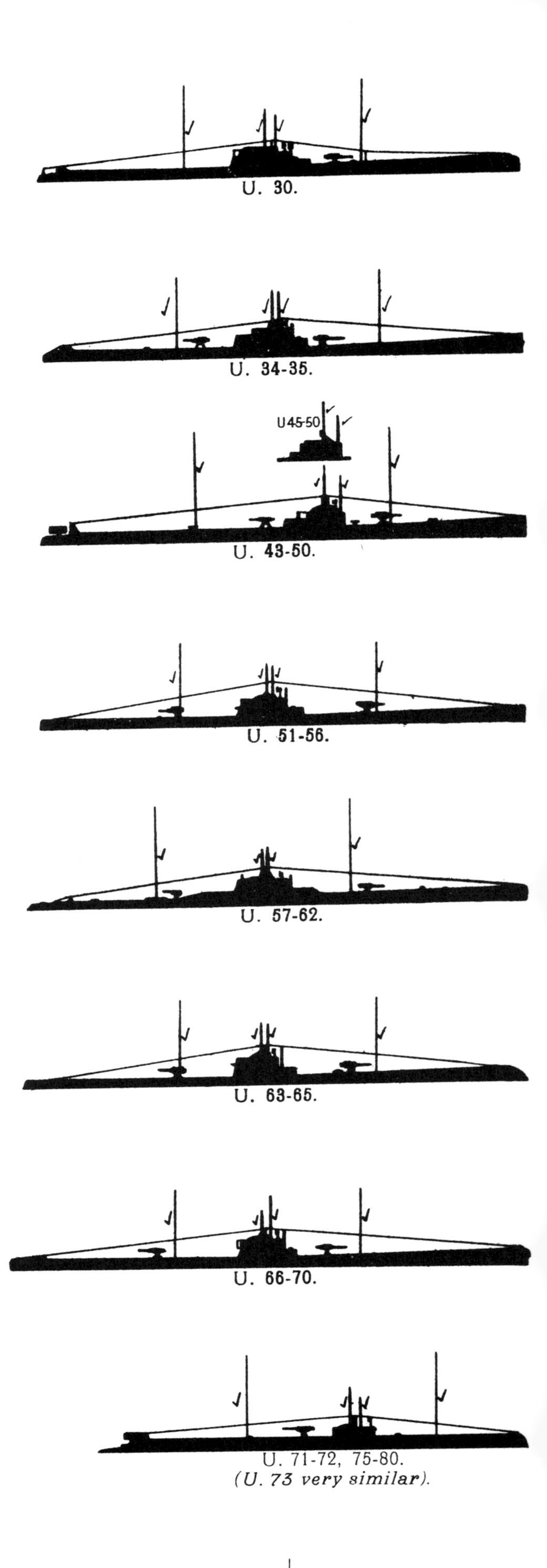

U. 30.
U. 34-35.
U45-50
U. 43-50.
U. 51-56.
U. 57-62.
U. 63-65.
U. 66-70.
U. 71-72, 75-80.
(U. 73 very similar).
U. 105-114.

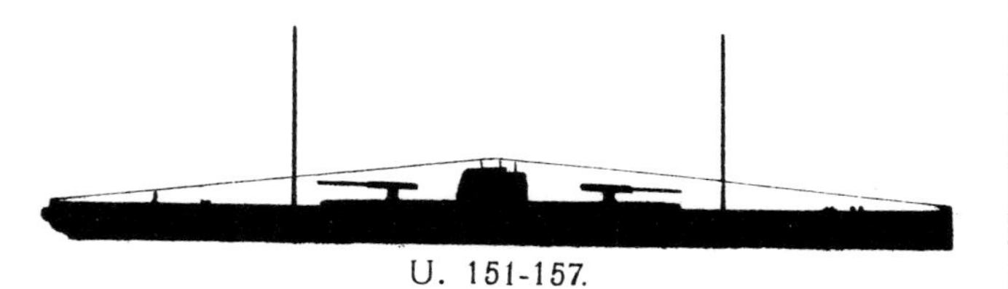

U. 151-157.

U.B. 2.

U.B. 7-8.

U.B. 9-17.

U.B. 18-23.

U.B. 24-29.

U.B. 30-41.

U.B. 42-47.

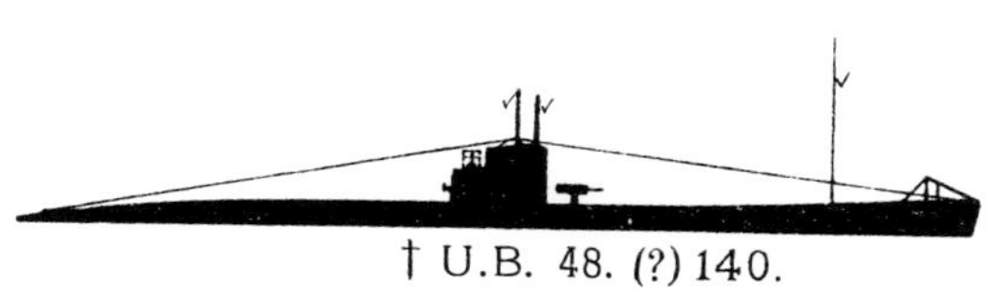

† U.B. 48. (?) 140.

†Note.—Many of these boats have a 4·1" gun in place of the 22pr. here shown. A few of these boats have a step on the fore side of the conning tower, and there may be other small differences between boats built in various yards.

U.C. 4.

U.C. 11-15.

*U.C. 16-79.

***Note.—Some of these Submarines have an overhanging stem and a net cutter as shown by pecked line; the conning towers also differ slightly.**

✓ **Disappearing.**

Ordnance Survey, May, 1918.

DETAILS OF

Part III. Section 5.

Sub-marines.

Official No.	Type. Hull.	Where Built.	Date of first Commissioning.	Normal Complement. (a) Officers. (b) Men.	Dimensions. (a) Length. (b) Breadth. (c) Draught.	Displacement. Surface. / Submerged.	Armament. (m. = machine gun ; S.T. = internal tube ; T. = external tube.)	Motive Power. Surface. / Submerged.
					Ft. Ins.	Tons.		
U. 157 - **U. 156** - **U. 155** -	Converted mercantile. Double hull.	Germania Yard, Kiel.	1917	(a) 12 (b) 65	(a) 213 3 (b) 29 2 (c) 14 9	1700 / 2100	**2**—5·9″; **1**—m.; **4** T. (19·7″); About **20** torpedoes.	Two Germania 6-cylinder, 4-cycle, Diesel engines. / Electric motors.
U. 154 - **U. 153** - **U. 152** - **U. 151** -	Converted mercantile. Double hull.	Germania Yard, Kiel.	1917	(a) 12 (b) 65	(a) 213 3 (b) 29 2 (c) 14 9	1700 / 2100	**2**—5·9″; **2**—22-pr.; **1**—m.; **4** bow S.T. (19·7″); **2** broadside S.T. (19·7″). About **20** torpedoes. Mines also are carried.	Two Germania 6-cylinder, 4-cycle, Diesel engines. / Electric motors.
U. 150 - **U. 149** - **U. 148** - **U. 147** - **U. 146** - **U. 145** - **U. 144** - **U. 143** -	Cruiser. Double hull.	Weser Yard, Bremen.	—	100*	(a) 360 0* (b) 36 0* (c) 20 0*	3150 / 4000 *	—	Two 10-cylinder, 4-cycle, M.A.N. Diesel engines / Electric motors.
U. 142 - **U. 141** - **U. 140** - **U. 139** -	Cruiser. Double hull.	Germania Yard, Kiel.	—	(a) 12 (b) 86	(a) 344 0* (b) 34 0* (c) 20 0*	2800 / 3500 *	**2**—5·9″ & **2**—4·1″; *or* **4**—4·7″; **1** *or* **2**—m.; **4** bow S.T. (19·7″); **2** broadside S.T. (19·7″); **2** stern S.T. (19·7″). About **16–20** torpedoes.	Two 6-cylinder, 4-cycle, M.A.N. Diesel engines. / Electric motors.

* Approximate only.

SUBMARINES.

Part III. Section 5. Sub-marines.

Brake H.P. Surface. / Sub-merged (normal).	No. of Pro-pellers	Speed. Surface. / Submerged.	Endurance (approximate). Surface. / Submerged.	Fuel and Fuel Stowage.	Batteries. No. of Cells. Type.	Periscopes.	Remarks.
		Knots.	Miles.	Tons.			
1,200 / 1,200*	2	11·5 / 8	17,000 at 6 kts. / —	Heavy oil, 250* (normal stowage; can be much increased by using ballast tanks for extra fuel).	330 to 360 Tudor lead cells.	2, 1 passing into conning tower, 1 through upper deck immediately before conning tower.	*See* Plates 1 and 11 and Silhouettes. *U. 155* is the former "Deutschland." *U. 156* was originally christened "Oldenburg," *U. 157* "Sachsen."
1,200 / 1,200*	2	11·5 / 8	17,000 at 6 kts. / —	Heavy oil, 250* (normal stowage; can be much increased by using ballast tanks for extra fuel).	330 to 360 Tudor lead cells.	2, 1 passing into conning tower, 1 through upper deck immediately before conning tower.	—
7,000 / 3,500*	2	18·5 / 9·5*	30,000 at 6 kts. / 90 at 3 kts.	Heavy oil 450*	330 to 360 Tudor lead cells.	3	Surface B.H.P. includes additional Augsburg 6-cylinder 4-cycle Diesel engine of 1,000 H.P. for charging. *None of these vessels are yet ready (April 1918), but the first of them may be completed shortly.*
4,250 / 3,000*	2	16 / 9*	25,000 at 6 kts. / 90 at 3 kts.	Heavy oil, 400*	330 to 360 Tudor lead cells.	3	*See* Plate 11. Surface B.H.P. includes additional Augsburg 6-cylinder, 4-cycle Diesel engine of 550 H.P. for charging. *None of these vessels are yet ready (April 1918), but two of them are reported to be doing trials.*

O (33) AS 4924—11 Pk 2972 5000 5/18 E & S

Part III. Section 5.

Submarines.

Official No.	Type. Hull.	Where Built.	Date of first Commissioning.	Normal Complement. (a) Officers. (b) Men.	Dimensions. (a) Length. (b) Breadth. (c) Draught.	Displacement. Surface. / Submerged.	Armament. (m. = machine gun; S.T. = internal tube; T. = external tube.)	Motive Power. Surface. / Submerged.
					Ft. Ins.	Tons.		
U. 138 **U. 137** **U. 136** **U. 135** **U. 134** **U. 133**	**Ocean-going. Double hull.**	—	—	—	—	—	—	—
U. 132 **U. 131** **U. 130** **U. 129** **U. 128** **U. 127**	**Ocean-going. Double hull.**	—	—	—	—	—	—	—
U. 126 **U. 125** **U. 124** **U. 123** **U. 122** **U. 121**	**Ocean-going. Double hull.**	—	—	—	—	—	—	—
U. 120 **U. 119** **U. 118** **U. 117** **U. 116** **U. 115**	Ocean-going. Double hull.	Imperial Yard, Danzig.	—	—	—	—	—	—

Brake H.P. Surface. / Submerged. (normal).	No. of Propellers.	Speed. Surface. / Submerged.	Endurance (approximate). Surface. / Submerged.	Fuel and Fuel Stowage.	Batteries. No. of Cells. Type.	Periscopes.	Remarks.
		Knots.	Miles.	Tons.			
—	—	—	—	—	—	—	Projected, but probably not yet laid down (April 1918).
—	—	—	—	—	—	—	One or two of these boats may be completed in latter half of 1918.
—	—	—	—	—	—	—	Some of these boats will be completed in latter half of 1918.
—	—	—	—	—	—	—	Some of these boats will be completed in latter half of 1918.

A 3

Part III. Section 5.

Sub-marines.

Official No.	Type. Hull.	Where Built.	Date of first Commissioning.	Normal Complement. (a) Officers. (b) Men.	Dimensions. (a) Length. (b) Breadth. (c) Draught.	Displacement. Surface. / Submerged.	Armament. (m. = machine gun; S.T. = internal tube; T. = external tube.)	Motive Power. Surface. / Submerged.
					Ft. Ins	Tons.		
U. 114, U. 113, U. 112, U. 111, U. 108, U. 107, U. 105	Ocean-going. Double hull.	Germania Yard, Kiel.	1917–18	(a) 7 (b) 31	(a) 220 0* (b) 22 0* (c) 12 0*	850 / 1,200*	**1**—4·1″; **1**—22-pr.; **1**—m.; **4** bow S.T. (19·7″); **2** stern S.T. (19·7″); **12** torpedoes.	Two Augsburg, 6-cylinder, 4-cycle, Diesel engines. / Electric motors.
U. 102, U. 101, U. 100	Ocean-going. Double hull.	Weser Yard, Bremen.	1917	(a) 7 (b) 30	(a) 220 0* (b) 22 0* (c) 12 0*	850 / 1,200*	**1**—4·1″; **1**—22-pr.; **1**—m.; **2** bow S.T. (19·7″); **2** stern S.T. (19·7″); **11** torpedoes.	Two M.A.N. 8-cylinder, 2-cycle, Diesel engines. / A.E.G. motors.
U. 98, U. 97, U. 96, U. 94	Ocean-going. Double hull.	Germania Yard, Kiel.	1916–17	(a) 7 (b) 30	(a) 220 0* (b) 22 0* (c) 12 0*	850 / 1,200*	**1**—4·1″; **1**—22-pr.; **1**—m.; **4** bow S.T. (19·7″); **2** stern S.T. (19·7″); **12** torpedoes.	Two Germania, 6-cylinder, 2-cycle, Diesel engines. / Siemens-Schuckert motors.
U. 92, U. 91, U. 90, U. 89	Ocean-going. Double hull.	Imperial Dockyard, Danzig.	1917	(a) 7 (b) 30	(a) 220 0* (b) 22 0* (c) 12 0*	850 / 1,200*	**2**—4·1″; **1**—m.; **4** bow S.T. (19·7″); **2** stern S.T. (19·7″); **12** torpedoes.	Two M.A.N. 6-cylinder, 4-cycle, Diesel engines. / Electric motors.
U. 86, U. 82	Ocean-going. Double hull.	Germania Yard, Kiel.	1916	(a) 7 (b) 30	(a)220 0* (b) 22 0* (c) 12 0*	850 / 1200*	**1**—4·1″; **1**—22-pr.; **1**—m.; **4** bow S.T. (19·7″); **2** stern S.T. (19·7″); **12** torpedoes.	Two Germania, 6-cylinder, 2-cycle, Diesel engines. / Electric motors.

* Approximate only.

Part III. Section 5.

Sub-marines.

Brake H.P. Surface. / Sub-merged (normal).	No. of Pro-pellers.	Speed. Surface. / Submerged.	Endurance (approximate). Surface. / Submerged.	Fuel and Fuel Stowage.	Batteries. No. of Cells. Type.	Periscopes.	Remarks.
		Knots.	Miles.	Tons.			
2,400 / 2,000 *	2	16·5 / 9·5 *	7,000 at 10 kts. / 100 at 3 kts.	Heavy oil. 105* (including stowage in ballast tanks usually appropriated for fuel on long cruises).	210 Tudor lead cells.	3, 2 passing into conning tower, 1 through fore part of fairwater into control room.	Full breadth superstructure with rounded sides. *See* Plate 2 and silhouettes.
2,600 / 2,100 *	2	17·0 / 9·5 *	7,400 at 10 kts. / 100 at 3 kts.	Heavy oil. 120 (including stowage in ballast tanks usually appropriated for fuel on long cruises).	240 Tudor lead cells.	3, 2 passing into conning tower, 1 through fore part of fairwater into control room.	Narrow superstructure with vertical sides. Bow slightly rounded. *See* Plate 2.
2,400 / 2,000 *	2	16·5 / 9·5	8,000 at 10 kts. / 100 at 3 kts.	Heavy oil. 120* (including stowage in ballast tanks usually appropriated for fuel on long cruises).	About 220 Tudor lead cells.	3, 2 passing into conning tower, 1 through fore part of fairwater into control room.	Full breadth superstructure with rounded sides.
2,400 / 2,000 *	2	15·5 / 9·0	7,000 at 10 kts. / 100 at 3 kts.	Heavy oil. 125* (including stowage in ballast tanks usually appropriated for fuel on long cruises).	About 220 Tudor lead cells.	2, 1 passing into conning tower, 1 through fore part of fairwater into control room.	Full breadth superstructure with rounded sides; upper vertical rudder.
2,400 / 2,000	2	16·5 / 9·0	7,800 at 10 kts. / 110 at 3 kts.	Heavy oil, 122 (including stowage in 4 ballast tanks usually appropriated for fuel on long cruises).	About 220 Tudor lead cells.	3, 2 passing into conning tower, 1 through fore part of fairwater into control room.	Full breadth superstructure with rounded sides.

A 4

Part III.
Section 5.

Sub-marines.

Official No.	Type. Hull.	Where Built.	Date of first Commissioning.	Normal Complement. (a) Officers. (b) Men.	Dimensions. (a) Length. (b) Breadth. (c) Draught.	Displacement. Surface. / Submerged.	Armament. (m. = machine gun ; S.T. = internal tube ; T. = external tube.)	Motive Power. Surface. / Submerged.
					Ft. Ins.	Tons.		
U. 80 **U. 79** **U. 78** **U. 73** **U. 72** **U. 71**	Ocean-going minelayers. Single hull with saddle tanks.	**Vulcan Yard, Hamburg.** (U. 80, 79, 78) **Imperial Dockyard, Danzig.** (U. 73, 72) **Vulcan Yard, Hamburg.** (U. 71)	1916	(a) 7 (b) 25	(a) 185 0 (b) 21 6*	650 ———* 800	**1**—4·1″ *or* 22-pr. ; **1**—m. ; **1** bow T. (19·7″) ; **1** stern T. (19·7″) ; **36** mines ; **2** torpedoes.	Two 6-cylinder, 2-cycle, Diesel engines (various makes). ——— Electric motors.
U. 70 **U. 67**	Ocean-going. Double hull.	Germania Yard, Kiel.	1915	(a) 7 (b) 29	(a) 225 0 (b) 21 0* (c) 12 0*	850 ———* 1200	**1**—4·1″ ; **1**—22-pr. ; **1**—m. ; **4** bow S.T. (17·7″) ; **1** stern S.T. (17·7″) ; **12** torpedoes.	Two Germania 6-cylinder, 2-cycle, Diesel engines. ——— Electric motors.
U. 65 **U. 64** **U. 63**	Ocean-going Double hull.	Germania Yard, Kiel.	1916	(a) 7 (b) 28	(a) 220 0 (b) 21 0* (c) 12 0*	850 ———* 1150	**1**—4·1″ ; **1**—22-pr. ; **1**—m. ; **2** bow S.T. (19·7″) ; **2** stern S.T. (19·7″) ; **12** torpedoes.	Two Germania 6-cylinder, 2-cycle, Diesel engines of Russian design. ——— Electric motors.
U. 62 **U. 61** **U. 60** **U. 57**	Ocean-going. Double hull.	Weser Yard, Bremen.	1916	(a) 7 (b) 28	(a) 217 6 (b) 20 8 (c) 11 6	820 ———* 1150	**1**—4·1″ ; **1**—22-pr. ; **1**—m. ; **2** bow S.T. (19·7″) ; **2** stern S.T. (19·7″) ; **11** torpedoes.	Two M.A.N. 8-cylinder, 2-cycle, Diesel engines. ——— Electric motors.
U. 55 **U. 54** **U. 53** **U. 52**	Ocean-going. Double hull.	Germania Yard, Kiel.	1916	(a) 7 (b) 28	(a) 212 0 (b) 20 6* (c) 12 0*	800 ——— 1120	**1**—4·1″ ; **1**—22-pr. ; **1**—m. ; **2** bow S.T. (19·7″) ; **2** stern S.T. (19·7″) ; **10** torpedoes.	Two M.A.N. 6-cylinder, 4-cycle, Diesel engines. ——— A.E.G. motors.

* Approximate only.

Brake H.P. Surface. / Sub-merged (normal).	No. of Pro-pellers.	Speed. Surface. / Submerged.	Endurance (approximate). Surface. / Submerged	Fuel and Fuel Stowage.	Batteries. No. of Cells. Type.	Periscopes.	Remarks.
		Knots.	Miles.	Tons.			
900 / 600 *	2	10·0 / 7·0	6,300 at 6 kts. / —	Heavy oil, 75* (including stowage in ballast tanks usually appropriated for fuel on long cruises).	Tudor lead cells.	2, 1 passing into conning tower, 1 through fore part of fairwater into control room.	Rounded bow. Gun abaft conning tower. Conspicuous upper rudder. *See* p. 61 and Silhouettes.
2,300 / 1,800 *	2	17·0 / 9·5	11,000 at 10 kts / 100 at 3·5 kts.	Heavy oil, 150* (including stowage in ballast tanks usually appropriated for fuel on long cruises).	220 Tudor lead cells.	3, 2 passing into conning tower, 1 passing through fore part of fairwater into control room.	Originally building for Austria-Hungary. A very successful class. Narrow superstructure with vertical sides. Straight stem. *See* Plate 12 and Silhouettes.
2,300 / 1,800 *	2	16·0 / 9·0	7,800 at 10 kts. / 100 at 3 kts.	Heavy oil, 125* (including stowage in ballast tanks usually appropriated for fuel on long cruises).	About 220 Tudor lead cells.	3, 2 passing into conning tower, 1 passing through fore part of fairwater into control room.	Full breadth superstructure with rounded sides. Bow noticeably rounded. No upper rudder. *See* Silhouettes.
2,100 / 1,700 *	2	15·0 / 9·0	7,000 at 10 kts. / 90 at 3 kts.	Heavy oil, 123 (including stowage in ballast tanks usually appropriated for fuel on long cruises).	About 220 Tudor lead cells.	3, 2 passing into conning tower, 1 passing through fore part of fairwater into control room.	Narrow superstructure with vertical sides. Bow slightly rounded. *See* Plate 13 and Silhouettes.
2,400 / 1,800 *	2	15·0 / 9·0	5,500 at 10 kts. / 90 at 3 kts.	Heavy oil, 110* (including stowage in ballast tanks usually appropriated for fuel on long cruises).	About 220 Tudor lead cells.	3, passing into conning tower.	Narrow superstructure with vertical sides. Straight stem. *See* Plates 3 and 4 and Silhouettes.

* Approximate only.

Part III.
Section 5.

Sub-marines.

Official No.	Type. Hull.	Where Built.	Date of first Commissioning.	Normal Complement. (a) Officers. (b) Men.	Dimensions. (a) Length. (b) Breadth. (c) Draught.	Displacement. Surface. / Submerged.	Armament. (m. = machine gun; S.T. = internal tube; T. = external tube.)	Motive Power. Surface. / Submerged.
					Ft. Ins.	Tons.		
U. 47, **U. 46**, **U. 43**	Ocean-going. Double hull.	Imperial Dockyard, Danzig.	1915–16	(a) 7 (b) 30	(a)213 3 (b) 21 4* (c) 11 10*	850 / 1150*	**1**—4·1"; **1**—22-pr.; **1**—m.; **4** bow S.T. (19·7"); **2** stern S.T. (19·7"); **10** torpedoes.	Two M.A.N. 6-cylinder, 4-cycle, Diesel engines. / A.E.G. motors.
U. 39, **U. 38**, **U. 35**, **U. 34**, **U. 33**, **U. 32**	Ocean-going. Double hull.	Germania Yard, Kiel.	1914–15	(a) 7 (b) 28	(a)210 0 (b) 20 6* (c) 12 0*	800 / 1100*	**1**—4·1"; **1**—22-pr.; **1**—m.; **2** bow S.T. (19·7"); **2** stern S.T. (19·7"); **8** torpedoes.	Two Germania 6-cylinder, 2-cycle, Diesel engines. / A.E.G. motors.
U. 30	Ocean-going. Double hull.	Imperial Dockyard, Danzig.	1914	(a) 7 (b) 28	(a)210 0 (b) 20 6* (c) 12 0*	800 / 1100*	**1**—4·1"; **1**—m.; **2** bow S.T. (19·7"); **2** stern S.T. (19·7"); **8** torpedoes.	Two M.A.N. 6-cylinder, 4-cycle, Diesel engines. / A.E.G. motors.
U. 25, **U. 24**	Ocean-going. Double hull.	Germania Yard, Kiel.	9.5.14, 6.12.13	(a) 7 (b) 26	(a)210 0 (b) 20 6* (c) 12 0*	800 / 1100*	**2**—22-prs.; **1**—m.; **2** bow S.T. (19·7"); **2** stern S.T. (19·7"); **8** torpedoes.	Two Germania 6-cylinder, 2-cycle, Diesel engines. / A.E.G. motors.
U. 22, **U. 21**	Ocean-going. Double hull.	Imperial Dockyard, Danzig.	25.11.13, 22.10.13	(a) 7 (b) 28	(a)208 0 (b) 20 0* (c) 12 0*	720 / 1000*	**2**—22-prs.; **1**—m.; **2** bow S.T. (19·7"); **2** stern S.T. (19·7"); **8** torpedoes.	Two M.A.N. 6-cylinder, 4-cycle, Diesel engine / A.E.G. motors.
U. 19	Ocean-going. Double hull.	Imperial Dockyard, Danzig.	6.7.13	(a) 7 (b) 28	(a)208 0 (b) 20 0* (c) 12 0*	720 / 1000*	**1**—4·1"; **1**—m.; **2** bow S.T. (19·7"); **2** stern S.T. (19·7"); **8** torpedoes.	Two M.A.N. 6-cylinder, 4-cycle, Diesel engines. / A.E.G. motors.

* Approximate only.

Brake H.P. Surface. / Sub-merged (normal).	No. of Pro-pellers.	Speed. Surface. / Submerged.	Endurance (approximate). Surface. / Submerged.	Fuel and Fuel Stowage.	Batteries. No. of Cells. Type.	Periscopes.	Remarks.
		Knots.	Miles.	Tons.			
2,000 / 1,600 *	2	14·5 / 9·0	7,600 at 10 kts. / 100 at 3 kts.	Heavy oil, 140* (including stowage in ballast tanks usually appropriated for fuel on long cruises).	220 Tudor lead cells.	2 passing into conning tower.	Full breadth superstructure with rounded sides. Straight stem. *See* Silhouettes.
1,700 / 1,400 *	2	15·5 / 9·5	7,500 at 10 kts. / 100 at 3 kts.	Heavy oil, 100* (including stowage in ballast tanks usually appropriated for fuel on long cruises).	About 220 Tudor lead cells.	3, 2 passing into conning tower, 1 through fore-part of fairwater into control room.	Narrow superstructure with vertical sides. Straight stem. *See* Plates 5, 6, and 7 and Silhouettes.
2,000 / 1,600 *	2	15·5 / 9·0	7,000 at 10 kts. / 100 at 3 kts.	Heavy oil, 100* (including stowage in ballast tanks usually appropriated for fuel on long cruises).	About 220 Tudor lead cells.	3, 2 passing into conning tower, 1 through fore-part of fairwater into control room.	Sunk in Borkum Roads 22.6.15, subsequently salved and repaired, but is now employed only for training purposes. Narrow superstructure with vertical sides. Slightly rounded bow. *See* Silhouettes.
1,700 / 1,400 *	2	15·0 / 9·0	6,500 at 10 kts. / 100 at 3 kts.	Heavy oil, 100* (including stowage in ballast tanks usually appropriated for fuel on long cruises.	About 220 Tudor lead cells.	3, 2 passing into conning tower, 1 through forepart of fairwater into control room.	*See* Plate 7 and Silhouettes. Closely resemble *U. 32–39*, but the armament differs.
1,700 / 1,400 *	2	15·0 / 9·0	5,600 at 10 kts. / 100 at 3 kts.	}			
1,700 / 1,400 *	2	15·0 / 9·0	5,600 at 10 kts. / 100 at 3 kts.	Heavy oil, 75* (including stowage in ballast tanks usually appropriated for fuel on long cruises).	About 220 Tudor lead cells.	3, 2 passing into conning tower, 1 through fore-part of fairwater into control room.	Narrow superstructure with vertical sides. Straight stem. *See* Silhouettes. *U. 21* has been one of the most successful of German submarines. She was the first submarine to be sent to the Irish Sea and again the first to be sent to the Mediterranean. She is still (April 1918) a very effective unit.

B 2

Part III. Section 5.

Sub-marines.

Official No.	Type. Hull.	Where Built.	Date of first Commissioning.	Normal Complement. (a) Officers. (b) Men.	Dimensions. (a) Length. (b) Breadth. (c) Draught.	Displacement. Surface. / Submerged.	Armament. (m. = machine gun ; S.T. = internal tube ; T. = external tube.)	Motive Power. Surface. / Submerged.
					Ft. Ins.	Tons.		
U. 17 -	Instructional. Double hull.	Imperial Dockyard, Danzig.	3.11.12	(a) 5 (b) 27	(a)202 0 (b) 19 6* (c) 11 6*	650 / 900*	**1**—4-pr.; **1**—m.; **2** bow S.T. (17·7"); **2** stern S.T. (17·7"); **8** torpedoes.	Four 8-cylinder, 2-cycle, Körting paraffin engines. / A.E.G. motors.
U. 16 -	Instructional. Double hull.	Germania Yard, Kiel.	28.12.11	(a) 5 (b) 27	(a)187 0 (b) 19 0* (c) 11 6*	600 / 800*	**1**—4-pr.; **1**—m.; **2** bow S.T. (17·7"); **2** stern S.T. (17·7"); **8** torpedoes.	Two 8-cylinder, two 6-cylinder, 2-cycle, Körting paraffin engines. / A.E.G. Motors.
U. 9 -	Instructional. Double hull.	Imperial Dockyard, Danzig.	18.4.10	(a) 5 (b) 23	(a)183 0* (b) 19 0* (c) 11 6*	550 / 700*	**1**—4-pr.; **1**—m.; **2** bow S.T. (17·7"); **2** stern S.T. (17·7"); **6** torpedoes.	Two 8-cylinder, two 6-cylinder, 2-cycle, Körting paraffin engines. / A.E.G. Motors.
U. 4 - / **U. 3** -	Instructional. Double hull.	Imperial Dockyard, Danzig.	1.7.09 / 29.5.09	(a) 3 (b) 20	(a)166 0 (b) 18 0* (c) 10 6*	400 / 500*	**1**—4-pr.; **2** bow S.T. (17·7"); **2** stern S.T. (17·7").	Two 8-cylinder, 2-cycle, Körting paraffin engines. / A.E.G. Motors.
U. 2 -	Instructional. Double hull.	Imperial Dockyard, Danzig.	18.7.08	(a) 3 (b) 18	(a)141 9 (b) 12 3* (c) 9 7*	233 / 295*	**2** bow S.T. (17·7"); **2** stern S.T. (17·7").	Two 8-cylinder, 2-cycle, Daimler paraffin engines. / A.E.G. Motors.
U. 1 -	Instructional. Double hull.	Germania Yard, Kiel.	14.12.06	(a) 3 (b) 18	(a)138 9 (b) 11 10 (c) 10 12	197 / 236	**2** S.T. (17·7"); **3** torpedoes.	Two 6-cylinder, 2-cycle, Körting paraffin engines. / A.E.G. Motors.
U.A. -	Instructional. Double hull.	Germania Yard, Kiel.	–.8.14	(a) 5 (b) 21	(a)150 0 (b) 16 6 (c)	246 / 332	**1**—4-pr.; **2** bow S.T. (17·7"); **1** deck T. (17·7"); **4** torpedoes.	Two Germania 6-cylinder Diesel engines. / A.E.G. Motors.

* Approximate only.

Part III, Section 5.

Sub-marines.

Brake H.P. Surface. / Sub-merged (normal).	No. of Pro-pellers.	Speed. Surface. / Submerged.	Endurance (approximate). Surface. / Submerged.	Fuel and Fuel Stowage.	Batteries. No. of Cells. Type.	Periscopes.	Remarks.
1,200 / 1,000 *	2	Knots. 15·5 / 9·0	Miles. 4,000 at 10 kts. / —	Tons. Paraffin, 60*	Tudor lead cells.	3 (control room periscope is very short).	*See* Silhouettes.
1100 / 900 *	2	15·5 / 9	4,000 at 10 kts. / —	Paraffin, 60*	Tudor lead cells.	3 (control room periscope is very short).	*See* Plate 7 and Silhouettes.
1000 / 700 *	2	13 / 9	3,500 at 10 kts. / —	Paraffin, 50*	Tudor lead cells.	3 (control room periscope is very short).	*See* Silhouettes.
600 / 400 *	2	12·5 / 8·5	3,500 at 10 kts. / —	Paraffin, 47	Tudor lead cells.	3 (control room periscope is very short).	*See* Silhouettes.
700 / 400 *	2	13·5 / 9·5	— / —	Paraffin, 41	Tudor lead cells.	3 (control room periscope is very short).	*See* Silhouettes.
450 / 300	2	10·9 / 8·7	— / —	Paraffin, 35	Tudor lead cells.	3, 2 passing into conning tower, one through forward hatch for trimming purposes.	*See* Silhouettes. The forerunner of all "U" boats. Built from designs of d'Equivilley-Montjustin.
900 / 650 *	2	14 / 10	— / —	Heavy oil.	Tudor lead cells.	3, 2 passing into conning tower.	*See* Silhouettes. Built for the Norwegian Navy, but taken over by Germany before delivery 7th August 1914.

B 3

Part III. Section 5.

Sub-marines.

Official No.	Type. Hull.	Where Built.	Date of first Commissioning.	Normal Complement. (a) Officers. (b) Men.	Dimensions. (a) Length. (b) Breadth. (c) Draught.	Displacement. Surface. / Submerged.	Armament. (m. = machine gun ; S.T. = internal tube ; T. = external tube.)	Motive Power. Surface. / Submerged.
					Ft. Ins.	Tons.		
U.B. 140 U.B. 139 U.B. 138 U.B. 137 U.B. 136 U.B. 135 U.B. 134 U.B. 133 U.B. 132 U.B. 131 U.B. 130 U.B. 129	Smaller Ocean-going. Double hull.	—	—	—	—	—	—	—
U.B. 128 U.B. 127 U.B. 126 U.B. 125 U.B. 124 U.B. 123 U.B. 122 U.B. 121 U.B. 120 U.B. 119 U.B. 118 U.B. 117	Smaller Ocean-going. Double hull.	Weser Yard, Bremen.	1918	(a) 7 (b) 28	(a)180 9 (b) 19 6 (c) 12 0	502 723	**1**—4·1″ *or* 22 pr. ; **1**—m. ; **4** bow S.T. (19·7″) ; **1** stern S.T (19·7″) ; **10** torpedoes.	Two 6-cylinder, 4-cycle, Diesel engines. Electric motors.
U.B. 116 U.B. 115 U.B. 114 U.B. 113 U.B. 112 U.B. 111	Smaller Ocean-going. Double hull.	Blohm & Voss Works, Hamburg.	1918	(a) 7 (b) 28	(a)180 9 (b) 19 6 (c) 12 0	502 723	**1**—4·1″ *or* 22-pr. ; **1**—m. ; **4** bow S.T. (19·7″) ; **1** stern S.T. (19·7″) ; **10** torpedoes.	Two 6-cylinder, 4-cycle, Diesel engines. Electric motors.

Brake H.P. Surface. / Sub-merged (normal).	No. of Pro-pellers.	Speed. Surface. / Submerged.	Endurance (approximate). Surface. / Submerged.	Fuel and Fuel Stowage.	Batteries No. of Cells. Type.	Periscopes.	Remarks.
		Knots.	Miles.	Tons.			
—	—	—	—	—	—	—	These boats may be ready shortly (April 1918).
1,060 / 1,000	2	13·4 / 7·8	8,300 at 8 kts. / 70 at 3 kts.	Heavy oil, 84†	124 Tudor lead cells.	2, passing into conning tower.	See p. 63, Plate 8, and Silhouettes.
1,060 / 1,000	2	13·4 / 7·8	8,300 at 8 kts. / 70 at 3 kts.	Heavy oil, 84†	124 Tudor lead cells.	2, passing into conning tower.	

† Including stowage in six ballast tanks, usually appropriated for fuel on long cruises.

Part III.
Section 5.

Sub-marines.

Official No.	Type. Hull.	Where Built.	Date of first Commissioning.	Normal Complement. (a) Officers. (b) Men.	Dimensions. (a) Length. (b) Breadth. (c) Draught.	Displacement. Surface. / Submerged.	Armament. (m. = machine gun; S.T. = internal tube; T. = external tube.)	Motive Power. Surface. / Submerged.
					Ft. Ins.	**Tons.**		
U.B. 110 **U.B. 109** **U.B. 108** **U.B. 107** **U.B. 106** **U.B. 105** **U.B. 104** **U.B. 103** **U.B. 102**	Smaller Ocean-going. Double hull.	Germania Yard, Kiel.	1917–18	(a) 7 (b) 28	(a) 180 9 (b) 19 6 (c) 12 0	502 723	**1**—4·1″ *or* 22-pr.; **1**—m.; **4** bow S.T. (19·7″); **1** stern S.T. (19·7″); **10** torpedoes.	Two 6-cylinder, 4-cycle, Diesel engines. Electric motors.
U.B. 101 **U.B. 100** **U.B. 99** **U.B. 98** **U.B. 97** **U.B. 96**	Smaller Ocean-going. Double hull.	Blohm & Voss Works Hamburg.	1917–18	(a) 7 (b) 28	(a) 180 9 (b) 19 6 (c) 12 0	502 723	**1**—4·1″ *or* 22-pr.; **1**—m.; **4** bow S.T. (19·7″); **1** stern S.T. (19·7″); **10** torpedoes.	Two 6-cylinder, 4-cycle, Diesel engines. Electric motors.
U.B. 95 **U.B. 94** **U.B. 93** **U.B. 92** **U.B. 91** **U.B. 90**	Smaller Ocean-going. Double hull.	Vulcan Yard, Hamburg.	1918	(a) 7 (b) 28	(a) 180 9 (b) 19 6 (c) 12 0	502 723	**1**—4·1″ *or* 22-pr.; **1**—m.; **4** bow S.T. (19·7″); **1** stern S.T. (19·7″); **10** torpedoes.	Two 6-cylinder, 4-cycle, Diesel engines. Electric motors.
U.B. 89 **U.B. 88** **U.B. 87** **U.B. 86** **U.B. 84**	Smaller Ocean-going. Double hull.	Weser Yard, Bremen.	1917–18	(a) 7 (b) 28	(a) 180 9 (b) 19 6 (c) 12 0	502 723	**1**—4·1″ *or* **1**—22-pr.; **1**—m.; **4** bow S.T. (19·7″); **1** stern S.T. (19·7″); **10** torpedoes.	Two 6-cylinder, 4-cycle, Diesel engines. Electric motors.

Part III. Section 5.

Sub-marines.

Brake H.P. Surface. / Sub-merged (normal).	No. of Pro-pellers.	Speed. Surface / Submerged.	Endurance (approximate). Surface. / Submerged.	Fuel and Fuel Stowage.	Batteries. No. of Cells. Type.	Periscopes.	Remarks.
		Knots.	Miles.	Tons.			
1,060 / 1,000	2	13·4 / 7·8	8,300 at 8 kts. / 70 at 3 kts.	Heavy oil, 84†	124 Tudor lead cells.	2, passing into conning tower.	*See* p. 63, Plate 8, and Silhouettes.
1,060 / 1,000	2	13·4 / 7·8	8,300 at 8 kts. / 70 at 3 kts.	Heavy oil, 84†	124 Tudor lead cells.	2, passing into conning tower.	
1,060 / 1,000	2	13·4 / 7·8	8,300 at 8 kts. / 70 at 3 kts.	Heavy oil, 84†	124 Tudor lead cells.	2, passing into conning tower.	
1,060 / 1,000	2	13·4 / 7·8	8,300 at 8 kts. / 70 at 3 kts.	Heavy oil, 84†	124 Tudor lead cells.	2, passing into conning tower.	

† Including stowage in 6 ballast tanks, usually appropriated for fuel on long cruises.

O AS 4924—11 C

Part III.
Section 5.
Sub-marines.

Official No.	Type. Hull.	Where Built.	Date of first Commissioning.	Normal Complement. (a) Officers. (b) Men.	Dimensions. (a) Length. (b) Breadth. (c) Draught.	Displacement. Surface. / Submerged.	Armament. (m. = machine gun; S.T. = internal tube; T. = external tube.)	Motive Power. Station. / Submerged.
					Ft. Ins.	Tons.		
U.B. 83, U.B. 82, U.B. 80, U.B. 79, U.B. 78	Smaller Ocean-going. Double hull.	Weser Yard, Bremen.	1917	(a) 7 (b) 28	(a) 180 9 (b) 19 6 (c) 12 0	502 723	**1**—4·1". *or* 22-pr.; **1**—m.; **4** bow S.T. (19·7"); **1** stern S.T. (19·7"); **10** torpedoes;	Two Körting 6-cylinder, 4-cycle, Diesel engines. / Siemens-Schuckert motors.
U.B. 77, U.B. 76, U.B. 74, U.B. 73	Smaller Ocean-going. Double hull.	Blohm & Voss Works, Hamburg.	1917	(a) 7 (b) 28	(a) 180 9 (b) 19 6 (c) 12 0	502 723	**1**—4·1" *or* 22-pr.; **1**—m.; **4** bow S.T (19·7"); **1** stern S.T. (19·7"); **10** torpedoes;	Two 6-cylinder, 4 cycle, Diesel engines. / Electric motors.
U.B. 71, U.B. 70, U.B. 68, U.B. 67	Smaller Ocean-going. Double hull.	Germania Yard, Kiel.	1917	(a) 7 (b) 28	(a) 180 9 (b) 19 6 (c) 12 0	502 723	**1**—4·1" *or* 22 pr.; **1**—m.; **4** bow S.T. (19·7"); **1** stern S.T. (19·7"); **10** torpedoes;	Two 6-cylinder, 4-cycle, Diesel engines. / Electric motors.
U.B. 65, U.B. 64, U.B. 62, U.B. 60	Smaller Ocean-going. Double hull.	Vulcan Yard, Hamburg.	1917	(a) 7 (b) 28	(a) 180 9 (b) 19 6 (c) 12 0	502 723	**1**—4·1" *or* 22-pr.; **1**—m.; **4** bow S.T. (19·7"); **1** stern S.T. (19·7"); **10** torpedoes.	Two 6-cylinder, 4 cycle, Diesel engines. / Electric motors.
U.B. 59, U.B. 57	Smaller Ocean-going. Double hull.	Weser Yard, Bremen.	1917	(a) 7 (b) 28	(a) 180 9 (b) 19 6 (c) 12 0	502 723	**1**—4·1" *or* 22-pr.; **1**—m.; **4** bow S.T. (19·7"); **1** stern S.T. (19·7"); **10** torpedoes.	Two Körting 6-cylinder, 4-cycle, Diesel engines. / Electric motors.

Brake H.P. Surface. / Sub-merged (normal).	No. of Pro-pellers.	Speed. Surface. / Submerged.	Endurance (approximate). Surface. / Submerged.	Fuel and Fuel Stowage.	Batteries. No. of Cells. Type.	Periscopes.	Remarks.
		Knots.	Miles.	Tons			
1,060 / 1,000	2	13·4 / 7·8	8,300 at 8 kts. / 70 at 3 kts.	Heavy oil, 84†	124 Tudor lead cells.	2, passing into conning tower.	
1,060 / 1,000	2	13·4 / 7·8	8,300 at 8 kts. / 70 at 3 kts.	Heavy oil, 84†	124 Tudor lead cells.	2, passing into conning tower.	
1,060 / 1,000	2	13·4 / 7·8	8,300 at 8 kts. / 70 at 3 kts.	Heavy oil, 84†	124 Tudor lead cells.	2, passing into conning tower.	*See* p. 63, Plate 8, and Silhouettes.
1,060 / 1,000	2	13·4 / 7·8	8,300 at 8 kts. / 70 at 3 kts.	Heavy oil. 84†	124 Tudor lead cells.	2, passing into conning tower	
1,060 / 1,000	2	13·4 / 7·8	5,700 at 8 kts. / 70 at 3 kts.	Heavy oil. 60‡	124 Tudor lead cells.	2, passing into conning tower.	

† Including stowage in 6 ballast tanks, usually appropriated for fuel on long cruises.
‡ Including stowage in 4 ballast tanks, usually appropriated for fuel on long cruises.

C 2

Part III.
Section 5.

Sub-marines.

Official No.	Type. Hull.	Where Built.	Date of first Commissioning.	Normal Complement. (a) Officers. (b) Men.	Dimensions. (a) Length. (b) Breadth. (c) Draught. Ft. Ins.	Displacement. Surface. / Submerged. Tons.	Armament. (m. = machine gun; S.T. = internal tube; T. = external tube.)	Motive Power. Surface. / Submerged.
U.B. 53 U.B. 52 U.B. 51 U.B. 50 U.B. 49 U.B. 48	Smaller Ocean-going. Double hull.	Blohm & Voss Works, Hamburg.	1917	(a) 7 (b) 28	(a) 180 9 (b) 19 6 (c) 12 0	502 723	**1**—4·1″ *or* 22-pr.; **1**—m.; **4** bow S.T. (19·7″); **1** stern S.T. (19·7″); **10** torpedoes.	Two Diesel engines. Electric motors.
U.B. 46 U.B. 45 U.B. 44 U.B. 42	Coastal. Single hull, with saddle tanks.	Weser Yard, Bremen.	1916	(a) 4 (b) 17	(a) 123 0 (b) 14 10 (c) 11 10	250 290	**1**—22-pr.; **1**—m.; **2** bow S.T. (19·7″), one above the other; **4** torpedoes.	Two 6-cylinder, 4-cycle, Diesel engines. Two Siemens-Schuckert motors, each with two armatures in tandem.
U.B. 40 U.B. 34 U.B. 33 U.B. 31 U.B. 30	Coastal. Single hull, with saddle tanks.	Blohm & Voss Works, Hamburg.	1916	(a) 4 (b) 17	(a) 123 0 (b) 14 10 (c) 11 10	250 290	**1**—22-pr.; **1**—m.; **2** bow S.T. (19·7″), one above the other; **4** torpedoes.	Two Körting 6-cylinder, 4-cycle, Diesel engines. Two Siemens-Schuckert motors, each with two armatures in tandem.
U.B. 28 U.B. 25 U.B. 24 U.B. 23 (*Interned in Spain, 30.7.17.*) U.B. 21	Coastal. Single hull, with saddle tanks. Coastal. Single hull, with saddle tanks.	Weser Yard, Bremen. Blohm & Voss Works, Hamburg.	1915–16	(a) 4 (b) 17	(a) 118 6 (b) 14 10 (c) 11 10	250 290	**1**—22-pr.; **1**—m.; **2** bow S.T. (19·7″), one above the other; **4** torpedoes.	Two 6-cylinder, 4-cycle, Diesel engines. Two Siemens-Schuckert motors, each with two armatures in tandem.

Brake H.P. Surface. / Sub-merged (normal).	No. of Pro-pellers.	Speed. Surface. / Submerged.	Endurance (approximate). Surface. / Submerged.	Fuel and Fuel Stowage.	Batteries. No. of Cells. Type.	Periscopes.	Remarks.
		Knots.	Miles.	Tons.			
1,060 / 1,000	2	13·4 / 7·8	5,700 at 8 kts. / 70 at 3 kts.	Heavy oil. 60†	124 Tudor lead cells.	2, passing into conning tower.	*See* p. 63, Plate 8, and Silhouettes.
285 / 250	2	8·5 / 6·0	4,500 at 5 kts. / 70 at 3 kts.	Heavy oil. 29‡	112 Tudor lead cells.	2, 1 passing into conning tower, 1 through fairwater just abaft it.	*See* p. 67, Plates 9 and 16, and Silhouettes.
285 / 250	2	8·5 / 6·0	4,500 at 5 kts. / 70 at 3 kts.	Heavy oil. 29‡	112 Tudor lead cells.	2, 1 passing into conning tower, 1 through fairwater jus abaft it	*See* p. 67, Plates 9 and 16, and Silhouettes.
285 / 250	2	8·5 / 6	4,500 at 5 kts. / 70 at 3 kts.	Heavy oil, 29.‡	112 Tudor lead cells.	2, 1 passing into conning tower, 1 through fairwater just abaft it.	*See* p. 67, Plates 9 and 16, and Silhouettes.

† Including stowage in 4 ballast tanks, usually appropriated for fuel on long cruises.
‡ Including stowage in 2 ballast tanks usually appropriated for fuel on long cruises.

C 3

Part III.
Section 5.

Sub-marines.

Official No.	Type. Hull.	Where Built.	Date of first Commissioning.	Normal Complement. (a) Officers. (b) Men.	Dimensions. (a) Length. (b) Breadth. (c) Draught.	Displacement. Surface. / Submerged.	Armament. (m. = machine gun; S.T. = internal tube; T. = external tube.)	Motive Power. Surface. / Submerged.
					Ft. Ins.	Tons.		
U.B. 17 U.B. 16 U.B. 14 U.B. 12 U.B. 11 U.B. 10 U.B. 9	Small coastal. Single hull.	Weser Yard, Bremen.	1915 (all U.B. boats)	(a) 1 (b) 15	(a) 90 0 (b) 10 0 (c) 9 9	125 ——— 140	**1**—1-pr. *or* **1**—m.; **2** bow S.T. (17·7") fitted abreast; **2** torpedoes.	One Benz 4-cylinder, Diesel engine. ——— Electric motor; two armatures fitted in tandem on one shaft.
U.B. 8 U.B. 6 (*Interned in Holland, 16.3.17.*) U.B. 5 U.B. 2	Small coastal. Single hull.	Germania Yard, Kiel.						
U.C. 120 U.C. 119 U.C. 118 U.C. 117 U.C. 116 U.C. 115 U.C. 114 U.C. 113 U.C. 112 U.C. 111	Minelaying. Double hull.	—	—	—	—	—	—	—
U.C. 110 U.C. 109 U.C. 108 U.C. 107 U.C. 106 U.C. 105 U.C. 104 U.C. 103 U.C. 102 U.C. 101	Minelaying. Double Hull.	—	—	—	—	—	—	—

Brake H.P. Surface. / Sub-merged. (normal).	No. of Pro-pellers.	Speed. Surface. / Submerged.	Endurance (approximate). Surface. / Submerged.	Fuel and Fuel Stowage.	Batteries. No. of Cells. Type.	Periscopes.	Remarks.
		Knots.	Miles.	Tons.			
60 / 120	1	8·4† / 5·0	1,200 at 5 kts / 50 at 2·5 kts.	Heavy oil, 2·4.	112 Tudor lead cells.	1 passing into conning tower.	See p. 73, Plate 9, and Silhouettes.
—	—	—	—	—	—	—	Believed to be under construction (April 1918) and to represent an improvement on U.C. 16–79 (see pp. 24–29).
—	—	—	—	—	—	—	

† With motors for a short period, but only 6·5 knots with oil engine.

C 4

Part III.
Section 5.

Sub-marines.

Official No.	Type. Hull.	Where Built.	Date of first Commissioning.	Normal Complement. (a) Officers. (b) Men.	Dimensions. (a) Length. (b) Breadth. (c) Draught.	Displacement. Surface. / Submerged.	Armament. (m. = machine gun ; S.T. = internal tube ; T. = external tube.)	Motive Power. Surface. / Submerged.
					Ft. Ins.	Tons.		
U.C. 100, U.C. 99, U.C. 98, U.C. 97, U.C. 96, U.C. 95, U.C. 94, U.C. 93, U.C. 92, U.C. 91, U.C. 90, U.C. 89, U.C. 88, U.C. 87, U.C. 86, U.C. 85, U.C. 84, U.C. 83, U.C. 82, U.C. 81, U.C. 80	Minelaying. Double Hull.	—	—	—	—	—	—	—
U.C. 78, U.C. 77, U.C. 75, U.C. 74	Minelaying. Double hull.	Vulcan Yard, Hamburg.	1916–17	(a) 4 (b) 25	(a) 162 3 (b) 17 1 (c) 12 0	395 475	**1**—22-pr. ; **1**—m. ; **2** bow T. (19·7″) ; **1** stern S.T. (19·7″) ; **5** torpedoes ; **18** mines.	Two 6-cylinder, 4-cycle, Diesel engines. Electric motors.
U.C. 73, U.C. 71, U.C. 70, U.C. 67	Minelaying. Double hull.	Germania Yard, Kiel.	1916–17	(a) 4 (b) 25	(a) 161 10 (b) 17 1 (c) 11 8	400 490	**1**—22-pr. ; **1**—m. ; **2** bow T. (19·7″) ; **1** stern S.T. (19·7″) ; **5** torpedoes ; **18** mines.	Two M.A.N. 6-cylinder, 4-cycle, Diesel engines. Electric motors.

Part III
Section 5.

Sub-marines.

Brake H.P. Surface. / Sub-merged (normal).	No. of Pro-peliers.	Speed. Surface. / Submerged.	Endurance (approximate). Surface. / Submerged.	Fuel and Fuel Stowage.	Batteries. No. of Cells. Type.	Periscopes.	Remarks.
		Knots.	Miles.	Tons.			
—	—	—	—	—	—	—	Believed to be under construction (April 1918) and to represent an improvement on U.C. 16–79 (*see* below).
600 / 460	2	12·0 / 7·0	6,000 at 8 kts. / 80 at 3 kts.	Heavy oil, 40†	124 Tudor lead cells.	2 passing into conning tower.	*See* p. 75, Plates 10 and 17, and Silhouettes. The dimensions here given correspond to the original design of these boats, *i.e.*, with a rounded bow. The boats which have been given a ship bow are about 3 feet longer.
600 / 460	2	12·0 / 7·0	6,000 at 8 kts. / 80 at 3 kts.	Heavy oil, 40†	124 Tudor lead cells.	2 passing into conning tower.	(see above)

† Normal stowage; can be considerably increased by using ballast tanks for extra fuel.

O AS 4924—11 D

Part III. Section 5.

Sub-marines.

Official No.	Type. Hull.	Where Built.	Date of first Commissioning.	Normal Complement. (a) Officers. (b) Men.	Dimensions. (a) Length. (b) Breadth. (c) Draught.	Displacement. Surface. / Submerged.	Armament. (m. = machine gun; S.T. = internal tube; T. = external tube.)	Motive Power. Surface. / Submerged.
					Ft. Ins.	Tons.		
U.C. 64	Minelaying. Double hull.	Weser Yard, Bremen.	1916–17 (U.C. 64, 60, 59, 58, 56)					
U.C. 60, **U.C. 59**, **U.C. 58**, **U.C. 56**	Minelaying, Double hull.	Imperial Dockyard, Danzig.		(a) 4 (b) 25	(a) 161 10 (b) 17 1 (c) 11 8	400 / 490	**1**—22-pr.; **1**—m.; **2** bow T. (19·7″); **1** stern S.T (19·7″); **5** torpedoes; **18** mines.	Two 6-cylinder, 4-cycle, Diesel engines. / Electric motors.
U.C. 54, **U.C. 53**, **U.C. 52**, **U.C. 49**, **U.C. 48** (*Interned in Spain 24.3.18.*)	Minelaying. Double hull.	Weser Yard, Bremen.	1916–17	(a) 4 (b) 25	(a) 161 10 (b) 17 1 (c) 11 8	400 / 490	**1**—22-pr.; **1**—m.; **2** bow T. (19·7″); **1** stern S.T. (19·7″); **5** torpedoes; **18** mines.	Two 6-cylinder, 4-cycle, Diesel engines. / Electric motors.
U.C. 40	Minelaying. Double hull.	Vulcan Yard, Hamburg.	1916	(a. 4 (b) 25	(a) 162 3 (b) 17 1 (c) 12 0	395 / 475	**1**—22-pr.; **1**—m.; **2** bow T. (19·7″); **1** stern S.T. (19·7″); **5** torpedoes; **18** mines.	Two 6-cylinder, 4-cycle, Diesel engines. / Electric motors.
U.C. 37, **U.C. 35**, **U.C. 34**	Minelaying. Double hull.	Blohm & Voss Works, Hamburg.	1916	(a) 4 (b) 19	(a) 161 10 (b) 17 1 (c) 11 8	410 / 500	**1**—22 pr.; **1**—m.; **2** bow T. (19·7″); **1** stern S.T. (19·7″); **5** torpedoes; **18** mines. (U.C. 37–25)	Two Körting 6-cylinder, 4-cycle, Diesel engines. / Electric motors. (U.C. 37–25)
U.C. 31, **U.C. 28**, **U.C. 27**, **U.C. 25**	Minelaying. Double hull.	Vulcan Yard, Hamburg.	1916	(a) 4 (b) 19	(a) 162 3 (b) 17 1 (c) 12 0	395 / 475		

Brake H.P. Surface. / Sub-merged (normal).	No. of Pro-pellers.	Speed. Surface. / Submerged.	Endurance (approximate). Surface. / Submerged.	Fuel and Fuel Stowage.	Batteries. No. of Cells. Type.	Periscopes.	Remarks.
		Knots.	Miles.	Tons.	Tons.		
600 / 460	2	12·0 / 7·0	6,000 at 8 kts. / 80 at 3 kts.	Heavy oil, 40†	124 Tudor lead cells.	2 passing into conning tower.	See p. 75, Plates 10 and 17, and Silhouettes. The dimensions here given correspond to the original design of these boats, *i.e.*, with a rounded bow. The boats which have been given a ship bow are about 3 feet longer.
520 / 460	2	11·5 / 6·8	6,000 at 8 kts. / 80 at 3 kts.	Heavy oil, 40†	124 Tudor lead cells.	2 passing into conning tower.	
520 / 460	2	11·5 / 6·8	6,000 at 8 kts. / 80 at 3 kts.	Heavy oil, 40†	124 Tudor lead cells.	2 passing into conning tower.	
520 / 460	2	11·5 / 6·8	6,000 at 8 kts. / 80 at 3 kts.	Heavy oil, 40†	124 Tudor lead cells.	2 passing into conning tower.	
500 / 460	2	11·5 / 6·8	6,000 at 8 kts. / 80 at 3 kts.	Heavy oil, 40†	124 Tudor lead cells.	2 passing into conning tower.	

† Normal stowage ; can be considerably increased by using ballast tanks for extra fuel.

D 2

Part III. Section 5.

Sub-marines.

Official No.	Type. Hull.	Where Built.	Date of first Commissioning.	Normal Complement. (a) Officers. (b) Men.	Dimensions. (a) Length. (b) Breadth. (c) Draught.	Displacement. Surface. / Submerged.	Armament. (m. = machine gun; S.T. = internal tube; T. = external tube.)	Motive Power. Surface. / Submerged.
					Ft. Ins.	Tons.		
U.C. 23 **U.C. 22** **U.C. 20** **U.C. 17**	**Minelaying.** **Double hull.**	**Blohm & Voss Works, Hamburg.**	1916	(a) 4 (b) 19	(a)161 10 (b) 17 1 (c) 11 8	410 500	**1**--22-pr.; **1**--m.; **2** bow T. (19·7″); **1** stern S.T. (19·7″); **5** torpedoes; **18** mines.	**Two 6-cylinder, 4-cycle, Diesel engines.** **Electric motors.**
U.C. 11	**Small Minelaying.** **Single hull.**	**Weser Yard, Bremen.**	1915	(a) 2 (b) 13	(a)110 3 (b) 10 4 (c) 9 7	179 202	**1**—4 pr. *or* **1** m.; **12** mines.	**One 4-cylinder, 4-cycle, Diesel engine.** **Siemens Schuckert motor, 2 armatures in tandem.**
U.C. 4	**Small Minelaying.** **Single hull.**	**Vulcan Yard, Hamburg.**	1915	(a) 3 (b) 15	(a)111 6 (b) 10 4 (c) 9 1	177 194	**1**—4 pr. *or* **1** m.; **12** mines.	**One Benz 6-cylinder, 4-cycle, Diesel engine.** **Electric motor, 2 armatures in tandem.**

Brake H.P. Surface. / Sub-merged. (normal).	No. of Pro-pellers.	Speed. Surface. / Submerged.	Endurance (approximate). Surface. / Submerged.	Fuel and Fuel Stowage.	Batteries. No. of Cells. Type.	Periscopes.	Remarks.
		Knots.	Miles.	Tons			
500 / 460	2	11·5 / 6·8	6,000 at 8 kts. / 80 at 3 kts.	Heavy oil, 40‡	124 Tudor lead cells.	2 passing into conning tower.	*See* p. 75, Plates 10 and 17, and Silhouettes. The dimensions here given correspond to the original design of these boats, *i.e.*, with a rounded bow. The boats which have been given a ship bow are about 3 feet longer.
80 / 155	1	7·5† / 5·0	1,100 at 4 kts. / 50 at 2·5 kts.	Heavy oil, 2·8	112 Tudor lead cells.	1 passing into conning tower.	*See* p. 84, Plates 10 and 18, and Silhouettes.
80 / 155	1	7·5† / 5·0	900 at 4 kts. / 50 at 2·5 kts.	Heavy oil, 2·3	220 Tudor lead cells.	1 passing into conning tower.	(see above)

† With motors, for a short period; but only 6 knots with oil engines.
‡ Normal stowage; can be considerably increased by using ballast tanks for extra fuel.

D 3

NOTE.

The following pages contain separate descriptions of the different types and classes of German submarines. The "U" ocean-going type, as the most important and most representative, is dealt with at greater length than the others. Information applicable to all types, e.g., that given under the headings "Demolition," "Navigation Apparatus," "Submersion," "Emersion," and "Trimming," will be found in the description of the "U" boats only, and this portion of the pamphlet should therefore be read by all officers who wish to acquire a general knowledge of German submarines.

Part III.
Section 5.

Sub-
marines.

DESCRIPTION OF SUBMARINES.

I.—"U." CONVERTED MERCANTILE TYPE.

U. 151–157.

(*See* Plates 1 and 11.)

GENERAL REMARKS.

Eight submarines of the "*Deutschland*" type were originally built. All of them were completed at Germania Yard, Kiel, though one or two of the hulls were built at Flensburg.

The "*Bremen*" having been lost, seven of these vessels remain. They have now all been converted into war vessels, receiving the numbers *U. 151–157*. The "*Deutschland*" was the first one converted and was given the number "*U. 155*."

It seems probable that this class of vessel may also be used for importing contraband cargo, as a large quantity could still be carried. The original cargo-carrying capacity was about 800 tons.

RECOGNITION.

Shape.

See Silhouette and Plate 1.

It should be noted that the converted mercantile submarines fall into two distinct classes as regards armament (*see under* Guns, p. 33, and Torpedo Tubes and Torpedoes, p. 34), and probably the two classes differ in other particulars also, though no details of such differences are yet available.

The bow of these vessels resembles that of a turret steamer.

The tops of the tanks are visible externally and form a deck at a lower level on either side of a central superstructure.

Colour.

A light French grey, except the deck, which is painted black.

When working in southern waters these submarines may adopt any light shades of blue, grey, or green, according to colour of water and conditions of light prevailing in the cruising locality.

DETAILS.

Batteries.

Large battery capacity, 330–360 cells, instead of the usual 210–240 cells of ordinary "U." boats. The accumulators are placed below compartments 4, 5, and 7 (*see* Internal Arrangements. p. 34).

Boats.

Two are carried, one of which is fitted with a detachable motor, for removal when diving.

Cable Cutting.

These submarines are specially equipped for cutting submarine cables. The apparatus is fitted over the stern; *see* Plate 1 (Stern View).

Complement.

A *Korvettenkapitän* or *Kapitänleutnant* in command, 5 other executive officers (including 2 prize officers), 1 engineer officer, 1 surgeon; about 70 warrant officers, petty officers, and men, including two prize crews of about 10 men each.

Deck and Superstructure.

A central, free-flooding superstructure is fitted, forming an upper deck. It can be used for storing cargo or loot which is impervious to the action of salt water.

Guard rails are fitted for part of the length of the superstructure.

Dimensions.

See tabulated details, p. 2.

Diving Capabilities.

The diving qualities of these boats were considerably improved when they were converted into war vessels, the vents having been made much larger. Originally the boats took 2½ minutes to dive to periscope draught; they now take about 1½ minutes only.

They are always somewhat down by the head, because the larger tanks are situated forward, and they are reported to take up an inclination of 20° bow down when diving.

The hulls are tested for a depth of 197 feet, but these submarines can dive to 300 feet for a short period in case of emergency.

Endurance.

One of these submarines has made a cruise lasting four months, and there is no reason why vessels of this class should not remain at sea for even longer periods, the only limitation being the provision and fresh-water supply.

Engines (Main).

Two 4-cycle, 6-cylinder reversible Diesel engines of the air injection type, of 600 brake H.P. each, which is a very low power in proportion to displacement.

Fresh-Water Supply.

A filter is provided and there is a very large stowage of fresh water. In addition, electrical evaporating plant is reported to be fitted. Each man is allowed about ½ gallon of water per day for all purposes.

Fuel.

Heavy oil, specific gravity ·87. Normal stowage about 250 tons. It is reported that 500–600 tons can be carried if necessary.

Guns, &c.

U. 151–154.—Two 5·9-inch guns are mounted in the centre line on platforms, one before and one abaft the conning tower. The type of gun is not yet known. It is, however, evident from photographs that the length of the gun is at least 40 calibres. The projectile is about 100 lbs. in weight, and the range of the gun is probably at least 15,000 yards.

Two 22-prs. are placed out of the centre line on opposite sides of the upper deck, the port gun just before and the starboard gun just abaft the conning tower. These are not A.A. guns; they are probably carried to save expenditure of heavy ammunition where this is unnecessary, and also for range-finding purposes. One machine gun is carried.

U. 155–157 carry two 5·9-inch guns, as above, but no 22-prs.

One or more additional light Q.F. guns may be carried for mounting in captured merchant vessels when these are used for scouting or decoy purposes or as a temporary base.

Ammunition Supply.—*U. 155* carried as much as 1,400 rounds of 5·9-inch ammunition on her first cruise. In *U. 151–154*, however, carrying 22-pr. guns, the stowage of 5·9-inch ammunition is presumably less.

The ready ammunition is stowed on deck in covered watertight lockers, which are fitted round the guns.

Rangefinder.—A rangefinder, probably 10-ft. base, is carried, and can be mounted on fore end of conning tower.

Hull.

These submarines are of the double hull type.

The length of the pressure hull is 164 feet; diameter, 21 feet 4 inches.

Internal Arrangements.

The sub-division in *U. 151–154* is stated to be as follows, commencing from forward (*see* Sketch drawn from Memory, Plate 11):—

(1) Forward torpedo room, containing torpedoes and torpedo compensating tanks.
(2) Crew's living quarters.
(3) *Port Side.*—Officers' cabins.
    Starboard Side.—Commanding officer's cabin and ward room.
(4*a*) Crew space on port and starboard sides. W/T cabinet on starboard side.
(4*b*) *Port Side.*—Chart room and seamen warrant officers' mess.
    Starboard Side.—Artificer engineers' mess.
(5) Control room.
(6) Crew space.
(7) Main engine room.
(8) Motor room.
(9) Tiller flat.

The accommodation for the crew is excellent.

In *U. 155–157* the sub-division is less elaborate.

Masts.

Two hinged W/T masts, about 50 feet high, which can be raised and lowered by motor power from inside the boat. It is reported that a crow's nest has been fitted on one of them.

Mines.

U. 151–154 are known to carry mines, which are ejected from the torpedo tubes. So far, only moored mines have been discovered. These are 19·7 ins. in diameter and 4 ft. 9 ins. in length, and are provided with four firing horns.

Periscopes.

Two periscopes are fitted, one in conning tower and one in control room, the latter periscope passing through the upper deck immediately before the conning tower, out of the centre line, on the starboard side.

Speed.

On surface, 11–12 knots.
Submerged, 8 knots.

Tanks.

Five external ballast tanks are fitted on either side. In normal trim the tops of these tanks are about level with the surface of the water.

The former cargo spaces, extending right down to the keel between the outer and inner hulls, have been converted into oil-fuel tanks.

Torpedo Tubes and Torpedoes.

In *U. 151–154* six 19·7-in. internal tubes are fitted, viz., four bow tubes and two broadside tubes. The angle at which the latter are set is not known.

Stern tubes cannot be fitted, as the engine compartment extends right aft.

In *U. 155*, and probably also in *U. 156* and *U. 157*, only four (?) 19·7-in. external frame tubes are fitted. The torpedoes are ejected from these by means of plungers operated by compressed air. The frames are fitted on each bow and quarter and angled out at about 35° from the fore-and-aft line.

About 20 torpedoes are carried, the majority being 17·7-in., which are used against merchant vessels, guide rails being then inserted in the tubes.

Wireless Telegraphy.

A 2-kilowatt set is fitted in the submarine, and a portable set is carried, which can be rigged up in a few hours in a captured merchant vessel if she is to be used for scouting or decoy purposes or as a temporary base.

II.—"U." CRUISER TYPE.
U. 139-142 and U. 143-150

GENERAL REMARKS.

Very little is known at present about these vessels.

None of them are yet at sea (April 1918). They are expected to be based on Kiel, forming with the converted mercantile submarines a Cruiser Submarine Flotilla.

Two distinct classes are known to be under construction, viz., *U. 139–142* and *U. 143–150.*

The two classes probably resemble one another in their general characteristics, but *U. 143–150* may be slightly larger than *U. 139–142*; they have considerably more powerful engines, and their surface speed should be appreciably higher (probably 18–19 knots).

It is reported that the vessels may be fitted with armour protection in the form of a pent-house roof, placed above the pressure hull, and running the whole length of it. Some of the main pipes and leads would be placed under this protection.

The vessels will probably have an overhanging bow. The freeboard will not be much greater than that of ordinary "U." boats.

DETAILS.
U. 143-150.

Completing at Weser Yard, Bremen.

The exact number of the vessels of this series is uncertain.

Dimensions.

See tabulated details, p. 2.

Engines.

Two M.A.N. 4-cycle, 10-cylinder reversible Diesel engines of the air injection type, of 3,000 brake H.P. each, for propulsion; one Augsburg 4-cycle, 6-cylinder Diesel engine of the air injection type, of 1,000 brake H.P., for charging.

It has been stated that the 3,000-H.P. engines made a successful 14 days' continuous run at the Augsburg Works.

The charging engine is reported to be placed on the centre line of the boat, abaft the main engines, and to be coupled direct to an electric generator. It is used to keep the batteries constantly charged up when the boat is proceeding on the surface, and can also serve as a very economical drive at low speeds, generating current for driving the main motors, which would then be used for propulsion. Under these latter conditions the boat will be in the best state for a quick dive, since the motors are ready for instant use and no unclutching or shutting off of the main Diesel engines is necessary. In view of the fact that such high-power Diesel engines have never before been fitted in submarines, it is conceivable that this charging engine of normal power is fitted partly as a stand-by in case of serious defects developing in the main Diesel engines, and partly to be used instead of the latter when cruising in order to economise fuel and personnel.

Speed.

On surface—maximum, probably about 18½ knots.
Submerged— „ „ „ 9½ knots.

DETAILS.
U. 139-142.

Completing at Germania Yard, Kiel. The first of these vessels was launched about the end of October 1917, and two of them are reported now to be doing trials (April 1918).

Accommodation.

Reported to be very good.

Batteries.

Probably 330–360 cells are carried. The batteries are installed below compartments 3 and 7 (*see* Internal Arrangements, p. 37).

Part III. Section 5.

Sub-marines.

Complement.

Estimated to be about 100 officers and men.

Conning Tower.

The conning tower is reported to be circular in section, about 5 feet in diameter, and made to raise and lower, working up and down in a gland, by means of a reversible electric winch with wires and pulleys, in the same way as the periscopes are operated. The idea of this may be to obtain all-round gunfire, but is more probably to reduce the visible portion of the boat as much as possible, the conning tower being the most conspicuous part of a submarine on the surface.

A low bridge is said to be fitted, sinking into the superstructure when the conning tower is housed. The whole report, however, lacks confirmation.

Control Room.

According to the same report, the control room is divided unequally into two portions by two fore and aft watertight bulkheads with a passage between them. The port control room is the larger, extending past the centre line of the boat, in order to take the conning tower and periscopes.

The control room is said to be divided in this way in order to prevent the boat being put out of action, should one side be hit and consequently flooded. If the port side is flooded, all the periscopes are put out of action, but the boat can be manœuvred from the starboard control room, which contains depth-gauges and hydroplane and steering pedestals, similarly to the port control room.

Dimensions.

See tabulated details, p. 2).

Endurance.

These vessels will be capable of keeping the sea for five or six months, and will have a radius of action of at least 25,000 miles at 6 knots, using the charging engine for driving the main motors, as described under U. 143–150 on p. 35. Electrical evaporating plant is reported to be fitted.

Engines.

Two M.A.N. 4-cycle, 6-cylinder reversible Diesel engines of the air-injection type, of 1,850 brake H.P. each, for propulsion, and one Augsburg 6-cylinder engine of the air-injection type of 550 brake H.P. for charging. For remarks regarding the charging engine, *see* under U. 143–150 on p. 35.

Fuel.

Heavy oil of specific gravity 0·87, stored in tanks which are all fitted externally (*i.e.*, outside the pressure hull). Normal stowage probably at least 400 tons.

Guns, &c.

Either two 5·9-inch and two 4·1-inch, all mounted in the centre line, *or* four 4·7-inch, all mounted in the centre line.

One *or* two machine guns, fitted for mounting on conning tower.

Details of 5·9-inch and 4·7-inch guns are not yet known. For probable details of 4·1-inch guns and ammunition supply, *see* under "U." boats (p. 47).

Rangefinder.—A rangefinder, probably 10-ft. base, is carried.

Hull.

The length of the pressure hull is stated to be about 266 feet. It is constructed by Krupp of special nickel steel, in the interests of which, according to report, the nickel coinage was withdrawn in Germany. This will tend to keep down the weight of the hull.

The pressure hull of these vessels is reported to have been made considerably thicker than that of the ordinary "U." boats in order to enable them to dive to a greater depth. (For details in the case of "U." boats, *see* p. 49.)

The increase in the weight of plating will diminish the surface buoyancy and tend to low freeboard.

The pent-house roof over the above-water part of the pressure hull, referred to in General Remarks, p. 35, is said to cover the oil-fuel suction pipes (on tank tops) and the circulating-water compensating pipes to the oil-fuel tanks.

Internal Arrangements.

The sub-division is reported to be as follows, commencing from forward (*see* Sketch drawn from Memory, Plate 11) :—

(1) Fore torpedo room and tubes crews' mess.
(2) Seamen's and daymen's mess.
(3) *Port side.*—Seamen petty officers' mess.
Starboard side.—Technical petty officers' mess.
(4) *Port side.*—Bath and W.C. ; W/T cabin ; 2nd officer's cabin ; ward room and chief engineer's cabin.
Starboard side.—3rd officer's cabin ; commanding officer's night cabin ; commanding officer's day cabin ; and executive officer's cabin.
(5) Control rooms—port and starboard.
(6) Broadside torpedo room.
(7) *Port side.*—Senior artificer-engineer's cabin ; navigating warrant officer's cabin ; warrant officer's mess.
Starboard side.—2nd engineer's cabin ; bath and W.C. ; artificer-engineers' cabins.
(8) Main engine room.
(9) Motor room.
(10) After torpedo room and stokers' mess.

The charging engine is fitted between compartments 8 and 9.

A passage in the centre line of the boat leads from the seamen's and daymen's mess to the broadside torpedo room, and from there to the main engine room.

Machinery.

For main engines *see* Engines, p. 36.

Periscopes.

Three are fitted.

Speed.

On surface—maximum probably about 16 knots.
Submerged— ,, ,, ,, 9 knots.

Steering and Hydroplane Gear.

It is probable that a small upper vertical rudder will be fitted, working on the same shaft as the lower vertical rudder, which is of the balanced type. The foremost hydroplanes will probably be fitted well forward, just above the waterline (in view of the high surface speed) ; and the after hydroplanes low down, right aft, immediately in line with and abaft the propellers.

Torpedo Tubes and Torpedoes.

Eight 19·7-inch torpedo tubes, viz. :—

Four bow, two broadside, and two stern tubes, all inboard and fixed.

About 16–20 torpedoes are carried.

III. "U." OCEAN-GOING TYPE.

U. 19–138.

(With the exception of U. 71–80, described separately on page 61.)

(*See* Plates 2–7, 12, and 13.)

GENERAL REMARKS.

These vessels are all built with two hulls—a partial outer hull, which is given a ship form, and an inner cylindrical pressure hull. Damage to the outer hull alone will not appreciably impair the diving qualities of the submarine, and certainly will not disable her. Should the fuel tanks, which are situated between the inner and the outer hull, be penetrated, oil will, of course, appear on the surface. In addition, an arrangement is fitted for ejecting oil in case of accident, to mark the position of the

E 3

submarine. German submarines are instructed to use this arrangement if it appears advisable, in order to mislead and delay an enemy. Oil seen on the surface must, therefore, never be accepted as proof of a submarine having been sunk, or even damaged.

RECOGNITION.

Shape.

See Silhouettes.

In general, these vessels are flushed-deck fore and aft.

To the eye :—

The top edge of the superstructure forms a continuous and nearly straight line from bow to stern, with slight sheer from the stem head aft. The conning tower is generally situated midway between bow and stern and, except in *U. 43–50*, has a fairwater stepped before it. (N.B.—This is not the case in British submarines, where the conning tower is generally well forward of the midship section, or, in some few cases, well abaft it, and is almost invariably vertical on its forward side.) The guns are mounted well clear of the conning tower as a rule, and, on account of their size and position, they are very conspicuous.

Smoke.

German submarines have never been observed to emit smoke from engine exhaust, except when starting main engines on coming to the surface, and then only in small quantities and for a short period. British submarines, on the other hand, at times make considerable exhaust smoke, and on many occasions smoke has established their identity as British.

Colour.

A light French-grey colour is nearly always adopted for the hull of submarines operating in the Atlantic and the North Sea. The deck is painted black or left unpainted, and the conning tower is painted grey or dark grey.

Submarines operating in the Mediterranean have their hull painted grey, and the deck usually dark blue.

In the past, black and white chequers have occasionally been used, and waves have been painted on the grey.

The periscopes have sometimes been painted green or in waved lines of white and grey, but are usually a light grey colour in northern latitudes. Silver-bronze has been used in the Mediterranean, where it merges with the colour of the water.

Some submarines have had a bow wave painted on, to give a false impression of very high speed. At one time it was the fashion in the Flanders flotilla to paint an eye on either bow for luck, after the style of a Chinese junk.

Camouflaging the conning tower by painting it a very light grey with black border, or in diagonal grey and white stripes, or in black and white chequers, is still sometimes practised, but is falling into increasing disfavour.

Recognition Marks.

All submarines operating from the German Bight are fitted with a special recognition mark, which is to be displayed when they anticipate sighting German aircraft. This consists of a white ring, about 12 inches broad and about 4 feet in diameter, painted half upon each of two plates of sheet-iron, which are hinged together. One of these plates is secured flat upon the deck forward, the other one being folded down on top of it when in harbour or when it is desired not to display the recognition mark, and opened out at other times.

It has also been stated that submerged submarines, which wish to signal to their own aircraft in the vicinity, eject coloured water, each colour having a different signification ; but this report is unconfirmed.

DETAILS.

Accommodation.

Owing to the large complements and the supernumeraries carried, the accommodation in German submarines is extremely cramped.

In the older boats there is only one bunk between every two members of the crew (petty officers excepted), but a few hammocks are provided in addition. In the newer boats, however, more bunks are fitted.

The crew work in two watches, six hours on and six off.

There are two water-closets, one or both of which can be used when submerged.

Steam heaters are fitted and are provided with supply and exhaust connections on upper deck for use in harbour. The engine exhaust gases are also used for heating, and in some cases electric radiators are fitted. An electric cooking stove is always fitted.

Accumulators.

See Batteries, page 40.

Air Service.

A number of steel bottles, containing compressed air at high pressure, are carried to enable ballast and other tanks to be blown rapidly. The bottles are made of 6 per cent. nickel steel and are ·47 inch thick. They are tinned inside and out, and coated with four coats of tar varnish. The containers inside the bottles are made of Siemens-Martin mild steel. The test pressure is 3,200 lbs. per square inch and the air is carried at a pressure of 2,275 to 2,348 lbs. per square inch.

Apparently one standard size of bottle is carried, about 7 feet in length and 1½ feet in diameter, capacity about 12·4 cubic feet. The bottles are usually fitted in pairs inside the superstructure, but in some cases they are also carried inside the pressure hull.

In *U. 43–50* 31 air bottles are carried, capacity of each about 12·4 cubic feet.

Two air bottles are fitted to the engines for blast and starting purposes.

The low-pressure air main is tested to a pressure of about 170 lbs. per square inch.

Owing to shortage of copper, both the H.P. and the L.P. air lines are now made of iron or steel.

Ammunition Supply.

See under Guns, page 47.

Anchor Gear.

A stockless anchor is usually carried in a hawse pipe in forward superstructure, stowing vertically. A capstan, which can be unshipped, is fitted on the upper deck right forward, for warping, &c., in the forward superstructure. The cable is self-stowing. The capstan and cable-holder can be worked either electrically or by hand from inside the pressure hull.

German submarines have no statical weight or method of remaining stopped submerged, except on the bottom (*see*, however, under "Trimming," page 59).

Ballast Tanks.

Are fitted externally, *i.e.*, between the inner and outer hulls, on either side and at both ends of the boat. The foremost and aftermost tanks extend right across the boat; the other tanks extend on each side down to the main keel. To increase the radius of action of the submarine, oil may be carried in some of the ballast tanks. As a rule, two on either side are fitted to be used in this manner (*see* Fuel Tanks, page 46). The more oil carried the less will be the normal reserve of buoyancy. But in case of accident the oil can, of course, be blown out in the same way as the water.

All ballast tanks are fitted with large flooding flaps (*see* under Kingston valves, page 50), which are kept permanently open in most of the boats.

For typical arrangement of flooding, venting, and blowing leads of the ballast tanks, *see* Plate 19.

The capacity of the ballast tanks of *U. 81–86* class is as follows :—

	Tons.*
No. 1 tank (aft)	13·10
No. 2 starboard tank	28·38
No. 2 port tank	28·38
No. 3 starboard tank	20·57
No. 3 port tank	20·57
No. 4 tank (forward)	26·48

Thus, in *U. 81–86*, total capacity of ballast tanks (six in number) = 137·5 tons.*
In *U. 57–62* there are 10 ballast tanks with a total capacity of 164·3 tons.*

* Fresh-water tons.

Batteries.

All boats carry two batteries of about 110 cells each, which are placed in compartments below the living quarters, as a rule both before and abaft the control room. Airtight flooring with traps in it is fitted over these compartments. Very elaborate ventilation arrangements are fitted to get rid of the hydrogen produced when charging the accumulators, a separate ventilation pipe being led to each cell.

The size of cell varies in different boats, but only one size is fitted in any one boat. The most recent pattern is that known as the "funnel" cell, dimensions estimated to be 43–47 inches in height and 20 inches by 17·5 inches in section. Each cell contains about 52 plates of the grid type.

The amount of deposit in the bottom of the cell is measured whenever the boat is in dockyard hands. When it is found that with the boat down by the bow at an angle of 45° or 50°, this deposit would touch the base of the plates at the forward end of the cell, the cells are removed and the deposit cleaned out. This should not be necessary more than about once in three years.

The life of a battery is reckoned as five years.

Charging from a shore station with a current of 1,000 ampères takes 18–20 hours to complete. Charging with the boat's own main engines can be completed in 10 hours. The voltage of the battery when fully charged is 305. The motors, however, are run at 220 volts, the light at 200 volts.

In order to obtain the maximum economy in ampère hours from the batteries, when proceeding at various speeds submerged, German submarines are fitted with "grouping switches," by means of which the two batteries can be put either in series or in parallel, and also the two armatures of each tandem motor either in series or in parallel with each other. The port and starboard main motors are independent of one another, and boats can proceed submerged with either or both running.

Table No. 1 shows the "grouping" arrangements adopted in *U. 57–62* for various speeds with both motors :—

TABLE No. 1.

Speed.	Knots Submerged.	Total Ampères.	Grouping.	
			Two Armatures.	Two Batteries.
Utmost speed	8·5	4,800	Parallel	Series
Full speed	8·0	4,000	,,	,,
Three-quarter speed	7·3	2,800	,,	,,
Half speed	6·0	1,500	,,	,,
Slow speed	5·1	800	,,	Parallel
Dead slow	3·5	230	Series	,,

Table No. 2 contains complete figures of the performance of the batteries and main motors of *U. 57–62* class.

TABLE No. 2.

Speed Submerged.	Revolutions per Minute.	Knots Submerged.	H.P. Two Motors.	Hours.*	Ampère Hours.*	Endurance.*
Utmost speed	350	8·5	1,472	1	4,800	8·5
Full speed	325	8·0	1,206	1·5	6,000	12·0
Three-quarter speed	300	7·3	830	3	8,400	21·9
Half speed	250	6·0	440	6	9,000	36·0
Slow speed	210	5·1	232	12	9,600	61·2
Dead slow	150	3·5	130	23	10,580	80·5

* Not taking into account current expended for lights, gyro compass and auxiliary machinery.

Boats.

A small canvas boat, capable of carrying four men only, can be stowed bottom upwards on deck abaft the conning tower. Many commanding officers, however, dispense with it. In any case in which communication with a merchant vessel is necessary, the ship's boats are used in preference. Some of the early boats (*e.g.*, *U. 19*) carried a small steel boat in fore part of superstructure.

Bulkheads.

The majority of the bulkheads fitted inside the pressure hull are convexed for resisting pressure, and these are subjected to the same test as the pressure hull. They are therefore known as "pressure bulkheads." The other bulkheads, which are straight, are usually watertight and gastight.

All bulkheads are fitted with watertight doors, which are hand-operated only. The pressure bulkheads have circular hinged doors, about 2 feet 9 inches in diameter, with the lower edge about 18 inches above the deck. These doors are closed water and air-tight by means of a locking ring which is actuated by a lever fitted on the starboard side of the bulkhead.

In the other bulkheads, ordinary hinged watertight doors, about 5 feet in height, are fitted.

Outside the pressure hull, between each pair of fuel tanks and between the fuel tanks and the adjacent ballast tanks, a double bulkhead is fitted, *i.e.*, each fuel tank has its own separate bulkheads. Between these bulkheads there is a space about $\frac{7}{16}$ inch wide, extending down to the keel, where there is an opening for free flooding and draining.

Clutches.

Between the main engines and the main motors a Bamag clutch was formerly fitted, but in later boats this has been replaced by a clutch of the Lohmann type. Both types are friction clutches, the main difference between them being that whereas in the Bamag clutch the friction surfaces are parallel to the shaft, serrated, and metal to metal, in the Lohmann clutch the contact discs are at right angles to the shaft and are sometimes covered with a thin layer of camel-hair felt.

In both types the clutch can be worked by hand or by compressed air, but the former method is usually employed and is preferred. If required, the clutches can be operated when the engines are moving.

In the Lohmann clutch the relative positions of the surfaces can be regulated by means of adjusting nuts, worked together by a circular rack on the engine side; but the wear is very slight. For details of this clutch, *see* Plate 14.

Compasses.

Gyro compasses are supplied by Anschütz. In the latest type the whole master compass is enclosed in a gastight binnacle, which is filled with hydrogen. The gyros revolve at 30,000 revolutions per minute instead of 20,000 (as formerly) without undue heating, and a considerable increase in "directive force" is thereby attained; on the other hand, the compass is inaccessible when running.

A fourth gyro is attached to the frame above the card and controls the rolling gimbal. This was fitted to counteract the very quick rolling period of submarines on the surface; it increases the rolling period of the frame and desynchronises it from that of the boat. It is probable that, before this fourth gyro was introduced, the frame fouled the floating system and caused the gyros to wander on the boat taking a sudden and heavy roll.

The gyro generator is usually situated in or just abaft the control room, the master compass near it, and receivers are usually fitted for bridge and in conning tower and control room. A portable magnetic compass is supplied as a reserve fitting.

The gyro compasses have given a good deal of trouble, and, in consequence of this, magnetic diving compasses have been introduced, having been copied from the British service. These compasses are mounted in a pressure-proof and watertight binnacle on fore end of bridge and reflect the card image down a tube on to a screen inside the conning tower.

Compensating Tanks.

See "Tanks," page 56.

Complement.

Very large complements in proportion to their size are borne in German submarines. About 38 officers and men are carried in recent boats, with, in addition, four or five supernumerary ratings for training.

Conning Tower.

Is fitted nearly midway between bow and stern. Despite numerous reports to that effect it is not specially armoured, but it is constructed of $\frac{7}{8}$-inch steel plates,

Part III. Section 5.

Sub-marines.

stiffened externally by three H-girders, which pass right round it, and is enclosed by a light casing about $\frac{1}{8}$ inch thick, which forms the fairwater.

The top of the conning tower is domed, the hatch in it being fitted with four clips worked simultaneously by a partial turn of a central hand-wheel. In the floor of the conning tower a second hatch is invariably fitted, also capable of being closed by hand very rapidly.

A light bridge is built over the conning tower, forming a surface steering and look-out position. In recent submarines, when on the surface, with tanks fully blown, the bridge is about 8–10 feet above water level.

The modern conning tower is roughly oval in section, but somewhat pointed forward, the length fore and aft being about $10\frac{1}{2}$ feet and the maximum breadth about $6\frac{1}{2}$ feet. In older submarines the conning tower is circular in section. The external appearance of the conning tower, surrounded as it is by the fairwater, varies greatly in different classes. Types of conning towers are shown on Plates 4 and 6.

Four or six circular or oval scuttles, provided with internal hinged deadlights, are usually fitted in the sides of the conning tower. Embrasures are cut in the fairwater to allow for looking nearly ahead or astern through the scuttles. Port and starboard lights are fitted in the fairwater.

The conning tower contains depth gauge, clinometer, engine telegraphs, torpedo-firing pistols, gyro receiver and steering wheel. Communication from the conning tower is by voice-pipe.

There are watertight wells fitted from the floor of the conning tower to the keel of the boat, into which the conning tower periscopes house. The wells have no opening inside the pressure hull; consequently, should the conning tower be damaged, no water can find its way in through this channel.

Deck and Superstructure.

In all "U." boats, with the exception of *U. 43–50*, *63–65*, *87–92*, *93–98*, and *105--114*, the upper deck is formed by—

(1) A superstructure built over the pressure hull, about one-third of the width of the vessel, merging into the outer hull forward and extending to the rudder post aft;

(2) A deck at a lower level on either side, formed by the tops of the external tanks; *see* Sketch below.

In this type of boat the frames of the pressure hull are fitted internally.

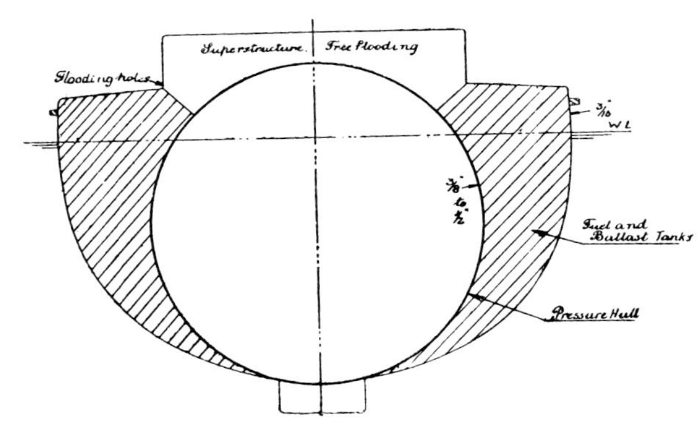

The superstructure is open to the sea through a number of flooding holes at the level of the deck on either side, and contains fuel and air-supply deck connections, engine muffler boxes, sounding machine drum, and various safety arrangements, such as telephone-and-light buoy, lifting shackles and divers' connections. The space above the pressure hull enclosed by the superstructure is, in addition, filled for the greater part of the boat's length with various kinds of stores not affected by sea-water or pressure, including rubber hoses and cordage, which may considerably reduce the effect produced by a bursting shell. Light frames, angles, and supports, fitted in the superstructure and tanks, will also tend to localise the damage from shell fire. The superstructure consequently affords considerable protection.

Some of the early submarines (*e.g.*, "*U. 19*") had a small steel boat stowed in fore part of superstructure. They also had their gun fitted to house in the super-structure, but this practice has now been abandoned.

In recent submarines of this type, with tanks fully blown, the top of the superstructure is 4 to 5 feet above the waterline. The tops of the tanks are then 1 to 2 feet above the waterline. Circular manholes are fitted in tops of the tanks.

In *U. 43–50, 63–65, 87–92, 93–98,* and *105–114,* the superstructure is rounded over to meet the outer hull, and the tops of the tanks are not visible; *see* Sketch below. In these boats the frames of the pressure hull are fitted externally. The superstructure is open to the sea through a number of flooding holes on either side, just below the rubbing strake.

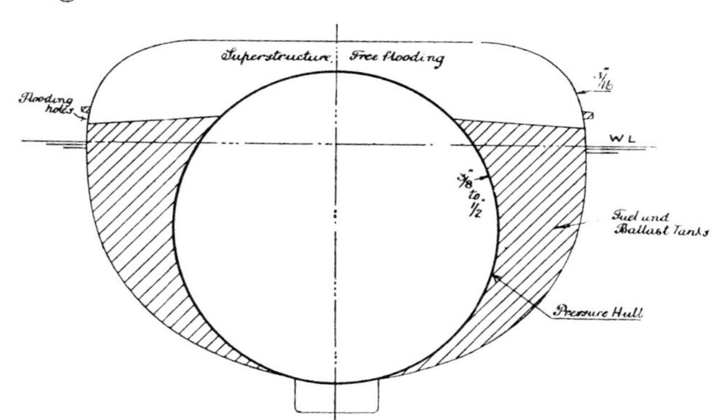

The height of the superstructure, with tanks fully blown, is about the same as in other "U." boats (*see* above), but the tops of the tanks are never more than about 1 foot above the waterline. When the reserve fuel tanks are filled with oil fuel, *i.e.*, nearly always on the *outward* voyage (*see under* Fuel Tanks, page 46), the tops of the tanks are below the surface of the water.

In *U. 43–50* class (and possibly in *U. 87–92*) the superstructure contains a reserve oil-fuel tank, and also a large air trap, in addition to the fittings mentioned above. The reserve fuel tank is situated immediately abaft the conning tower. The air trap is fitted in the bow superstructure, extending right across the boat. A water-tight manhole, kept closed, is fitted in the top of it, and there are the usual flooding holes on either side of the boat. An air cushion is thus formed when the boat is on the surface, giving her good buoyancy forward. On diving, a vent valve is opened from the fore torpedo room, allowing the trap to be flooded.

Originally, in *U. 43–50* the whole superstructure was divided up into watertight tanks; these were used as additional ballast tanks and served to give the boat a large reserve of buoyancy, which might have proved useful in the event of damage to the main ballast tanks. In practice, however, it was found that two systems of tanks caused delay in diving and unduly complicated the process. The whole of the superstructure was accordingly made free-flooding, with the exception of the reserve oil-fuel tank and the air trap already mentioned, which thus are really survivals of the former arrangement.

Demolition.

For demolition purposes submarines are supplied with cylindrical explosive charges about 4½ inches in diameter and 7½ inches in length. These charges are fitted with a time fuse, about 7 feet long, which burns for from 5 to 10 minutes, being ignited by means of a wooden firing pistol. The latter consists of a knobbed plunger held by a safety pin; the fuse is lit by forcing the plunger home on to a small igniting charge. About 50 charges are usually carried.

When used to destroy steamers, &c., the charges are placed outboard, lashed together in groups of two or three, abreast engine-room and cargo holds. They are lowered down over the ship's side till well below the waterline, and then fired by a single fuse.

These explosive charges are used for scuttling the submarine or for destroying important documents and machinery in the event of impending capture. In this case charges are usually attached to one of the torpedoes in the bow compartment and to one of those in the stern compartment, the confidential books being previously placed in the vicinity. The detonation of one torpedo ensures that of all the others stowed in the compartment. As a rule, charges are also placed amidships in the control room or wireless cabinet to wreck important machinery, sound-signalling apparatus, &c.

F 2

Depth when Submerged.

In most modern "U." boats, periscope draught, reckoned from top prism to keel, is 45 feet.

In bad weather or fog, boats usually proceed at depths between 66 and 131 feet, reckoned to keel.

All boats are tested for a depth of 197 feet, but will dive to 295 feet in case of serious emergency. For lying on the bottom they do not ordinarily choose localities in which the depth exceeds 164 feet.

Diving Capabilities.

In areas in which patrol craft are frequently met with, a system often followed in "U." boats is to run one Diesel engine for both propulsion and charging (fuel and air supply adjusted for a higher number of revolutions than are actually required), whilst the opposite main motor is run at such a speed as exactly to expend the current which is being generated; the number of revolutions maintained on each side being adjusted until the needle of the ammeter stands at zero. They are thus ready for instant diving, and it is not necessary first to start the motors, as described under Submersion (page 55), and in this manner the electricity expended is constantly replaced.

As a rule the ballast tanks are not flooded before the order to dive is given. It has been found in practice that if they are flooded, the boat is sluggish in answering her helm, and in a seaway she may become extremely unmanageable and make an involuntary dive. These disadvantages are considered to more than outweigh the additional protection against gunfire which would be obtained by bringing the pressure hull below the surface of the water. Some commanding officers probably do resort to the practice of flooding their tanks beforehand, but the balance of opinion in the German submarine service appears to be against this. With modern Kingstons and venting arrangements, no appreciable time is saved by flooding the tanks in advance.

In most of the boats it is now the custom to keep the Kingstons permanently open and to hold the boat up on the vents. When the order to dive is given, all that has to be done, so far as the tanks are concerned, is to open four master vent valves. Only half the crew need therefore be turned out for diving.

In the classes *U. 43–50* and *U. 87–92*, both built at Danzig, this cannot be done, because if any appreciable amount of water enters the tanks on either side due to the vents leaking, the boats immediately begin to roll violently.

In an efficient "U." boat, proceeding with both oil engines, on the order to dive being given, periscope draught (*see* above) can be reached in one minute thirty seconds. If proceeding with one oil engine and one electric motor, as described above, the time required is appreciably less. If the submarine is going deeper, the further dive will be at the rate of about a foot a second.

Electric Lighting and Power.

See under Batteries (page 40), Motors (page 51), and Machinery (page 50).

Emersion.

German submarines come to the surface by two methods :—

(1) By means of the diving rudders and hydroplanes.
(2) By blowing certain tanks with compressed air, first from the high-pressure, then from the low-pressure air service.

As a rule, both methods are employed simultaneously.

Directly the conning tower is out of the water, the high-pressure air is shut off, and the low-pressure turbo-blower is started and the remaining tanks blown with this, the compressed air in high-pressure air service and bottles thus being saved.

Endurance on Surface.

For approximate figures for the various classes, *see* pages 5–13. These figures represent ordinary cruising with both engines. They can be exceeded by using one engine only.

The usual duration of a long cruise is about four to five weeks, and it is not considered that this period can be greatly exceeded by a "U." boat working round the British Isles, as it is necessary to keep the larger part of the crew almost continually below. When in the "danger area" the men are never allowed to go on the upper deck except on duty. They are allowed to take short spells in turn on the conning tower if conditions permit, but it is not unusual for an engine-room rating to remain continuously below from the time the boat leaves her home port until she returns. All hatches, except the conning tower hatch, are usually kept closed.

Endurance Submerged.

For figures in the case of *U. 57–62* class, *see* under Batteries, Table No. 2, page 40.

The following have been given as the extreme periods for which modern "U." boats can remain submerged in case of absolute necessity:—

(*a*) If stationary, 5 to 6 days (assuming that the crew are not doing any hard work).

(*b*) At dead slow speed, *i.e.*, about 2½ knots, 42 hours (batteries then being exhausted).

(*c*) At slow speed, *i.e.*, about 4 knots, 23 hours (batteries then being exhausted).

(*d*) At utmost speed, *i.e.*, about 9 knots, 2 hours (batteries then being exhausted).

In practice, however, no submarine will voluntarily run her batteries right down, as this injures the accumulators.

Experience shows that it is not at all customary for boats to remain submerged for long periods when on active service, except during very bad weather; about 12 hours is the usual maximum.

Engines (Main).

Both 2-cycle and 4-cycle engines are fitted in "U." boats. The 4-cycle engines were originally all made at Augsburg, the 2-cycle engines by Krupp (Germania type), and later at Nürnberg also. The Nürnberg engines had eight cylinders; all the others have six. Now, however, various other firms are turning out both 4-cycle and 2-cycle engines of the same standard types.

The Germania 2-cycle engines have proved very successful and give higher speeds than 4-cycle engines. *U. 57–62* are the only boats fitted with the Nürnberg 2-cycle engines, which apparently have not been very successful.

Engineers often prefer the 4-cycle engine as being quieter and more reliable, but more 2-cycle engines have been fitted.

The following is believed to be the distribution of engines:—

4-cycle.	2-cycle.
U. 19–23.	*U. 24–27.*
U. 28–30.	*U. 31–41.*
U. 43–56.	*U. 57–86.*
U. 87–92.	*U. 93–104.*
U. 105–114.	(57 boats in all.)
(38 boats in all.)	

The 4-cycle cylinders are made of steel, the 2-cycle of bronze. Bronze is found to be more durable, but the supply of material is limited. Steel is not suitable for 2-cycle engines.

Air injection is used in all cases. Up to recently, in 2-cycle engines the pistons have been cooled with fresh water, but oil-cooling has now been introduced for them, as in the case of 4-cycle engines, since it was found that in the event of any small leakage, trouble was immediately caused by water mixing with the lubricating oil.

"*U. 81*" and "*U. 48*" may serve as examples of the two types. The revolutions and corresponding speeds of these boats are approximately as follows:—

——	"U. 81" (2-cycle).		"U. 48" (4-cycle).	
	Revs.	Knots.	Revs.	Knots.
Utmost speed	425	16·5	450	14·5
Full speed	360	14·0	400	13·0
Three-quarter speed	} 320	12·0	{ 350	11·3
Half speed			280	9·0
Slow speed	210	8·0	200	7·2
Dead slow	180	7·0	150	4·8
Slowest possible	—	—	80	2·5

Exhaust.

The conspicuous funnels fitted in early boats have all been removed and replaced by hinged funnels, which can be used when lying down on the deck, the exhaust passing into two fixed pipes which branch away to either side of the boat. Sometimes these funnels are tilted up at an angle of about 30° to the deck.

In all modern boats the exhaust is led out abreast of the engine room, on either side, just above the water line, and is practically invisible. A muffler box is fitted on the exhaust pipe in the superstructure, almost directly over each engine.

Upright funnels may, however, sometimes be used as a means of disguise or concealment, dense smoke being then purposely emitted.

Fresh Water.

The fresh-water supply for drinking purposes is stated to be worked out on a basis of half a gallon per man per day.

In all but the most recent boats the fresh-water supply has proved very restricted in view of the augmented complements carried and long cruises undertaken. Washing is consequently reduced to a minimum or entirely dispensed with. The fresh-water tanks are usually situated under the conning tower, "ready-use" drinking tanks being placed in living spaces.

U. 43–50 carry 7 tons, *U. 57–62* carry 6¾ tons, *U. 105–114* are reported to carry 12 tons. *U. 57–62* carry, in addition, about 1 ton of distilled water for topping up batteries and for use in H.P. air compressors. Topping up the batteries, however, is generally done in harbour.

No distilling plant of any kind is carried.

Fuel.

From "*U. 19*" onwards all German submarines use, when possible, heavy oil of a specific gravity of 0·87. They are, however, not dependent on this, but can use any oil which is not too viscuous and which has a flash-point over 100° Fahr.

The oil fuel usually employed is Galician oil, which is passed through a refining process, the paraffin wax and asphaltum or creosote being removed. In the absence of Galician oil, a first distillate of tar oil has proved almost as satisfactory for the 4-cycle engines, though the heating value is lower; with 2-cycle engines it is not so satisfactory.

In *U. 43–50*, at 14·5 knots (maximum speed), about 100 gallons of oil fuel are consumed per hour; at 11 knots the consumption drops to about 58 gallons; at 9 knots (cruising speed) to about 37 gallons. In *U. 66–70* the consumption at 12 knots (cruising speed) is about 60 gallons an hour; at 8 knots it is about 20 gallons an hour. In *U. 105–114* the consumption at 16 knots (maximum speed) is about 110 gallons per hour.

Fuel Tanks.

A number of the external tanks between the outer and inner hulls are appropriated for oil fuel. These tanks are compensated automatically, all oil used being replaced by circulating water from the main engines, which is led into the oil fuel tanks through an open funnel fitted against the outside of the conning tower (*see* Plate 20). Consequently, when the boat dives, the pressure inside and outside the oil fuel tanks is levelled off through this open compensating water pipe, which branches to the bottom of each external oil fuel tank. Arrangements are made so that only about nine-tenths of the fuel can be withdrawn from each tank, in order to prevent water finding its way to the engines. These tanks are not fitted with Kingstons

To provide against surging, in some boats the fuel tanks are fitted with half bulkheads, whilst in others each of the tanks has two or three full-size bulkheads with syphons. With the latter arrangement the oil is less likely to mix with the water and there is less danger of losing oil when the boat is out of the horizontal.

Four fuel tanks are usually fitted on either side of the boat, two forward and two aft, but in some classes (*e.g.*, *U. 66–70*) there are only two fuel tanks on either side. In addition, in all classes, two or four of the external ballast tanks are always used for carrying extra oil fuel when the submarine is proceeding on a long-distance cruise (*e.g.*, from a North Sea port into the Atlantic). They are specially fitted for the purpose and have the automatic compensating arrangements.

The oil fuel in these reserve tanks is always expended first, *i.e.*, on the outward voyage, the port foremost and starboard after tanks being emptied together, and *vice versâ*. After this the regular oil fuel tanks are used in pairs; they are usually of smaller dimensions than the ballast tanks, consequently opposite tanks can be emptied simultaneously without much effect on the trim.

The use of warm water for compensation of fuel expended prevents coagulation of oil, which would occur if cold water were used.

In *U. 43–50*, and possibly in *U. 87–92* also, an additional fuel tank is fitted in the superstructure, immediately abaft the conning tower.

The capacity of the main fuel tanks in *U. 81–86* is as follows :—

Starboard.		Port.	
No. 1 tank	8·5 tons.	No. 1 tank	8·5 tons.
No. 2 tank	12·8 "	No. 2 tank	12·8 "
No. 3 tank	10·9 "	No. 3 tank	10·7 "
No. 4 tank	7·5 "	No. 4 tank	7·6 "
	39·7 "		39·8 "

In addition, 42·5 tons* can be stowed in two of the external ballast tanks, making a total of 122 tons* (excluding ready-use tanks).

In *U. 57–62* the normal oil fuel stowage is 73·2 tons,* the total capacity, including reserve tanks, 123 tons*; for details *see* Plate 13.

Funnels.

See under Exhaust, p. 46.

Guns, &c.

Modern "U." boats carry one of four armaments :—

(*a*) One 4·1-inch and one 22-pr. gun
(*b*) Two 4·1-inch guns
(*c*) Two 22-pr. guns
(*d*) One 4·1 inch gun
} and one machine gun (sometimes two).

(*a*) is the most usual armament in modern boats, but frequently a new boat makes her first cruise or two with a lesser armament, the construction of boats being ahead of the construction and supply of guns.

The 4·1-inch is believed to be an L/45 gun. Two patterns are supplied, one on a low destroyer mounting with a maximum range of 10,000 yards, the other on a higher mounting, specially designed for submarines and ranging up to 13,500 yards. The latter mounting gives a maximum elevation of about 45°, and to that extent only can the gun be used for anti-aircraft purposes. The guns are fitted with telescopic sights (*see* below) for both gunlayer and trainer, but in fair weather they are usually laid and trained by the gunlayer single-handed, by means of a breast-piece.

The modern 22-pr. has a maximum range of 12,400 yards, but some submarines have guns of older pattern.

Both Krupp and Ehrhardt 4·1-inch and 22-pr. guns and mountings are fitted.

Percussion firing only is used.

In some of the early boats the gun housed in the superstructure; in all modern boats the gun or guns (except machine guns) remain in position when the submarine dives. All the guns are fitted with an expanding muzzle tampion and with a special breech tampion, shaped like the base of a cartridge case, which has a leather seating and is backed by a Belleville spring washer; this breech tampion is held in place by the breech block.

Each gun is mounted on a slightly raised platform, which extends beyond the superstructure on either side and is provided with guard rails.

The guns are not fitted with shields.

The machine gun can usually be mounted either on the after end of the fairwater or on the bridge, the bridge mounting serving also for the portable searchlight; as a rule, the after periscope pedestal is utilised for the bridge mounting. The machine gun is used for sinking mines or engaging aircraft, but some commanding officers have returned it to store as a useless encumbrance.

Sights.

The 4·1-inch guns are fitted with two telescopic prismatic sights, manufactured by Zeiss—one direct, the other angled at about 80°. Both telescopes are fitted with

* Tons of oil fuel, specific gravity ·87.

Part III. Section 5.

Sub-marines.

cross wires, which can be illuminated for night work. In the eyepiece, light filters are fitted, of varying density of colour.

These sights are pressure proof and can be left shipped if the boat has to dive hastily, but they are ordinarily kept stowed in the control room.

All German guns are also fitted with open sights of the "H" and "Barleycorn" patterns, for use in emergency.

Ammunition Supply.

Boats mounting two 22-prs. can carry about 750 rounds; those with one 4·1-inch gun, about 400 rounds; whilst those with the more usual mixed calibre armament carry about 210 rounds of 4·1-inch and 170 rounds of 22-pr.

The position of the magazines varies in different classes. In *U. 43–50* ammunition is stowed below the crew space immediately abaft the control room, in the warrant officers' mess forward, and in the wing passage on the port side leading from the control room to the fore torpedo room. In *U. 66–70* it is stowed in two magazines, under the forward and after torpedo rooms respectively.

A ready supply of ammunition is kept on deck, being stowed in lockers fitted near the gun in the superstructure or in one or both ends of the fairwater. In boats carrying mixed-calibre armament, this consists of 12 rounds of 4·1-inch and eight of 22-pr. Each round is stowed nose downwards in a cylindrical pressure-proof holder with a watertight bung, these holders fitting tightly into recesses in the lockers. As each round is taken from its holder, the bung is left open to allow it to fill with water, thus practically compensating for the loss of weight when the boat dives.

The ammunition is got up by whip (or passed up by hand) through the ordinary hatchways or, in bad weather, through the conning tower. In some boats the ammunition is transported from the conning tower to the gun by means of a traveller running on the jumping wires.

Hatches.

Access to the inner hull is usually given through:—

1.—(*a*) Circular hatch in top of conning tower.
(*b*) Inner hatch on the pressure hull at base of conning tower, fitted to enable the boat to maintain her buoyancy should the conning tower be holed or shot away.

2. A circular hatch trunked up to the level of the superstructure deck before the conning tower.

3. A similar hatch abaft the conning tower, sometimes trunked up through afterpart of fairwater.

4 and 5.—Torpedo hatches fitted in superstructure over foremost and after torpedo rooms. These also form access hatches for the crew.

All hatches are domed, and are secured by four clips, worked simultaneously by a partial turn of a hand wheel fitted in the centre. No automatic closing device is fitted.

Horse-Power.

	Maximum Speed.	Corresponding brake H.P. (2 engines).
"U. 81" - - -	16·5 knots.	2,400
"U. 48" - - -	14·5 „	2,000

The brake H.P. of "U." boats has increased from about 1,700 in the case of *U. 19–26*, which were completed shortly before the war, to about 2,400 (1,200 each engine) in the most recent boats, *U. 81–114*.

Hull (outer).

The plating of the outer hull is about $\frac{3}{16}$ inch thick. Its form is chosen solely with regard to seagoing qualities and speed on surface. The space between the inner and outer hulls is subdivided by transverse bulkheads to form external ballast, regulator, oil-fuel and lubricating oil tanks.

The outer hull must, therefore, be regarded primarily as a receptacle for water-ballast and oil fuel. If it is holed, this will merely entail, at most, a loss of oil fuel unless the pressure hull also is penetrated. It supplies a great factor of safety against collision or gunfire, particularly in view of the cushion of water or oil between the two hulls.

The outer hull, as a whole, is not required to resist pressure, as, when the boat is submerged, the Kingston valves at the bottom of the external ballast tanks are open, and thus equilibrium is maintained between the pressure inside and outside the tanks. In the case of the oil-fuel tanks a similar equilibrium is ensured by the automatic compensating arrangements described on page 46.

The regulator tanks and external lubricating oil tanks, however, are pressure-proof. (For tests, *see* Plate 13.)

A strong plate keel is fitted, usually filled with lead pigs. In the earlier boats, parts of this keel (generally about 12 tons) were detachable, forming safety weights; but this system has now been abandoned.

In the majority of "U." boats, bilge keels, about 2 feet in depth, are fitted low down in way of the regulator tanks, *i.e.*, abreast the conning tower.

Hull (pressure).

Is of circular section, cylindrical in form amidships, tapering towards the extremities, and with blunt ends. In most modern "U." boats it is 164 feet in length and 13 feet 8 inches in diameter. *U. 57–62* form an exception, the pressure hull being enlarged both longitudinally and transversely in way of the Diesel engine room (*see* Plate 13).

All the machinery and complete accommodation for the crew is contained within the pressure hull. It is tested for a depth of 197 feet, but cases are known of boats having dived to 295 feet without injury.

The plating of the pressure hull is from $\frac{3}{8}$ inch to $\frac{1}{2}$ inch thick. The frames are of bulb angle bar, rather heavier when they are externally fitted than when they are internal.

In the classes *U. 43–50*, *63–65*, *87–92*, *93–98*, and *105–114*, in which the superstructure is rounded off on either side to meet the outer hull and the tops of the tanks are completely hidden, the frames of the pressure hull are fitted externally; in all other classes, where the tops of the tanks are visible and form a deck at a lower level on either side of the superstructure, they are fitted internally.

Hydrophones.

See Sound-Signalling Apparatus, page 54.

Hydroplanes.

See Steering and Hydroplane Gear, page 55.

Internal Arrangements.

The usual system of subdivision is as follows, starting from forward:—

(1) *Fore Torpedo Room*, with the inner lengths of the bow tubes projecting through fore end of pressure hull. It contains spare torpedoes and torpedo compensating tanks. The after part of this compartment usually forms living quarters for the torpedo-tubes' crews, &c. On the after side it is closed by a pressure bulkhead.

(2) *Commanding Officer's Cabin*, separated in recent submarines by a watertight bulkhead from

(3) *Living Quarters* for officers, and sometimes also for men, the latter divided off by a light bulkhead. This compartment extends aft as far as the control-room, where another pressure bulkhead is fitted.

(4) *Control Room*, situated immediately below the conning tower, with a pressure bulkhead both on the fore side and on the after side. It contains ballast and trimming pumps, turbo-blower, periscope motors, diving and steering wheels, compasses, depth gauge, clinometer, vent and blow valves of the ballast tanks, &c.

Preserved provisions are kept in lockers in the control-room, and fresh-water tanks are situated below it.

(5) *Living Quarters* for petty officers and men, separated by a watertight bulkhead from

(6) The *Engine Room*, containing the Diesel engines, separated by a watertight bulkhead from

(7) The *Motor Room*, at the after end of which a pressure bulkhead is fitted.

(8) The *After Torpedo Room*, with the inner lengths of the stern tubes projecting through end of pressure hull. It contains spare torpedoes, torpedo compensating tanks, and berths for petty officers and men (occasionally for warrant officers).

Watertight doors are fitted in all bulkheads (*see under* Bulkheads, page 41.)

"Alarm" bells are fitted in all compartments.

In *U. 43–50* and *U. 87–92* wing passages are fitted inside the pressure hull: on the port side leading from the control-room past the officers' quarters to the fore torpedo room, and on the starboard side leading from the control-room past the men's quarters to the engine room. These passages, which are about 5 feet high and 18 inches wide, are closed at either end by a watertight door.

In *U. 66–70* the subdivision is even more elaborate than that described above, and a species of air lock formed by two bulkheads close together is fitted in the compartment before the control-room. For details, *see* Plate 12.

Jumping Wires.

Two jumping wires (*Minenabweiser*), about 1½ inches, are fitted to carry mines, hawsers, or nets clear of the conning tower. They run the whole length of the boat, extending from the bow to stanchions on either side of the bridge and thence to the stern. They are insulated for a portion of their length on either side before and abaft the conning tower, to enable them to be used as an auxiliary aerial when it is impossible or inadvisable to raise the W/T masts (*see* Wireless Telegraphy, page 60). In recent boats, two or three iron spreaders are fitted between these two wires at intervals, both before and abaft the conning tower. Between and over the spreaders three thin wires are stretched, evidently to increase the capacity of this auxiliary aerial.

Kingston Valves.

The external ballast tanks are fitted with "flooding flaps." These are square or rectangular in shape, and consist of flaps with a leather seating and fitted with a hinge. They open inboard through an arc of 90°.

In places where it is impracticable to fit a flap valve, and in regulator tanks and all internal tanks, a conical-shaped valve is fitted, which is also leather-seated and inboard opening.

These valves and flaps can all be operated either from the control room or at the valve itself.

Lubricating Oil.

Forced lubrication is employed, a special mineral oil being used, specific gravity ·91.

U. 43–50 carry about 3,600 gallons of lubricating oil; *U. 57–62* carry 3,323 gallons.

The main *Lubricating Oil Tanks* are generally built-up welded pressure tanks, fitted one on either side nearly abreast the conning tower between the pressure and the outer hull.

Machinery.

For main engines, *see* "Engines," page 45; for main motors, *see* "Motors," page 51.

Auxiliary Machinery.

Practically all electrical.

The following auxiliary machinery is fitted in *U. 43–50*:—

(1) In fore torpedo compartment:—
 1 mast-raising motor.
 1 capstan motor.

(2) In control room:—
 1 turbo-blower for blowing ballast tanks.
 1 ballast pump for pumping out any compartment, including regulator tanks.
 1 trimming pump, also available as auxiliary ballast pump.
 2 periscope motors.
 1 steering motor for vertical rudder.
 1 motor for foremost hydroplanes.

1 motor for diving rudder.
2 transformers for order transmitters.
1 transformer for gyro compass.
2 fans for ventilation purposes, and for removing the gases generated by the battery.
Compressed air installation for W.C.

(3) In petty officers' mess:—
1 hand pump for pumping drinking and washing water from the various fresh-water tanks.

(4) In main engine room:—
2 circulating water pumps } driven off main engines.
2 lubricating oil pumps }
2 oil fuel pumps }
1 pump for taking in lubricating oil.
1 oil fuel pump for pumping from the fuel tanks to the ready-use tanks.
1 hand pump for pumping lubricating oil from the lubricating oil ready-use tanks.

(5) In main motor room:—
2 high-pressure air compressors (driven off main engines).*
2 circulating water pumps.
2 fans for main motors.
2 air coolers for main motors.
1 transformer for W/T.
1 transformer for sound-signalling.
1 hand pump for thrust block lubrication.
1 hand pump for lubrication of main motor bearings.
1 pump for taking in oil fuel (fitted to be coupled to the motor of the port circulating water pump).

(6) In after torpedo compartment:—
1 auxiliary ballast pump.
1 auxiliary bilge pump serving W.C.
1 sounding machine motor.
Hand gear for vertical rudder.

Manœuvring Powers.

"U." boats are extremely handy and very quick on the helm when on the surface.

Their turning circle *submerged* is about 440 yards; rate of turning about 4 minutes for 90°, or 3 minutes if already under helm.

Masts.

See under Wireless Telegraphy, page 60.

Mines.

As a general rule are carried only by submarines of the class *U. 71–80* (*see* page 62), but the torpedo tubes of many, if not all, "U." boats are fitted to allow of mines being ejected from them. Both Leon and moored mines can be used.

Motors (Main).

Are mostly supplied by the Allgemeine Elektrizitätsgesellschaft (A.E.G.) and by Siemens-Schuckert. Various types of winding are in use. No essential change in the type of motor has been made from "*U. 19*" onwards, but the switchboard has been much simplified and improved.

In *U. 57–62* the main motors are shunt-wound and have separately excited reversible fields, operated for direction "Ahead" or "Astern" by means of a reversing switch. Ordinary "chopper switches," introducing resistances, are fitted for starting and stopping. Speeds are regulated by means of a rheostat in the shunt winding.

In all cases two armatures are fitted in tandem on each shaft. In *U. 66–70* the two armatures are built separately and connected by a fixed coupling.

For details of "grouping," &c., *see under* Batteries, page 40.

* In newer boats only one of these is driven off main engine.

Part III. Section 5.

Sub-marines.

The following are approximately the number of revolutions and corresponding speeds made by the main motors of "*U. 81*" and "*U. 48*":—

—	*U. 81.*		*U. 48.*	
	Revolutions.	Knots. Submerged.	Revolutions.	Knots. Submerged.
Extra emergency speed	327	9·2	330	8·25
Utmost speed	297	8·65	300	7·7
Full speed	275	8·2	275	7·1
Three-quarter speed	240	7·3	240	6·25
Half-speed	180	5·6	180	4·8
Slow speed	125	4·0	150	4·0
Dead slow	90	3·0	90	2·5

When on the surface, the motors are available as generators for charging the batteries and are run as such by the main engines.

Navigation Apparatus.

There are two known types of apparatus lately fitted in "U." boats and designed for the purpose of detecting the presence of an electrical cable or cables laid along the bed of a channel which is swept through minefields or other areas dangerous to submarines. In both types of apparatus the principle is the same; they differ only in the method of application.

By means of suitable gear, sensitive to electrical vibration and mounted outside the boat, the vibrations produced by an alternating current passed through the submarine cable are transmitted to a telephone circuit and thence to the ear of the operator in the W/T room inside the boat. It is thus possible to locate and dive along the cable and hence to keep accurately in the channel.

In the first type of apparatus, which appears now to have been abandoned:—

Two lengths of wire are fitted horizontally along the sides of the boat, about 18 inches apart, the lower wire being flush with the water line.

In the second type of apparatus, fitted more recently:—

Two hinged rectangular frames, each carrying about 50 turns of insulated wire, are fitted in a casing, just before or abaft the conning tower inside the superstructure. The frames are made of angle bar, and, when in use and erected, one is in the longitudinal and one in the transverse plane. From one corner of each frame a twin lead, connected to each end of the coil and suitably insulated, is led through glands in the hull to a 3-valve amplifier in the W/T cabinet. The amplifier is connected to a telephone headpiece, as used with the sound signalling apparatus and W/T. The frames are hinged and stow flat, one above the other, when not in use.

Net Cutter.

A good many "U." boats now carry net-cutters, the upper one being a species of straight saw edge, about 10 feet in length, slanting upwards and aft from the stem-head and supported by iron stays, whilst the lower one, though of similar type, is curved and secured direct to the rounded portion of the stem under the boat. The saw edges are formed by 10 separate knives or teeth, about 2 inches wide and 1½ inches in depth, set in between two narrow iron plates.

Oil.

See Fuel (page 46) and Lubricating Oil (page 50).

Periscopes.

Three periscopes are fitted in all boats except *U. 43–50* and *87–92*, which have two only. These periscopes are fitted for use either from the conning tower or the control room, never from both.

When only two periscopes are fitted, the foremost one is used from the control room; it passes through the fore part of the conning tower fairwater and is fitted for use against aircraft. By the movement of a lever the top prism can be directed upwards, so as to enable the sky to be searched, instead of the horizon.

In boats fitted with three periscopes, the foremost one again is used from control room, passing through fore part of fairwater, but usually only the centre periscope, *i.e.*, the foremost of the two in conning tower, is fitted for use against aircraft. In *U. 105–114*, both the centre and the foremost periscopes are so fitted.

The periscopes house into circular wells in the tanks under the control room, which extend to the keel. For conning tower periscopes this well is trunked up to the floor of the conning tower.

In *U. 57–62, 66–70, 105–114*, and possibly other classes, in the case of the principal periscope (the after one), a platform, on which the observer stands, ascends and descends with the periscope inside the trunk passing through the control room. *U. 43–50* were originally so fitted, but the arrangement was discarded. In these boats the trunk is now subdivided to form fresh-water tanks.

A small motor is fitted for raising each periscope, and alternative hand-gear is provided. Hauldowns are always fitted to periscopes as a safety arrangement in case the gland is too tight or freezes over. The periscopes are revolved by hand.

All "U" boats are supplied with an instrument known as the Goertz range estimator, which can be attached to the periscope in place of the ordinary eye-piece when making a torpedo attack. The use of the instrument depends upon knowledge of the masthead height or some other dimension of the target, two glass prisms being separated by means of a thumb-screw until the required cut is obtained (as with a sextant). The range can then be read off directly from a scale of curves fitted on the casing of the instrument.

The periscope tube is made in two parts, the upper portion being usually of brass and about 4 feet long and of from 2 to 3 inches external diameter; whilst the lower portion is usually of nickel steel and about 20 feet in length, with an external diameter of 5·9 inches. The periscopes are believed all to be manufactured by Goerz or Zeiss.

The field of view is about 40°.

German submarines do not possess a periscope specially adapted for night work, but light filters are provided for the periscopes.

A flashing lamp is usually fitted on one of the periscopes for signalling purposes.

A dummy periscope has occasionally been towed some distance astern of a submerged submarine to mislead the vessel attacked.

Prize Crews.

Submarines making extended cruises at one time carried a prize crew of one or two officers and from four to seven men, in addition to complement. These officers and men were used to man captured vessels, which were then employed as decoys or for housing crews of ships sunk. With the introduction of "unrestricted" warfare their purpose has disappeared.

Propellers.

All modern boats have two propellers, three-bladed; diameter, about 4 feet 3 inches.

Horizontal perforated steel plates are fitted on either quarter, about 1½ feet above the waterline, as propeller and hydroplane guards.

Range-Finder.

Some later submarines carry a range-finder, about 3 feet in length, similar to Barr & Stroud type. When in use, it is mounted on a portable tripod stand on the bridge. In addition, there are range-finding attachments for the periscopes (*see* above).

Regulator Tanks.

See under Tanks, page 56.

Safety and Salvage Arrangements.

Are of a most detailed nature, and include a telephone-and-light buoy (*see* page 56), lifting hooks, and divers' connections on pressure hull, a large supply of air-purifying cartridges fitted for individual use, safety masks and jackets, &c. Electric torches are provided.

The lifebuoys carried are of horse-shoe shape, usually painted red and very conspicuous.

All "U." boats have two pairs, and some four pairs, of salvage buoys, which are fitted in the centre line forward and aft. The buoys of each pair are connected to each other by means of a wire hawser, which passes through a sheave on the pressure hull. Should the submarine sink, the buoys are released and float to the surface. The salvage ship "Vulcan," on her arrival, places herself over the buoys and attaches to one of them a species of large grapnel with which she is fitted. She then heaves in the other buoy, and the arrangement is such that the grapnel is drawn down into a recess in a strong saddle built on to the crown of the pressure hull of the submarine. This grapnel is fitted with three steel toes, compressed outwards by means of a strong spring, which afford the necessary grip.

A number of spherical cork buoys are carried by submarines for the purpose of assisting the crew to reach the surface from depths of about 200 feet or less. A buoy of this type is attached to the end of a rope and allowed to float to the surface, the other end of the rope being secured to the hatch from which the crew escape. The idea is to allow the crew to ascend to the surface slowly, holding on to the rope, thus avoiding serious injury from rising too rapidly.

Searchlight.

In all recent submarines a portable searchlight is fitted for mounting on the bridge. It is 11·8 inches in diameter and can be run either from a portable battery or from the main batteries.

It stows in the engine room.

Smoke Boxes.

Smoke boxes (*Nebel Unterwasser Bomben or N.U.B.*) are carried for producing a smoke screen. They are small metal cylinders, painted green, with a tin spiral spring and fuse on top. They contain a liquid which burns the flesh or clothes if it comes into contact with them.

When adjusted and thrown overboard, they at once develop a white or greyish smoke. The smoke does not smell or taste and can be inhaled with impunity.

Sound-Signalling Apparatus.

Receiving Apparatus.

A hydrophone set, called by the Germans "U.G." (*Unterwasser-Geräusch-Empfänger*), is invariably fitted. It is supposed to pick up all sounds within a radius of about 10 nautical miles and to determine their approximate direction, character and cause. It is only reliable when used by operators having considerable experience of it, and seldom in practice at distances of over 5 miles. It is, however, reported that a new set having a greater range and reliability is now being introduced.

When listening, all noise must be stopped in the boat (*e.g.*, men talking or moving about, auxiliary machinery, &c.). For the *best* results, the boat must be lying on the bottom, with all machinery stopped, but it is possible to get fairly good results with the motors running at slow speeds.

The present standard installation consists of six microphones, a 50 to 60 milli-amperes battery, a switchboard, an adjustable resistance and a telephone receiver.

The six microphones are arranged in pairs, one pair being placed in one of the starboard midship ballast tanks, another pair in the corresponding port ballast tank, whilst the third pair is placed forward in the bow ballast tank. As this latter compartment extends right across the boat, the microphones in it cannot give directional results. Those in the midship ballast tanks, on the other hand, enable a rough direction to be obtained, the sound being most distinct when the vessel producing it is on the beam.

The two microphones of each pair are each placed inside a small tank filled with water, and the two small tanks are suspended one above the other in the midship ballast tanks, and abreast each other in the bow ballast tank. Of each pair of broadside microphones, one is tuned low and is used for the sounds of reciprocating engines and propellers; the other is of a higher pitch and detects the sounds produced by the running of turbines and electric motors.

The switchboard, battery and resistance are placed in the W/T room. The switchboard is so arranged that sounds can be received at will from either the port or the starboard side, or from the bow receivers, and from either microphone of each pair.

The adjustable resistance enables the strength of the current to be regulated and the sounds heard in the telephone to be damped or amplified as necessary.

In order to determine sounds at extreme ranges, the valve amplifier of the W/T apparatus can be inserted between the switchboard and telephone receiver. By this means the sound is greatly magnified; but absolute silence is essential if accurate results are to be obtained, as any sounds inside the boat will then likewise be magnified.

These hydrophones have proved very useful for locating British destroyers and other hunting craft. Their utility is much reduced if the hunting vessels do not approach singly, but in groups of not less than three.

As a rule the hydrophones are not used when the boat is on the surface.

Sending Apparatus.

Sending apparatus is fitted in most "U." boats, but not in all. No detailed description of it is available.

In modern boats, the signals are sent by means of a large diaphragm fitted externally, near the bow of the boat. On the inner side of the diaphragm there are two coils, through one of which a direct current is passed, whilst an alternating current is passed through the other. In general design the system appears to be very similar to that of the Fessenden transmitter. A musical note is produced.

In earlier boats a water syren may be fitted

Sounding Gear.

A sounding machine, worked by hand, is usually fitted in the fore torpedo compartment. A clockwork device, registering the depth, is fitted in the lead. The latter is led through a watertight tube which passes down through the pressure hull to the outer hull near the keel. The tube is closed, at top and bottom, by sluice valves. When the lead has been hove up, the lower sluice is closed and the tube drained, enabling the upper sluice to be opened, the lead removed, the depth read off, after which the registering device is reset to zero.

Speed.

Surface.—Maximum speed varies between about 14·5 knots in the case of *U. 43–50* and 17 knots in the case of *U. 66–70* and some of the most recent boats.

Cruising speed usually varies between 8 and 12 knots, according to the class of boat and the circumstances of the cruise.

Submerged.—Maximum speed lies between 8 and 9½ knots in all classes. Normal speed when attacking is about 4 knots.

See also under Engines (Main), page 45, and Motors (Main), page 51.

Steering and Hydroplane Gear.

In most modern boats a single vertical rudder of the balanced type is fitted. *U. 43–50,* and *87–92* (also *U. 71–80, see* page 61), are fitted in addition with a conspicuous upper vertical rudder, which increases the turning power when submerged, though it causes the boat to heel considerably when under helm. It is fitted on the same shaft as the lower rudder.

Hydroplanes are fitted both forward and aft near the extreme ends of the boat. The after hydroplanes, in modern boats, are always fitted in line with and immediately abaft the propellers. No housing gear is fitted.

Both vertical rudder and hydroplanes are worked by their own electric motors; alternative hand gear is fitted for use in emergency.

In addition to the steering positions on bridge and in conning tower and control room, a reserve hand steering position is usually fitted right aft. The hydroplanes are operated from the control room.

Good hydroplane guards are fitted, the after pair consisting of horizontal perforated steel plates which serve also as propeller guards.

Submersion.

See also Diving Capabilities, page 44.

The compensating and trimming tanks are always kept adjusted so that, for the boat to submerge, only the external ballast tanks have to be filled. At the same time the main motors are started ahead to force the boat under by means of the hydroplanes.

The procedure, on the alarm being rung, is at once to open the flooding flaps (or Kingstons) of the ballast tanks, and to change over from main engines to main motors. Indicators in the control room show the engineer when all the flooding flaps have been opened, and he then reports to conning tower.

G 4

On the order being given, an engine-room P.O. opens the four master vent valves worked from the control room. At the same time red lights are shown in the foremost and after compartments, in order that the men stationed there may help to open these valves should any difficulty be experienced.

As soon as the Officer of the Watch is relieved by the Commanding Officer in the conning tower, he goes into the control room and controls the depth steering.

To assist the dive, the trimming pump is started and water run from the after to the foremost trimming tank. Directly the boat's bow dips, the connection is reversed, the pump still being kept running, so that the water is pumped back again to the after trimming tank. Boats dive as a rule with a forward inclination of about 15°.

When a submarine dives quickly, a cloud of steam and black smoke may be observed, which is apt to convey a false impression that the submarine has been hit. The black smoke is caused by fuel oil being discharged unconsumed into the exhaust pipe, the heat in which, combined with the absence of air, causes partial combustion. The white portion of the cloud is steam, due to the hot exhaust pipe coming into contact with the sea water. In addition, the venting of the tanks during submerging might possibly be mistaken for the fall of a shell.

Superstructure.

See Deck and Superstructure, page 42.

Tanks.

See also under Ballast Tanks, Fuel Tanks, Lubricating Oil, Torpedo Tubes and Torpedoes.

(*a*) *Compensating Tanks.*

Are fitted inside the pressure hull below the control room and serve to compensate for the weight of stores consumed, sea water being admitted to them as necessary. As the fuel expended is automatically replaced in the fuel tanks by sea water (*see* page 46), compensation in this case is only necessary for the difference in specific gravity. The compensating tanks are also used to compensate for variations in the specific gravity of the sea-water.

(*b*) *Regulator Tanks.*

Usually fitted externally, abreast the conning tower on either side, and situated so that their centres of gravity are in the same transverse section as the centre of buoyancy of the boat when trimmed down. They can thus be flooded or blown without altering the fore and aft trim.

The ordinary ballast tanks are of such capacity that, when they are flooded, the boat with normal equipment will lie submerged approximately to base of conning tower. Regulator tanks are fitted in the first place to cancel the buoyancy of the conning tower, but they also serve to compensate for expended ammunition, provisions, and stores. Over and above the total maximum amount of compensation required, they should have a reserve of buoyancy of 6 to 7 tons each. They are always kept adjusted to compensate for buoyancy of conning tower and amount of stores, &c., expended, so as to be in constant readiness for diving.

They have also been used latterly as anti-rolling tanks; the connection between the two, which leads through the pressure hull, being opened. This procedure, it is stated, has proved effective in reducing rolling.

In *U 99–104* the regulator tanks, four in number, are fitted internally under the control room.

(*c*) *Trimming Tanks.*

Are situated near the extreme ends of the boat, usually inside the pressure hull, and are used for adjusting the longitudinal trim.

Telephone and Light Buoy.

German submarines were all formerly supplied with a telephone-and-light buoy, but apparently this fitting is now being abandoned as liable to betray the position of the submarine should it be accidentally released by the explosion of a depth charge.

Plate 22, Figs. 1 and 2, shows the type of buoy formerly carried by all "U." boats. The buoy, which is countersunk into the superstructure and held in place by a locking pin, is filled with compressed air. A steel spring fitted to its lower end is maintained in a compressed condition by the locking pin, which, when withdrawn, gives the buoy a propelling force to clear the boat and rise to the surface.

The telephone set consists of a transmitter and receiver, connected to about 6 or 8 feet of flexible 3-core cable, at the end of which is a heavy brass connector. This set is contained in a watertight recess in the top of the buoy. In the centre of the top of the buoy is a locking wheel, securing a pin which protects the contact pieces for the brass connector of the telephone. When the telephone is required for use, the locking wheel is revolved, the pin removed and the connector inserted and locked by the wheel.

A watertight electric light with a U link is fitted on top of the buoy, and on the side of the light fitting is an ebonite push-piece. Presumably this light can be operated as a flashing lamp from inside the submarine, and the push-piece enables answering signals to be sent in the event of a failure of the telephone. The U link ensures that the light shall be constantly on, when not in use for signalling.

Plate 22, Figs. 3 and 4, show a simpler form of buoy (recovered about October 1917). It is not fitted with a telephone attachment, but there is a plug fitting in the lamp circuit where a telegraph or telephone instrument could be plugged up.

Thrust Block.

In recent boats the old type of collar-and-shoe thrust block has been superseded by a ball thrust (*see* Plate 15). The details are roughly as follows :—

A sleeve with a heavy collar at each end is shrunk on to the shaft. Into the annular space between these collars are fitted two sets of ball races. Each set consists of two rings with the balls mounted between them.

The thrust casing is of stiff construction and made in two halves, being bolted together through a horizontal flanged joint, and has a "T"-shaped ring in halves let into it. This ring, which is rigidly secured to the thrust casing, is interposed between the two sets of ball races and takes the ahead and astern thrust transmitted through the shafting.

Torpedo Tubes and Torpedoes.

"U." boats are fitted with either four or two bow tubes and two stern tubes, except class *U. 66–70*, which have four bow and only one stern tube. (For mine-laying type, *U. 71–80*, *see* page 62.)

Submarines up to "*U. 39*" and the classes *U. 51–56*, *57–62*, *63–65*, and *99–104*, all have two bow and two stern tubes fitted abreast in each case.

U. 43–50, *81–86*, *87–92*, *93–98*, and *105–114* have four bow and two stern tubes: bow tubes fitted thus : o o / o o ; stern tubes fitted abreast.

No broadside tubes and no side-loading tubes are fitted.

Only 19·7-inch tubes are mounted in "U." boats (from *U. 19* onwards, with the exception of *U. 66–70*, which were originally designed for Austria-Hungary, and probably have 17·7-inch tubes). All boats, however, now carry a proportion of 17·7-inch torpedoes for use against merchant vessels, guide rails being inserted in the 19·7-inch tubes when these smaller torpedoes are used.

The number of torpedoes normally carried varies from 8 to 12, including those in the tubes. The spare torpedoes are divided between the two torpedo rooms (bow and stern compartments) and stowed on the deck or against the sides of the boat or overhead. Additional torpedoes can, however, be stowed in the living quarters, and a case is known of a "U." boat having carried 16 torpedoes in all.

All 19·7-inch torpedoes are fitted with gyroscopes capable of being angled in 15 steps up to 90° either way.

The charge of the 19·7-inch torpedo is between 360 and 440 lbs. of high explosive, varying with the type ; that of the 17·7-inch torpedo, between 300 and 350 lbs.

Part III. Section 5.

Submarines.

Types of Torpedoes.

The types of torpedoes carried in German submarines are as follows :—

Type.	Size and Mark of Torpedo.	Range and Speed.
Fresh-water heaters	19·7-inch G/7, G/7*, and G/7**	9,900-11,000 yards at 28½ knots ; 4,000–5,500 yards at 35 knots.
	17·7-inch Fiume G/125	6,000 yards at 24 knots. 2,200 " 36 "
Salt-water heaters	19·7-inch G/6 D	7,500 " 30 " 3,800 " 35 "
	17·7-inch C/06 D	6,500 " 26½ " 3,300 " 36 "
	17·7-inch C/03 D	5,000 " 26½ " 2,200 " 36 "
Dry heaters	19·7-inch G/6 A.V.	5,500 " 27 " 2,200 " 38 "
	17·7-inch C/06 A.V.	3,900 " 27 " 2,200 " 32 "
	17·7-inch C/03 A.V.	3,400 " 26½ " 1,400 " 36 "
Cold torpedoes	19·7-inch K. III.	1,300–1,600 yards at 32–29 knots.

"U." boats usually carry a majority of G/6 A.V. or K. III. torpedoes and only a few fresh-water heaters.

High-speed setting is almost always used.

Cold torpedoes are being manufactured in considerable quantities, and are popular on account of their simplicity. The much more visible track made by these weapons is, however, a disadvantage.

Torpedo net cutters of an explosive type can be used with all types except fresh-water heaters.

Details of Tubes.

The torpedo tubes mounted in the "U." boats are internal tubes, *i.e.*, tubes whose rear portion projects into the interior of the submarine through the pressure hull, to which they are rigidly bolted by flanges.

The tube mountings (viz., side stop, tripper, firing valves, &c.) are readily accessible for inspection and test.

The tubes are closed at their outboard end by hinged caps with rubber seatings, operated from the interior of the boat by shafting and worm or bevel gearing.

These torpedo tubes in operation do not differ very greatly from the upper deck tube of a destroyer, but they are fitted for air impulse only and carry more fittings.

The following features are worth noting :—

(*a*) In connection with the air impulse, a *swirl reducer* is fitted, which prevents unnecessary air escaping after the torpedo has been fired and thus tends to eliminate that violent "swirl" or "splash" which is generally observable on the surface of the water on a torpedo being fired from a submerged submarine.

The firing reservoirs must be exhausted between each shot.

(*b*) A *silencer* is fitted on the exhaust pipe of reservoirs to prevent noise when the above takes place.

(*c*) No "outboard vents" at all appear to be fitted on the torpedo tubes. The tube is flooded after replacing the firing gear by opening an outboard drain.

(*d*) The torpedo tube *firing gear* can be operated electrically or by hand. When it is electrically operated, by pressing the torpedo tube firing pistol key in control room or conning tower, a small cartridge filled with powder is discharged, and the force of the explosion pushes in a spring plunger, which allows the firing bar to revolve and the tube to fire.

(*e*) Torpedoes can be set for depth and long or short range shot either before loading or when in the tubes. Angling can be effected *only* when the torpedo is in the tube.

To set depth, a key is inserted through a cock on the shell of the tube, engaging on depth-setting spindle of torpedo, and is screwed right or left as many turns as are requisite to adjust.

To angle gyroscope, a spring-loaded spindle (which is made watertight) operates through the shell of the tube and engages in angling socket of torpedo. An external indicator is mounted on the torpedo tube to record the amount the gyroscope has been angled right or left in degrees.

Torpedo Compensating Tanks.

These are located below the torpedo rooms inside the pressure hull, and serve to compensate for torpedoes fired, and also to flood the torpedo tubes before the caps are opened for firing, so that the diving trim shall not be disturbed on this being done.

Trimming.

The operation of trimming is usually carried out directly after submerging. As a general rule, no submarine will pick up an exact trim after flooding her ballast tanks and diving, unless she has already dived and trimmed a short time before. This is due to slight errors in compensation for fuel or stores consumed, small leaks, varying densities of the sea water, &c. Trimming consists in adjusting the buoyancy and the inclination of the boat (fore and aft and athwartships), so that she is enabled to manœuvre and keep a good depth-line submerged at the lowest possible speed and using the smallest amount of helm on the hydroplanes. As a general rule these conditions will be obtained when the buoyancy of the boat is approximately zero.

The tanks used for adjusting the trim are the compensating tanks, situated near the centre of gravity of the boat, and the trimming tanks, situated near the extreme ends.

In German submarines, by means of a roturbo pump (the trimming pump), water can be pumped from the after trimming tank into the forward trimming tank, and *vice versâ*, till the desired inclination is obtained. Compressed air can be used in place of the trimming pump, if desired; for instance, when the submarine is hunted by hydrophone vessels and the noise of the pump might betray her.

On diving, water is usually pumped forward to assist in inclining the boat's head down, and, when the boat has dived, it is pumped aft again till the desired inclination is reached (*see* Submersion, page 55). As a general rule all submarines are trimmed horizontal when proceeding submerged, but in some cases a slight inclination of 1° to 3° bow down is used, when under helm, to check the tendency to rise above the desired depth-line and break surface.

Statical Diving.

A very well trimmed boat, when proceeding submerged, may be able to remain stopped for as much as 15 minutes at a time, but probably not for longer, except in the Baltic, Kattegat, and a few other localities where salt and fresh water meet in large volumes. The salt water here forms a layer under the fresh water, and submarines may be able to lie totally submerged on top of the salt-water layer, with all machinery stopped, a great saving in battery power being thus effected. In places where the fresh-water layer is only about 30 feet deep or less, a very efficient periscope watch can be kept in this manner, as, with the boat lying stopped, no "feather" or wash of any kind is made.

Turning Circle.

See Manœuvring Powers, page 51.

Ventilation.

(*a*) *When Submerged :—*

A great deal of attention is paid to air purification, which is very necessary in view of the large complements carried by German submarines. The system is to absorb the CO_2 by passing the air, by means of ventilating fans, through a number of purifiers containing fragments of caustic alkali.

See Plate 21 for typical arrangement of ventilation leads and air purifiers.

The air in the boat is driven through the purifiers every five or six hours when the boat is submerged.

In addition, cylinders of oxygen are carried in order to renew the supply of oxygen in the air. In *U. 43–50* ten such cylinders are carried, the capacity of each being about 5·1 cubic feet.

Apparatus for testing the condition of the air is also carried.

2

(*b*) *On Surface*:—

A fresh air trunk and an exhaust air trunk are led up inside conning tower fairwater. A vertical ventilation tube, communicating with the fresh air trunk, is sometimes fitted immediately before the conning tower. The main air supply, however, is through the conning tower hatch, which is usually opened immediately the boat comes to the surface. The Diesel engines suck air from the engine room and consequently from the conning tower and other openings for ventilation, thus adequately renewing the supply of air.

When charging battery the ventilating fans also suck air from the living spaces and discharge it overboard through an exhaust trunk together with the explosive gases drawn from the cells.

Vents.

All vents are hand-operated. Mechanical operating by compressed air was tried in 1908, but did not prove a success.

All tanks can be vented by means of four domes, each of which has a master vent valve operated from the control room or alternatively from the compartment next the dome. Sketch shows typical arrangement of vents and domes.

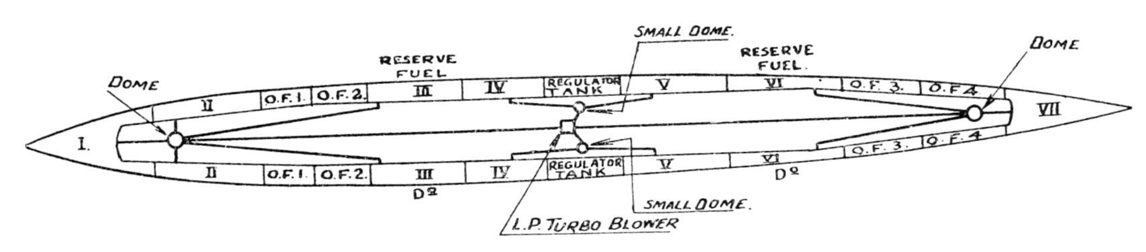

Inboard vents are fitted only to regulator tanks.

All vent piping is of large bore. The venting leads serve also as blowing leads for the L.P. turbo-blower.

In addition to the master vent valves, each tank has its own vent valve, situated in the vent pipe at the tank end between tank and dome.

In *U. 43–50* the master vent valves are 20 inches in diameter.

After diving, all vents are closed immediately if an attack by depth-charge is expected, whereas under normal conditions they are left open for about 10 minutes in order to ensure that no air cushions are formed.

Wireless Telegraphy.

Usually two masts, 30–40 feet high and 80–90 feet apart, are fitted to carry the main aerial; they are hinged at the heel and lie aft along the deck when not required. They are raised and lowered by wire purchases from inside the hull by hand or motor power. The aerial consists of two parts of bare stranded phosphor-bronze wire. Single feeders of similar wire lead down from the centre of the aerial.

When the mast aerial is to be used, these feeders are connected to the main feeder, which passes up through the deck insulator in fore part of fairwater. At other times the main feeder is kept connected to the jumping wires, which are fitted as an auxiliary aerial, as described on p. 50. Though frequently short-circuited by spray, the jumping wires offer the great advantage that the submarine is able to intercept wireless messages directly she comes to the surface, without putting up her masts.

Wireless communication is, however, practically impossible in heavy weather, as the masts cannot be got up for fear of their going over the side, and the seas breaking over the boat short-circuit the jumping wires.

The wireless room, as a rule, is usually just before the control room.

The wireless set usually fitted is a 1-kilowatt commercial quenched spark set of the Telefunken type, with perikon detector. It is only arranged to receive damped waves. Transmitting range, using mast aerial: by day 100 to 200 miles reliable, 300 to 700 miles extreme; by night, about double the distance. The use of the jumping wires in place of the mast aerial roughly halves the range.

The range of wave-length for transmitting is 980 to 2,690 feet; for receiving, 650 to 20,000 feet. German submarines generally transmit on a wave-length of 1,312 feet and receive on a longer wave-length. Three-valve amplifiers are usually carried.

It is reported that a few German submarines are now fitted experimentally with continuous-wave transmitter and receiver.

Submarines, as a rule, send as few wireless messages as possible in order not to betray their presence. According to prisoners, they generally only report important intelligence obtained in the course of their operations, and, when returning, their approximate time of arrival in home waters, together with any defects the boat may have developed, necessitating a premature return.

IV.—"U." MINELAYING TYPE.

U. 71–80.

GENERAL REMARKS.

These large minelaying submarines, sometimes known as "E" boats, form a distinct class. Their main distinguishing characteristics will, therefore, be described as far as information is available.

They have not proved a success, and the type is not being perpetuated

For minelaying methods, *see* under *U.C. 16–79* class, pp. 76 and 81.

RECOGNITION.

Shape.

See Silhouettes.

These boats have more freeboard than the ordinary "U." boats. The upper vertical rudder, which is fitted above the aftermost end of the superstructure, and the rounded bow are conspicuous marks by which they may be recognised; also the single gun, mounted *abaft* the conning tower.

Colour.

See "U." Boats, p. 38.

DETAILS.

Accommodation.

Officers and warrant officers are provided with bunks. There are, in addition, sufficient bunks for half the crew, and a few hammocks are carried.

Anchor Gear.

A stockless anchor is carried in a hawsepipe forward, and an electric capstan is fitted.

Compasses.

A magnetic diving compass is fitted on fore end of fairwater, and the reflector and screen below it in the control room. In addition, there is a gyro installation, consisting of a master compass in the mine room aft and repeaters in the control room and conning tower.

Complement.

Seven officers and about 25 petty officers and men.

Dimensions.

See tabulated details, p. 8.

Endurance).

Maximum surface endurance: about 6,300 miles at 6 knots.

The length of the cruises made by these boats varies very greatly, according to the locality in which they lay their mines, but the usual maximum duration of a cruise is four weeks. Owing to their low speed, their radius of action is small compared with that of ordinary "U." boats.

Engines.

Two-cycle Diesel engines, made by various firms, maximum brake H.P. about 900. Most of these engines, it is stated, give constant trouble.

Guns, &c.

These boats originally carried one 22-pr. gun, but in some of them this gun has been replaced by a 4·1-inch. The gun is mounted about 25 feet abaft the conning tower. In addition, a machine gun and some rifles are carried.

The magazine is forward under the crew space. The ammunition supply is by hand, through the conning tower. A ready supply of ammunition is usually kept on deck, in watertight holders, lashed near the gun.

Hatches.

There are three access hatches, one in dome of conning tower, one forward, above the warrant officer's mess, and one aft, leading to the mine room, abaft the gun. A special hatch is fitted aft to enable mines to be taken on board.

Hull.

The diameter of the pressure hull is 17 feet, *i.e.*, considerably greater than in the ordinary "U." boats. There is no outer hull, but saddle tanks are fitted. The ballast tanks are internal and the fuel tanks external.

Internal Arrangements.

The pressure hull is subdivided as follows:—

(1) *Bow compartment*, containing the men's living quarters right forward, and abaft these, separated by a curtain, the petty officers' mess. An ordinary watertight bulkhead separates this from

(2) The *warrant officers' mess*, a small compartment separated by a watertight bulkhead from

(3) The *officers' quarters*, separated by a pressure bulkhead from

(4) The *control room*, which again is separated by a pressure bulkhead from

(5) The *engine room*, separated by an ordinary watertight bulkhead from

(6) The *motor room*, abaft which, separated by another watertight bulkhead, is

(7) The *mine room*.

Watertight doors are fitted in all bulkheads.

Mines.

From 34 to 36 mines are stowed on board these submarines in the mine room, which takes up about two-thirds of the space between conning tower and stern, and extends right aft. The mines are stowed on their sides in three tiers on either side of the vessel, rails being fitted for each tier. At the fore end, next to the engine room, there is a small lift, operated electrically, by means of which the mines in the upper tiers are lowered on to the floor of the mine room. There they are placed on rails and transported to the two horizontal mine tubes fitted right aft, one on either side of the centre line. These tubes each take three mines at a time, and the mines are loaded into the tubes, sinker first. On each side of the tube a small hand-wheel is fitted, which, when turned, slowly ejects the mines at regular intervals by means of gearing. Safety arrangements are fitted in each tube to prevent the second mine dropping before the first mine is well clear, &c. As a rule, two men attend to each tube.

The advantage that this method of mine stowage possesses over that adopted in the "U.C." type of boat (*see* pp. 76 and 81) is that the mines are always accessible and protected from the action of the sea and weather; they can be set for any desired depth of water immediately before they are laid.

The mines are of the same type as those carried by "U.C. boats" (*see* p. 81).

Periscopes.

Two are fitted, one used from conning tower, the other from control room.

Speed.

Maximum, about 10 knots on surface and 7 knots submerged. Cruising speed, about 6 knots.

Torpedo Tubes and Torpedoes.

Only two tubes are fitted, both being 19·7-inch external tubes; a bow tube on the port and a stern tube on the starboard side. These tubes are probably identical with the external tubes fitted in U.C. 16–79 class (*see* p. 83). The torpedoes are fired from the conning tower or control room.

Only two torpedoes are carried, stowed in the tubes. The safety pins of the pistols are usually removed on leaving harbour, so that there is no need to touch the torpedoes again.

Wireless Telegraphy.

A silent cabinet is fitted on the port side of the engine room.

V.—EARLY "U." TYPE (used for Instructional Purposes).

U.A. and U. 1–18.

(*See* Plate 7.)

GENERAL REMARKS.

These boats are now all used for instructional purposes.

With the exception of "*U.A.*," which is a small modern submarine built for the Norwegian Navy but taken over by Germany on the outbreak of war, they are earlier editions of the regular ocean-going "U." boats (*see* p. 37).

The main point of difference is that these early "U." boats are fitted with Körting paraffin engines, which are greatly inferior to the Diesel engines of later classes, both as regards reliability and also as regards smoke and noise.

For details, *see* tabular statement (p. 12) ; for appearance, *see* Silhouettes.

VI.—"U.B." SMALL OCEAN-GOING AND COASTAL TYPE.

U.B. 48. — (?) 140.

(*See* Plate 8.)

GENERAL REMARKS.

These submarines, known as the B.B. or B² boats, are a development of the coastal classes *U.B. 1–17* and *18–47*. But, being provided with a double hull and a watertight hatch at base of conning tower, they are far less vulnerable to gunfire than these earlier classes. They must be regarded as small ocean-going submarines, several of them having proceeded independently to the Mediterranean *viâ* the Straits of Gibraltar. The exact number of boats of this series is uncertain, but there are certainly at least 75 of them, believed in all essentials to form one class and consequently the product of an advanced system of standardisation. As might be expected from this, they have been turned out by the different building yards in very rapid succession.

RECOGNITION.

Shape.

See Silhouettes.

The fairwater enclosing the conning tower, as a rule, has no step at either the forward or the after end. Round the fore part of the conning tower it is continued up to form a permanent bridge screen, giving the impression of a very high conning tower rising vertically from the deck on the fore side. This square, solid appearance of the conning tower is a noticeable characteristic (*see* Plate 8).* A ship bow with slight overhang also serves to distinguish this class of boat from *U.B. 18–47* class, in addition to the greater length.

The line of the superstructure is continuous fore and aft, with a slight sheer from the bow to the stern, the latter being very nearly awash in surface trim. A net cutter is generally mounted on the bow, stayed by heavy struts in an athwartship direction.

The gun is mounted immediately before the conning tower, the lower part of the fairwater, at its forward end, being recessed to allow of the recoil.

Colour.

As for "U." boats, *see* p. 38.

---

* In some boats of the class, however, the permanent bridge screen is fitted abaft the magnetic compass and steering pedestal at the fore end of the bridge, and in this case there is the appearance of a step at the forward end of the fairwater.

DETAILS.

Accommodation.

Twenty-six bunks were originally fitted, but five have since been removed to give more space for stowage of torpedoes in the fore torpedo room.

Air Service.

Ten bottles of equal size, holding about 13·4 cubic feet of compressed air each at 2,350 lbs. pressure per square inch, are carried: four forward and four aft in the superstructure, and two in the control room bilges.

Anchor Gear.

One stockless anchor is carried in a hawse pipe in the superstructure forward, stowing vertically. The cable is self-stowing.

Ballast Tanks.

There are nine ballast tanks fitted externally between pressure and outer hulls; the foremost tank extends right across the boat.

Battery.

Sixty-two lead cells in each of the two half-batteries, stowed in watertight compartments.

Performance of Battery and Main Motors (U.B. 62).

——	Revs.	Amps. per Side.	Amps., Field.	Knots Submerged.	Grouping.		Hours.*	Amp. Hours* (Total).	Endurance.*
					Armatures.	Half Batteries.			
									Miles.
Utmost speed	390	1,200	20	7·5	Parallel	Series	1·6	3,977	12·0
Full speed	290	750	30	6·0	,,	,,	3·2	5,140	19·2
Three-quarter speed	250	850	14	5·3	,,	Parallel	6	5,460	31·8
Half speed	210	520	20	4·5	,,	,,	9	5,094	40·5
Slow speed	160	200	34	3·3	,,	,,	19·5	5,577	66
Dead slow	100	50	20	1·8	Series	,,	60	6,960	108

* Allowing for an auxiliary consumption of 46 ampères per hour.

Boat.

A small collapsible boat is sometimes carried, lashed keel uppermost on after part of superstructure.

Bulkheads.

See Internal Arrangements, page 66. The bulkheads are all straight and water-tight only, none of them being convexed to resist pressure as in "U." boats.

Compasses.

An Anschütz gyro-compass installation is fitted, consisting of a master compass in the engine room and three receivers, one in the control room, one on the bridge, and one in the after torpedo room.

A magnetic diving compass is also fitted on the bridge.

Complement.

The complement of these boats is about 35 officers and men.

Conning Tower.

See under Shape, page 63.

The fairwater is about 10 feet high at the forward end, where it is continued up to form a permanent bridge screen.

The conning tower contains a glass gauge to show the level of the water outside, depth gauge, clinometer, engine room telegraphs, torpedo firing pistols, &c.

Deck and Superstructure.

The outer hull is rounded above the waterline and forms a narrow deck on either side of the superstructure.

The superstructure deck stands about 2 feet above the outer hull amidships and aft, and merges with it forward. It is open to the sea through numerous holes, about 6 inches in diameter, on either side.

Dimensions.

See tabulated details, pages 14–20.

Diving Capabilities.

These boats are reported to be exceptionally quick in submerging and to rise to the surface with equal rapidity. Apparently they go down at a very steep angle (approximately 20°). It is stated that they can dive to periscope draught, viz., 40 feet (reckoned to keel), in 40–50 seconds from full speed ahead with both oil engines. They are tested for a depth of 197 feet.

Electric Lighting and Power.

Cables are metal encased.

Switches.—There are three switchboards, one directly over the motors, one forward for lighting and power, and one aft for lighting only.

Fuzes.—The fuze boxes are placed close to the batteries.

Endurance.

Surface endurance with extra stowage of oil fuel, 5,700–8,300 miles at 8 knots. For submerged endurance, *see* under Battery, page 64.

Duration of cruises varies from about 14 to 24 days.

Engines (Main).

Two 6-cylinder, 4-cycle, Diesel reversible engines of the air injection type. Direction of revolution inwards. The bed plate is of cast steel. The engines develop 520 brake H.P. each at 450 revolutions. Forced lubrication is used. The pistons are oil-cooled.

The following table (for "*U.B. 55*") gives the consumption of oil fuel per hour, endurance and brake horse power at different speeds on the surface, assuming the initial amount of oil fuel carried to be 60 tons :—

Number of Engines.	Speed in Knots.	Revolutions per Minute.	B.H.P. Total.	Consumption of Oil Fuel per Hour (lbs.).	Consumption of Oil Fuel per B.H.P. per Hour (lbs.).	Endurance.*
						Miles.
2	13·4	455	1,260	609	·483	1,500
2	12·0	410	880	436	·495	3,180
2	10·0	330	490	251	·512	4,350
2	8·0	265	260	138	·531	5,720
1	10·0	415	610	283	·464	3,920
1	8·0	333	320	148	·463	5,400
1	6·0	255	130	69	·531	7,075

* Allowing ·4 ton of fuel per day for auxiliary purposes.

The coolers are of the tubular type; tubes are either of brass or copper, to withstand the action of salt water, and they are heavily tinned.

Lohmann friction clutches (*see* Plate 14), actuated by hand, are fitted between main engines and main motors.

Fresh Water.

There is stowage for 1,100 gallons of drinking water and 105 gallons of distilled water.

Fuel and Fuel Tanks.

Heavy oil is used, specific gravity 0·87.

Four oil fuel tanks are fitted externally between the pressure and outer hulls, one on either side abreast engine room, one on either side forward.

These tanks give a normal stowage of 37 tons, but, in addition, oil fuel can be stowed in four of the external ballast tanks in the earlier boats of this class and in six of them in later boats, increasing the stowage to 60 and 84 tons respectively.

Guns.

One 4·1-inch *or* one 22-pr. gun, close before the conning tower, and one machine gun. There are mountings for the latter on the forward and after parts of bridge.

All these boats will probably be armed with 4·1-inch guns, as the guns become available.

For details of guns, &c., *see under* "U." Boats, page 47.

Hatches.

A circular watertight hatch is fitted to the top of the conning tower, and another is fitted at the base of it on the pressure hull, so that the conning tower can be shut off in case of damage.

One access hatch is fitted aft, leading to galley flat and engine room. Above the fore torpedo room a special torpedo hatch is fitted, which also serves as an access hatch for the crew; a similar hatch is fitted above the stern torpedo compartment.

Part III.
Section 5.

Sub-
marines.

Hull.

These boats have a complete inner cylindrical pressure hull and a partial ship-shaped outer hull, like "U." boats; the sides of the outer hull are rounded and merge with the superstructure forward. Bilge keels are not fitted.

The length of the pressure hull is 131 feet 7 inches, diameter 12 feet 10 inches.

The extreme ends of the boat, beyond the pressure hull, are free-flooding.

Hydrophones.

See Sound-Signalling Apparatus, below.

Hydroplanes.

See Steering and Hydroplane Gear, page 67.

Internal Arrangements.

Reported to be as follows, commencing from forward :—

(1) *Fore torpedo room and crew space*, containing four bow torpedo tubes and spare torpedoes, separated by a watertight bulkhead from
(2) *Warrant Officers' quarters*, separated by a light bulkhead from
(3) *Officers' quarters*, with Commanding Officer's cabin curtained off;
(4) *Control room*, divided into two parts by a light bulkhead, fitted primarily to deaden the noise of the auxiliary machinery, which is placed abaft it. The fore part contains diving and steering pedestals. A W/T cabinet and a W.C. are fitted on the port side. The control room is enclosed by a watertight bulkhead at either end.
(5) *Crew space*, separated by a watertight bulkhead, fitted with an air-lock, from
(6) *Engine room* and *motor room*, containing the two Diesel engines, and abaft these the two electric motors. It is separated by a watertight bulkhead from
(7) *Galley flat*, containing the galley on starboard side and petty officers' mess on port side, separated by a watertight bulkhead from
(8) *After torpedo room*, containing the stern torpedo tube and steering and hydroplane actuating gear.

All watertight bulkheads are fitted with hinged doors, closed by two locking bars.

Jumping Wires.

As in "U." boats, *see* page 50.

Lubricating Oil.

2,120 gallons are carried, 1,900 gallons of which are stowed in external tanks.

Masts.

For W/T masts, *see* Wireless Telegraphy, page 67.

In addition, a small telescopic mast for signalling purposes is fitted at after end of conning tower.

Motors (Main).

Two tandem sets each developing 360 kilowatts at 400 revolutions. For details of performance, *see* under Battery, page 64.

Net Cutter.

As in "U." boats, *see* page 52.

Periscopes.

Two are carried, both fitted for use in the conning tower only. They are bi-focal, power × 1·5 and × 6. The foremost one is fitted for use against aircraft. The after one is fitted with an observer's platform, both periscope and platform moving up and down in a watertight trunk which passes through the auxiliary engine room.

The periscopes are raised by motor power with alternative hand gear.

Searchlight.

In some of the boats a small portable 12-inch searchlight is carried. The lighting current is derived from a portable 6-volt accumulator set.

The searchlight can be mounted on the machine-gun mounting on the bridge.

Sound-Signalling Apparatus.

Receiving Apparatus.—Probably similar to that in "U." boats, *see* page 54.

Sending Apparatus.—A number of boats of this class are fitted with sending gear, similar to that fitted in "U" boats and briefly described on page 55.

Speed.

See under Engines, page 65, and Battery, page 64.

The fastest boats of this series attain a speed of 14 knots on the surface and 9 knots submerged. They usually cruise at a speed of 8–10 knots.

Steering and Hydroplane Gear.

There are four steering positions, one on the bridge, one in the conning tower, one in the control-room, and one in the engine room. Two vertical balanced rudders are fitted.

Hydroplanes are fitted low down, forward and aft.

A separate motor is fitted for each pair of hydroplanes and for the vertical rudders.

Tanks.

See Ballast Tanks, page 64, and Fuel, page 65.

One regulator tank (capacity 14 tons) and one lubricating oil tank are fitted externally on either side amidships, between the pressure and the outer hull.

Torpedo Tubes and Torpedoes.

Five internal 19·7-inch tubes are fitted, viz., four bow tubes and one stern tube.

Ten torpedoes are carried, viz., five in the tubes, four stowed in the forward torpedo room, and one in the engine room between the two main engines.

For details of torpedoes and tubes, *see* under "U." boats, page 57.

Towing Arrangements.

A fairlead is fitted in the upper part of the stem, presumably for a towing pendant.

Ventilation.

The system is believed to be the same as in "U." boats, *see* page 59. Eight oxygen bottles are carried, five stowed in the forward and three in the after torpedo room.

Wireless Telegraphy.

Hinged W/T masts, about 30 feet in height, are fitted, but are rarely used. They can be raised from inside the hull by motor, or alternatively, by hand gear.

The jumping wires are fitted for use as an auxiliary aerial, and suffice for most purposes.

A silent cabinet is fitted in the control room.

U.B. 18–47.

(*See* Plates 9 and 16.)

GENERAL REMARKS.

This series of submarines, known as B.A. boats, forms practically one class, though there are a few points of difference between *U.B. 18–29* and *U.B. 30–47*, including a small difference in length.

These boats have only a single hull, but light external tanks are fitted. Owing to their single hull and the absence of a hatch at base of conning tower they are much more vulnerable to gunfire than the "U." or later "U.B." boats.

They are constructed in five sections so as to be transportable by rail. The only representatives of the class actually transported overland, however, were *U.B. 42–47*, all six of which were sent by rail to Pola and there assembled for service in the Mediterranean. (*U.B. 43* and *47* have since been sold to Austria-Hungary.)

RECOGNITION.

Shape.

See Silhouettes.

In addition to their small size, two very marked features which distinguish these craft are the bluff rounded bow (usually surmounted by a straight net cutter) and the sheer aft towards the stern, the latter being nearly awash in surface trim.

The gun is mounted close before the conning tower.

I 2

DETAILS.

Accommodation.

There are 12 bunks in forward compartment for a total complement of 4 officers and 17 men. Men and officers all live together.

A few cupboards are provided for provisions and mess utensils.

Cooking is done by an electric stove. Steam heaters, with supply and exhaust deck connections, are fitted in each compartment.

Water closets are fitted aft, one for use on the surface and one for use when submerged.

Air Service.

Four large air bottles are carried outside the pressure hull in the superstructure. Capacity of each bottle about 12·4 cubic feet.

Two small bottles are fitted abreast each engine for blast and starting purposes.

No L.P. blower is fitted in these boats.

The H.P. air main is not duplicated (*i.e.*, it is a single line, not a ring main); it is made of copper pipes and screwed unions with copper washers.

On coming to the surface, all tanks are blown with H.P. air. All blows are fitted in control room. The internal tanks are fitted with a relief valve, and the external with a differential gauge.

Anchor Gear.

One stockless anchor is carried in a hawse pipe in the forward superstructure, stowing vertically. When let go, it passes through an opening in the starboard hydroplane guard. The wire is on a drum in the superstructure, with an indicator inside the hull. The motor for working it is in the torpedo room forward.

Ballast Tanks.

There are four external ballast tanks, fitted amidships, two on either side. The combined capacity of these external ballast tanks is 15·6 tons. In addition, there are two internal ballast tanks (inside the pressure hull), capacity as follows:—

Forward tank - - - - - - - - - 12·1 tons.
After tank - - - - - - - - - - 10·1 „

The total ballast tank capacity is thus 37·8 tons.

Battery.

A battery of 112 lead cells is carried, stowed under metal screw-down covers in forward compartment amidships, and also in two rows at the sides, each row forming a settee. Ventilation pipes are fitted to each cell.

The cells are each about 39 inches high by 13 inches by 16 inches; weight about 816 lbs.

Each cell contains 20 positive and 21 negative plates. The positive plates are solid, the negative plates of the grid type.

Performance of Battery and Main Motors (U.B. 26).

Revs.	Knots, submerged.	H.P.	Total Amps.	Hours.*	Amp. Hours.*	Endurance.* Miles.
416	5·8	248	1,125	3·2	3,600	18
386	5·21	189	840	5	4,200	26
310	4·24	105	480	10	4,800	42
215	2·81	34	160	34	5,440	96
180	2·4	22	120	50	6,000	120
165	2·1	12	100	60	6,000	126

* Not taking into account current expended for lights, gyro compass, and auxiliary engines.

Grouping switches are fitted as in "U." boats (*see* page 40), but in the above table the system of grouping has been disregarded, as full details are not available.

Bulkheads.

See Internal Arrangements, p. 71.

Clutches.

Friction clutches of the "Bamag" or "Lohmann" type (*see* under "U." boats, p. 41) are fitted between the main engines and main motors. Star clutches are fitted between the motor shafts and the tail shafts.

Compasses.

An Anschütz gyro-compass installation is fitted, consisting of a master compass in the after part of machinery compartment, one receiver on bridge and one in control room. A magnetic diving compass is fitted in a pressure-tight binnacle on the fore part of the bridge, the card image being reflected on to a screen in the control room.

Complement.

The normal complement is 4 officers and 17 petty officers and men, but several supernumerary ratings are frequently carried for training.

Conning Tower.

Circular, about 4½ feet in diameter, the casting being in two parts connected by horizontal flanges. It is stiffened with H girders and surrounded by a fairwater of light plating. A light bridge is built over it, with surface steering position at fore end.

The conning tower contains a glass gauge tube to show level of water outside, depth gauge, clinometer, engine telegraphs, torpedo firing pistols, &c. A circular watertight hatch is fitted to top of conning tower as in ocean-going type.

There is a large opening in the hull under conning tower, and no hatch is fitted for closing it, consequently a single hole in the conning tower should be sufficient to prevent the boat diving.

Circular scuttles, 3 inches in diameter, and provided with hinged internal deadlights, are fitted in the sides of the conning tower.

Deck and Superstructure.

The superstructure stands about 3½ feet above the pressure hull, and is open to the sea through numerous small holes on either side. It contains fuel and air supply deck connections, air bottles, engine muffler boxes, sounding machine drum and various safety arrangements such as telephone and light buoy, lifting shackles and diver's connections. The plating is about ⅛ inch thick. The breadth at the ends is about 3 feet, but this increases in wake of the gun, conning tower, and engine hatch, so that amidships a fairly roomy upper deck is obtained. The deck is formed of wooden gratings and steel frames, and is generally portable throughout the whole length.

Dimensions.

See tabulated details, p. 20.

Diving Capabilities.

This type of boat is able to dive very quickly.

Two depth gauges are fitted, one graduated from 0 to 197 feet, and one of large scale from 0 to 82 feet.

The hull is tested for a depth of 164 feet, but probably these boats rarely go to a greater depth than 100 feet.

Periscope draught, from top prism to keel, is 38 feet.

Electric Lighting and Power.

All cables are braided only, no lead casing is used. Main fuzes are situated at the battery terminals with cables leading straight aft to the motors.

Open switchboard over each main motor.

Endurance.

Surface endurance 4,500 miles at 5 knots. For submerged endurance, *see under* Battery, p. 68.

The usual length of cruises is from about 7 to 14 days.

Engines (Main).

Two 6-cylinder, 4-cycle, reversible Diesel engines of air injection or blast type; cylinders, 8¼ inches diameter, 9 inches stroke. Direction of revolution—counter clockwise. Bedplate is of bronze. Forced lubrication is fitted. Engines develop 142 brake H.P. each at 478 revolutions.

No main bearings are fitted between the pairs of cylinders (which is similar to motor-car practice).

The compressors for air-starting and engine fuel spraying are of the two-stage tandem type, driven from cranks at the fore end of the crank shaft. The H.P. and L.P. coolers are vertical, and are embodied with the crank case of the compressors. The engines are connected to the main shafting by friction clutches, which are incorporated with the fly wheels.

The following table of surface speeds and horse powers with both Diesel engines (of *U.B. 26*) may be accepted as applying generally to this class of boat :—

——	Revs.	Knots.	Brake H.P.
Utmost speed - - - - - - - - -	478	8·5	290
Full speed - - - - - - - - -	380	6·9	140
Three-quarter speed - - - - - - - - -	300	5·7	90
Half speed - - - - - - - - -	240	4·75	70
Slow speed - - - - - - - - -	150	3·4	20
Dead slow - - - - - - - - -	100	2·6	10

Fresh Water.

There is stowage for 110 gallons of distilled water and 880 gallons of drinking water.

Fuel and Fuel Tanks.

Heavy oil is used, specific gravity 0·87.

There are four external oil fuel tanks, situated two before the external water ballast tanks, and two abaft them. These external fuel tanks hold 17·4 tons. In addition, there are two internal oil fuel tanks, holding 4·2 tons. Total fuel capacity is thus 21·6 tons (normal stowage).

If necessary, oil can also be carried in two of the external ballast tanks, giving an additional capacity of about 7 tons.

The consumption per hour at the full speed of 8½ knots is about 16 gallons, and at 5 knots about 5 gallons.

Guns, &c.

One 22-pr. gun is mounted just before the conning tower. A muzzle tampion of the expanding type is fitted.

A machine gun is carried, two positions being provided on the bridge.

For details of gun, &c., *see under* " U." boats, p. 47.

Ammunition Supply.

There is a small ready-use magazine in the superstructure on the fore side of the gun, the ready ammunition being stowed in pressure-proof holders, one round in each. The main supply of ammunition is stowed under the torpedo room forward, and is passed up by hand through the conning tower. The total stowage is about 120–150 rounds.

Hatches.

Two access hatches are fitted in pressure hull, trunked up to the level of the superstructure deck, one before the gun leading to the torpedo room, one just abaft the conning tower leading to the engine room. In addition, a torpedo hatch, set at about 25° from the horizontal, is fitted immediately before the foremost of these two hatches.

Hull.

The pressure hull is about 89½ feet long, with a maximum breadth of 12½ feet. The sections of the pressure hull are circles. The length and breadth beyond the pressure hull are obtained by building light steel casings at the ends to give the submarine ship-shaped extremities, and by fitting external saddle tanks, which carry water ballast amidships and oil fuel towards the ends. These casings and the tank-plates are about $\frac{3}{16}$ inch thick. The extremities of the boat, beyond the ends of the pressure hull, are free flooding.

The pressure-hull plating is $\frac{7}{16}$ inch thick, worked flush on the frames with double butt straps. The frames are 31 inches apart and of angle-bulb section 6 inches deep.

The pressure hull is built in five sections connected together by double butt straps, running round the boat and double riveted on each side of the joint. The middle section, about 29 feet long, is a cylinder. The other four sections are, roughly, frustrums of cones. The lines of the pressure hull are therefore all straight and the ends are blunt. The few internal tanks are placed so that they do not cross the junctions of the pressure hull. This permits all the sections to be built complete with their tanks, and the latter to be tested in the sections. The whole is then ready for transport, if necessary, by rail and for re-erection on arrival at its destination. This re-erection can easily be carried out in a fortnight, the various internal fittings being then quickly installed.

The superstructure, saddle tanks, external casings, and keel are put on after the sections of the hull have been connected up.

Hydrophones.

See Sound-Signalling Apparatus, p. 72.

Hydroplanes.

See Steering and Hydroplane Gear, p. 72.

Internal Arrangements.

See Plate 16.

Four main watertight bulkheads are fitted. Two of these are amidships, and form the ends of the control room, which is under the conning tower; the other two are near the bow and stern, the spaces between them and the ends of the pressure hull forming the trimming tanks.

Before the control room are living quarters and the torpedo room. Abaft the control room is the engine and motor room.

The fresh-water tanks and battery are in the forward compartment.

The control room, which is very cramped, contains the usual mechanism (*see* under " U." boats, p. 49.)

Jumping Wires.

As in " U." boats, *see* p. 50.

Keel.

The ballast keel runs nearly the whole length of the pressure hull; it is 4 feet wide and 12 inches deep, formed of strong transverse girders, with thin casing at the bottom and sides. Lead pigs are stowed along the keel.

The centre part of the keel originally formed a detachable safety weight, but has now been replaced in most of these boats by a non-detachable keel.

Machinery.

For main engines, *see* Engines, p. 69; for main motors, *see* Motors, below.

Auxiliary Machinery.

Includes two main ballast pumps, two ready-use fuel pumps, and two lubricating pumps.

A 4-stage, 2-crank compressor is driven by gearing from one of the main shafts; the inter-coolers are contained in a separate casing bolted to the hull. This compressor draws air from the boat and discharges to the H.P. system. A small 2-stage compressor, driven from the other main shaft, takes its suction from the air injection pumps' discharge at 900 lbs. per square inch, and delivers at 2,350 lbs. per square inch.

Masts.

See under Wireless Telegraphy, p. 73.

Mines.

Not carried by this class of submarine.

Motors (Main).

Made by Siemens-Schuckert; 125-H.P. each, at about 220 volts and 416 revolutions. Each has two armatures in tandem on one shaft, the commutators being at the forward and after ends.

For details of performance, *see* under Battery, p. 68.

Net Cutter.

Most of the boats of this class are fitted with the type of net cutter since adopted for " U." boats, *see* page 52.

Periscopes.

Two are carried, one in conning tower, one in fairwater just abaft it. The forward one, used either from conning tower or from control room, is bifocal, 1·5 and 6-powered. The tube is 2·4 inches in diameter at top, 5·9 inches at the bottom, and 21 feet 4 inches in length. The after one, used from the control room only, is 3·3 inches in diameter at top, 5·9 inches at bottom, and likewise 21 feet 4 inches in length.

A motor is fitted for raising each periscope. The drums on which the periscope-raising wires are wound can be worked by hand if necessary.

An electric flashing lamp is fitted on one of the periscopes, for night use on the surface.

Propellers.

Two : solid, 3-bladed, 3 feet 9½ inches in diameter, 2 feet 4 inches pitch, projected area 4½ square feet. A guard is fitted outside each, and is continued to support the outer ends of the after hydroplane pintle.

Safety and Salving Arrangements.

Generally as in " U." boats, *see* page 53.

Sound Signalling Apparatus.

In most, if not all, of these boats no *sending apparatus* is fitted.

Receiving Apparatus.

The following are the arrangements in " *U.B. 26* " and probably in other boats of the class.

Two distinct types of instruments are fitted, presumably for the reception of sounds of different character.

Set 1 consists of two small tanks, filled with water, and each containing a microphone, placed inside the pressure hull on either side of the boat forward and bolted inside the skin plating. The microphones are tuned for sounds of low pitch.

Set 2 consists of two microphones, enclosed in metal casings and suspended by brackets, one on either side of the boat, in one of the internal ballast tanks. These microphones are tuned for sounds of high pitch.

Both sets are wired up to a common switchboard, by means of which it is possible to connect a telephone head set to either the port or the starboard receiver of either of the above sets, or to both receivers of either set.

Sounding Gear.

A hydro-pneumatic sounding machine, worked by hand, is fitted, the drum being carried in the superstructure.

The lead and line pass through a watertight tube fitted in the foremost of the two internal ballast tanks. The line consists of a rubber tube with an outer covering of plaited jute, and the lead is hollowed to form an air chamber. When sounding, a sufficient air pressure is admitted to the tube to expel the water, and, when an equilibrium is reached, a gauge inside the hull registers the depth.

Speed.

Maximum speed is about 8·5 knots on the surface and 6 knots submerged.

Surface cruising speed is about 5 knots.

See also under Engines, p. 69, and Battery, p. 68.

Steering and Hydroplane Gear.

The rudder is 32 square feet in area, the after hydroplanes 16 square feet each, and the forward hydroplanes 15 square feet each. They are all worked by manual power only. There are three steering positions, one on the bridge, one in the control room, and one right aft by the master compass. The hydroplane pedestals are at the forward end of the control room. Elaborate guards are fitted to the hydroplanes.

Tanks.

See also Ballast Tanks, page 68, and Fuel Tanks, page 70.

The trimming tanks are situated at the extreme ends inside the pressure hull. Of the six main internal tanks (three on each side of the middle line), the two after ones, under the engine room, are oil-fuel tanks. The other four, under and on either side of the control room, are ballast and compensating tanks respectively. These compensating tanks fulfil the same purpose as the regulator tanks of the "U." boats, *i.e.*, they give the boat a good margin of buoyancy on coming to the surface for a rapid look round, and can be very quickly blown or flooded.

Torpedo Tubes and Torpedoes.

There are two 19·7-inch torpedo tubes, one above the other in the bow.

Six torpedoes are now usually carried, two in the tubes, one on either side abreast the tubes, and two on the middle line abaft the tubes, one of these being slung overhead.

For general description of torpedo tubes and types of torpedoes, *see* under "U." boats, page 58.

Ventilation.

As in "U." boats, *see* page 59.

Provision is made for long periods of submersion, and four oxygen cylinders are carried.

Wireless Telegraphy.

A small W/T set is fitted in a box, about 4 feet by 2 feet by 1½ feet, in the control room. There is no silent cabinet.

In the earlier boats of this class, two hinged masts were originally fitted, raised from inside the hull by screw gear and wire tackles, worked by electric motor or, alternatively, by hand. They were of steel tubing, 28 feet long, tapering from 5 inches to 2½ inches. The jumping wires were fitted for use as an auxiliary aerial.

In several, if not all, cases the masts have since been removed, and the jumping wires now form the only aerial.

U.B. 1—17.

(*See* Plate 9.)

GENERAL REMARKS.

These boats were constructed in sections, so as to be transportable by rail, and three of them, viz., *U.B. 7*, *8*, and *14*, were sent overland to the Mediterranean. They have only a single hull, and there are no external tanks. They are exceedingly vulnerable to gunfire.

In general appearance they resemble the small minelaying type, *U.C. 1–15* (*see* page 84), but their length is about 20 feet less.

This class of boat does not now possess much interest, and only general details are here given.

RECOGNITION.

Shape.

See Silhouettes.

Colour.

As for "U." boats, *see* page 38. The German cockade—black, white, and red—is sometimes painted on either bow, looking like an eye. In combination with the fore hydroplane guards, which resemble fins, this produces the general impression of a whale's head.

DETAILS.

Anchor Gear.

Believed to be similar to that in *U.C. 1–15*, *see* page 85.

The weight of the anchor is 3 cwt.

Battery.

112 Tudor lead cells are carried, stowed before and abaft the control room.

Dimensions of the cells: Length 10 inches, breadth 12½ inches, depth 24½ inches. Weight of a cell with its connections, 321 lbs.

Compasses.

A complete Anschütz gyro installation is fitted. A magnetic compass is also carried.

Conning Tower.

Similar to that in *U.C. 1–15, see* page 85. There is no means of shutting off the conning tower from the hull.

Dimensions.

See tabulated details, page 22.

Endurance.

The usual length of cruises is from 2 to 8 days. The theoretical surface endurance of these boats is 1,200 miles at 5 knots.

Engines (Main).

One reversible Diesel engine of the air-injection type, developing about 60 brake H.P. at 540 revolutions. Circulating pump capacity, 440 gallons per hour.

Fresh Water.

118 gallons are carried.

Fuel.

Normal stowage, 2·4 tons of heavy oil (specific gravity 0·87). This stowage cannot be much increased, as there are no external tanks. The consumption per hour at the full speed of $6\frac{1}{2}$ knots is about 3 gallons, and at 5 knots about $1\frac{2}{3}$ gallons.

Guns.

One 1-pr. or a machine gun is carried, the 1-pr. being mounted before the conning tower.

Hydrophones.

Believed to be similar to those in *U.B. 18–47* (*see under* Sound-Signalling Apparatus, page 72).

Hydroplanes.

Believed to be generally the same as in *U.C. 1–15* (*see* Steering and Hydrophone Gear, page 87).

Hull.

The pressure hull is built in three sections. Probably the two after sections resemble those of the class *U.C. 1–15, see* page 85; the foremost section is entirely different in internal arrangement to that of *U.C. 1–15*, but the bow is very similar in shape. Length of pressure hull, 77 feet; thickness of pressure hull plating, $\frac{11}{16}$ inch forward, $\frac{7}{16}$ inch aft. The frames are 30 inches apart. The length beyond the pressure hull is obtained by light steel casings of $\frac{1}{8}$-inch plate, built on so as to give the submarine ship-shaped ends.

Internal Arrangements.

The control room is amidships, below the conning tower. On the fore side are the living quarters, forward accumulators and torpedo room: on the after side the after accumulators, and engine and motor room. The ends of the submarine, beyond the pressure hull, form the trimming tanks.

Jumping Wires.

Similar to those in "U." boats, *see* page 50.

Keel.

The ballast keel runs the length of the pressure hull. The centre portion, 4 tons in weight, is detachable.

Mines.

Not carried by this class of submarine.

Motor (Main).

Two armatures, fitted in tandem, on the same shaft as, and abaft, the main engine; 120 H.P. at about 210 volts, and 630 revolutions.

Periscope.

One only is fitted, to be used alternatively from conning tower or control room. A motor of about 4 H.P. is fitted for raising it, with alternative hand gear.

Propeller.

One—3-bladed.

Safety and Salvage Arrangements.

Generally the same as in "U." boats, *see* page 53, though less elaborate.

Sound Signalling Apparatus.

Receiving Apparatus believed to be similar to that in *U.B. 18–47*, *see* page 72.

No sending apparatus is fitted.

Sounding Gear.

Similar to that in *U.B. 18–47*, *see* page 72.

Speed.

A maximum speed of 6·5 knots can be attained with the main engine. With the main motor, a speed of 8·4 knots can be attained on the surface, and a speed of 5·2 knots submerged (for one hour).

Steering and Hydroplane Gear.

Believed to be generally the same as in *U.C. 1–15*, *see* page 87.

Torpedo Tubes and Torpedoes.

Two 17·7-inch torpedo tubes, fitted abreast, in the bows.

Two torpedoes are carried in the tubes; no reserve of torpedoes.

For details of torpedoes and tubes, *see under* "U." boats, page 58.

Ventilation.

The same system as in "U." boats, *see* page 59.

Capacity of ventilating fan, 2,650 cubic feet per hour.

Two oxygen cylinders are carried, capacity 1·75 cubic feet each.

Wireless Telegraphy.

The arrangements are believed to be the same as in *U.C. 1–15*, *see* page 87.

VII.—"U.C." MINELAYING TYPE.

U.C. 80–120.

None of this series of submarines have yet been completed (April 1918). They are believed to be a development of the U.C. 16–79 class, carrying more mines and with greater engine power, but at present no details are known.

U.C. 16–79.

(*See* Plates 10 and 17.)

GENERAL REMARKS.

This class of submarine (called "C.A." boats) represents a very great advance on *U.C. 1–15* (*see* page 84) and in endurance and radius of action closely approaches the ocean-going type. The general idea of the *U.C. 1–15* class has been followed, but the single hull has been abandoned in favour of a double hull; the dimensions and engine-power have been considerably increased, and the interior is divided into 6, instead of 3, watertight compartments. The mine capacity has been improved, and torpedoes and a gun are carried in addition.

The losses among these vessels have been proportionately very high. This has been stated to be due to indifferent diving qualities and to the absence of a watertight hatch between conning tower and hull. But it must in fact be attributed mainly to

Part III. Section 5.

Sub-marines.

the very dangerous character of their work, which has to be performed in the vicinity of the coast and of harbour entrances.

There are slight differences between submarines built in different yards, but one general description will serve for the whole class.

MINELAYING METHODS.

Mines are usually laid at night, and either at high or low water. In moderate depths, high water is generally preferred, as giving greater depth for diving. Also, when a submarine is proceeding to a locality in which mines have already been laid, she will endeavour to arrive there at high water.

In this class of minelaying submarine the upper of the three mines in each tube can be adjusted at any time when the boat is on the surface; but the other two are not accessible, and the submarine commander must therefore decide before leaving harbour whether the mines are to be laid at high or low water.

In the Atlantic and North Sea mines are laid at any depth between 7 and 80 fathoms. In the Mediterranean they have been laid in depths up to 150 fathoms.

Whilst laying their mines these submarines may proceed either on the surface or submerged. When laying mines on the surface they usually proceed at half speed, about 6½ knots. When submerged, they generally lay their mines at full speed, also about 6½ knots, and maintain a depth of about 50 feet (reckoned to keel).

Mines were formerly laid in a straight line, about 55 yards apart, in groups of 4, 6, or 12. Now they are frequently laid on a zigzag course, particularly in narrow channels, and the distance has been increased in some cases up to 800 yards.

Submarines depend greatly upon light vessels, buoys and conspicuous landmarks for obtaining their position for minelaying.

It is customary for each of these boats to be assigned a particular minelaying area, which forms her primary objective often in cruise after cruise. This ensures local knowledge and discourages any tendency to carelessness in fixing and plotting the position of minefields.

Outward bound minelayers do not hesitate to attack merchant vessels which they encounter whilst proceeding to their proper objective, but they do not go out of their way to search for prey until they have laid their mines. Homeward-bound minelayers follow a trade route where possible; but after laying his mines, the submarine commander has considerable discretion as to his further operations.

RECOGNITION.

Shape.

See Silhouettes.

The high bow, surmounted by a net-cutter, and the conspicuous dip in the forward superstructure where the gun is mounted, just before the conning tower, render recognition easy. A good many of the earlier vessels of the class, *e.g.*, *U.C. 25–33* and *46–54*, originally had a rounded "whale" bow, but some of these have since had a ship bow built on. Most of the later boats were built with a ship bow in the first instance.

Colour.

See under "U." boats, page 38.

DETAILS.

Accommodation.

In some cases 23 bunks are fitted, in others there are only 18, supplemented by hammocks.

The living quarters are wood-lined.

An electric cooking stove is fitted on starboard side of mine-room, opposite the wireless cabinet. There are four electric heaters, and steam heaters are fitted in all compartments for use in harbour.

There are two water-closets, one forward on starboard side of mine-room, and one as a reserve fitting, in control room. They are merely curtained off, and, as a rule, are only used when submerged.

Air Service.

The general arrangement of the air service is a system of piping and distributor boxes, connecting the high pressure air bottles and the tanks and auxiliary machinery. The H.P. air pressure is 2,350 lbs. per square inch.

Air bottles, 10 in number, are stowed :—

Four in superstructure, immediately abaft conning tower ;
Three in engine-room ;
Three in control room.

Each bottle has a capacity of 12·36 cubic feet and is fitted with a stop valve.

These bottles are connected up in pairs, thus forming five groups, each of which is connected to the H.P. air distributor box in control room through a special stop valve with a pressure gauge on the bottle side.

This special stop valve can be set in three ways :—

(1) Wide open, for charging bottles.
(2) In non-return position, for discharging bottles (if a pipe then bursts, the air cannot escape from other groups).
(3) Closed.

From the H.P. air distributor, the air is taken to the L.P. air distributor through a pressure-reducing valve. L.P. air pressure is 176 lbs. per square inch.

Anchor Gear.

A mushroom-head anchor, worked by an electric windlass, is fitted right in the bow: weight, 660 lbs. The cable consists of about 55 fathoms of ·79-inch chain, and is self-stowing. The capstan motor is mounted on a platform inside the pressure hull before the foremost mine-tube. A capstan, which can be unshipped, is fitted on the upper deck right foreward, for warping, &c. When this is in use, the motor is unclutched from the cable-holder.

Ballast Tanks.

There are seven external ballast tanks situated between the outer and pressure hulls.

These tanks do not extend down to the keel, but to well beyond the turn of the bilge, where the tank bottom forms a narrow shelf like the under side of a bilge keel. The forward ballast tank (No. 4) extends right across the boat ; the others are on either side, port and starboard.

The tanks are all fitted with flooding flaps (*see under* "U." boats, pages 39 and 50).

The capacity of the ballast tank differs slightly in boats built in different yards.

Capacity of Ballast Tanks (U.C. 34–39).

No. 4 ballast		14·2 tons*	Total capacity, 80·6 tons.*
„ 3 „	starboard	11·7 „	
„ 3 „	port	11·7 „	
„ 2 „	starboard	11·9 „	
„ 2 „	port	11·9 „	
„ 1 „	starboard	9·6 „	
„ 1 „	port	9·6 „	

The regulator tanks (*see under* "Tanks," page 83) separate No. 3 from No. 2 ballast (port and starboard) tanks.

For details of flooding, venting and blowing arrangements *see* Plate 19.

Battery.

Consists of 124 cells divided into two half-batteries of 62 cells each. Dimensions of each cell: Length, 20 inches; breadth, 13½ inches, height, 28 inches ; weight, 776 lbs. Weight of complete battery, 43 tons.

* Fresh-water tons.

Part III. Section 5.

Submarines.

Performance of Battery and Main Motors (U.C. 39).

Revolutions.	Knots, Submerged.	H.P.	Total Amps.	Hours.*	Amp. Hours.*	Endurance,* Miles.
420	7·1	627	2,020	1	2,020	7·1
395	6·8	493	1,680	2	3,360	13·6
340	5·8	266	860	5	4,300	29·0
300	5·1	163	540	9	4,860	45·9
240	4·1	90	300	17	5,100	69·7
180	3·0	53	180	32	5,760	96·0
120	2·0	35	110	55	6,050	110·0
90	1·5	30	100	65	6,500	97·5

* Not taking into account current expended for lights, gyro compass, and auxiliary machinery.

Grouping switches are fitted as in "U." boats, *see* p. 40, but in the above table the system of grouping has been disregarded, as details are not available.

Boat.

A small canvas boat is sometimes carried stowed bottom up on deck abaft the conning tower.

Bulkheads.

There are five main bulkheads inside the pressure hull, of which two, viz., those at the forward and after ends of the control room, are convexed to resist pressure, whilst the others are watertight only. *See also under* Internal Arrangements, p. 80.

Compasses.

A magnetic diving compass is fitted on the fore end of the fairwater, reflecting to the control room.

In addition there is a gyro installation, the master compass being fitted in the control room, one repeater in the conning tower and another right aft in the torpedo room. It is found that the gyro compass cannot be relied upon in heavy weather.

Complement.

A crew of 29 officers and men is usually carried.

Conning Tower.

Circular in section, about 4½ feet in diameter, and 6 feet in height.

It is unarmoured, and there are no means of shutting it off from the control room below it. A grating is fitted over the opening in the hull.

A light bridge is built over the conning tower, fitted with a permanent weather screen on the fore side. The conning tower contains the usual fittings and instruments (*see under U.B. 18–47*, page 69).

Deck and Superstructure.

A superstructure, the top of which forms a deck, is carried above the pressure hull for the whole of the boat's length.

Forward it is raised to a height of about 4 feet to provide the necessary extra height for the mine tubes, and right in the bow it merges with the outer hull, thus giving a resemblance to the bow of a destroyer.

At the after end of the mine room, *i.e.*, just before the conning tower, the superstructure slopes down to a lower level and forms an emplacement for the gun. Abaft the conning tower the superstructure continues aft, at first parallel to the waterline, then quickly sloping down to the stern of the boat, where it is almost awash. To permit of free flooding and draining, limber holes are cut in the sides of the superstructure at intervals all the way along it, except right forward, where it is the full width of the boat.

Depth when submerged.

These submarines usually proceed at a depth of 66 to 81 feet when cruising submerged, and at a depth of about 33 feet when making an attack, the depth in each case being reckoned to the keel.

Periscope draught, reckoned from top prism to keel, is 39 feet.

The boats are tested for a depth of 197 feet, but will dive to greater depths in case of serious emergency.

Dimensions.

See tabulated details (pages 24–28).

Endurance.

The usual length of cruises is from one to three weeks, but these submarines are capable of keeping the sea for longer periods. One boat of the class is known to have spent 55 days continuously at sea, though this is exceptional.

With the normal stowage of about 40 tons of oil fuel, the surface endurance is 6,000 miles at a speed of 8 knots, but this figure can be considerably increased by using external ballast tanks for stowing additional fuel.

For submerged endurance, *see under* " Battery," page 78.

Engines (Main).

All the boats of this series are fitted with two sets of Diesel engines developing about 260 brake H.P. each at 450 revolutions per minute. The engines are of the reversible 6-cylinder, 4-cycle, air-injection type, made either by Körting, Benz, or the M.A.N. The cylinder dimensions are as follows :—

	Diameter.	Stroke.
Benz - - - -	10·24 inches.	12·6 inches.
Körting - - -	9·84 ,,	14·17 ,,

Each engine drives by means of gearing off its crank shaft a circulating water pump and either a two or a three-stage air injection pump, which delivers at 900 lbs. per square inch. Governors are fitted.

The following table of surface speeds and horse-powers with both Diesel engines may be accepted as applying generally to this class of boat :—

Speed.	Revolutions.	Knots.	Brake H.P.
Utmost speed - - - - - - - -	450	11·6	520
Full speed - - - - - - - - -	390	10·2	365
Three-quarter speed - - - - - - -	300	8·1	182
Half speed - - - - - - - - -	240	6·7	105
Slow speed - - - - - - - - -	200	5·5	60
Dead slow - - - - - - - - -	150	3·7	25

Fresh Water.

About 1,036 gallons of drinking water are carried.

Fuel and Fuel Tanks.

For details of oil fuel system, *see* Plate 20.

Heavy oil of specific gravity 0·87 is used.

Four external tanks are appropriated for oil fuel. The precise capacity varies in boats built in different yards, but for practical purposes the following may be accepted :—

Capacity of Fuel Tanks (U.C. 34–39).

Tank	Capacity	
No. 1 tank, starboard - - - -	10·4 tons.*	Total Capacity 39·9 tons.*
No. 1 ,, port - - - - -	10·4 ,,	
No. 2 ,, starboard - - - -	9·4 ,,	
No. 2 ,, port - - - - -	9·4 ,,	
Two ready-use tanks - - - - -	·3 ,,	

Additional fuel can be carried if required in external ballast tanks.

Guns, &c.

One 22-pr. gun is mounted just before the conning tower. A machine gun is carried and can be mounted either on the standard of the after periscope, or on a bracket extending from the after end of the conning tower fairwater.

For details of gun *see under* "U." boats (page 47).

* Tons of oil fuel, specific gravity ·87.

K 4

Ammunition Supply.

The main magazine is under the control room, and ammunition is passed up by hand through the conning tower.

A ready rack containing about 24 rounds is built into the after slope of the bow superstructure just before the gun. Each round stows in a pocket made watertight by means of a bung, which is left open when the round has been removed, in order to allow the pocket to fill on diving and thus compensate for the loss of weight.

Machine gun, rifle and pistol ammunition is stowed in lockers underneath the flat at the after end of the mine room.

Hatches.

There are three access hatches: one in dome of conning tower; one just abaft mine room, leading to warrant officers' mess; and one aft, leading to engine room. The after hatch also allows of the passage of torpedoes.

Hull.

These submarines have a partial double hull, the outer hull plating only extending to just beyond the turn of the bilge, and the bottom plating being that of the pressure hull. The latter is built in sections.

Length of pressure hull - - - - - - 141 feet.
Diameter of pressure hull - - - - - - 12 ,,

Internal Arrangements.

These boats are subdivided into six compartments by means of watertight bulkheads and doors, usually as follows, commencing from forward:—

(1) *Mine Room,* containing the six mine tubes, and on them the mine-releasing gear, the forward trimming tank (before the foremost mine tube), and the six mine-compensating tanks (three on either side). Lockers for machine-gun, rifle and pistol ammunition are fitted underneath the flat.

(2) *Officers' Quarters,* divided into two parts by a light transverse partition: the warrant officers' quarters on the fore side, and the officers' quarters on the after side. The forward battery is under the floor of this compartment.

(3) *Control Room,* divided into two parts by a light transverse non-watertight, but sound-proof, bulkhead, fitted for the purpose of deadening the noise of the auxiliary machinery, which is placed abaft it. The fore part of the control room contains periscopes, diving and steering wheels, gauges, &c. A watertight hatch gives access to the conning tower above. Under the floor of the control room there are lubricating oil tanks, a drinking water tank, the periscope wells, and the 22-pr. magazine.

(4) *Crew Space,* containing bunks and accommodation for the men. The after battery is under the floor of this compartment.

(5) *Engine and Motor Room,* containing the two Diesel engines and, abaft these, the two main motors. A lubricating oil tank is fitted in the bilge between the two engines.

(6) *Petty Officers' Mess* and *After Torpedo Room,* the two being separated by a light transverse partition. On the fore side of this partition there are bunks and accommodation for the petty officers; on the after side there is the stern torpedo tube, emergency steering position, steering and hydroplane actuating gear and indicators.

In some boats of this class the officers' and warrant officers' quarters are aft, and the men's quarters forward.

In the two control room bulkheads (which are pressure bulkheads), circular, hinged, watertight doors, about 2 feet 9 inches in diameter, are fitted. The other bulkheads have ordinary hinged watertight doors, about 5 feet in height by 1 foot 9 inches in breadth.

Jumping Wires.

Fitted as in "U." boats (*see* page 50), and forming an auxiliary aerial.

Lubricating Oil.

Forced lubrication is employed, a special mineral oil being used, specific gravity ·91.

Lubricating Oil Tanks.

Fitted inside the pressure hull, two below the control room, and one in the engine room. Capacity in *U.C. 34–39* as follows:—

Capacity of No. 1 tank (engine room) - - - - - -	468	gallons.
,, ,, No. 2 tank (control room, port) - - - -	532	,,
,, ,, No. 3 tank (control room, starboard) - - -	532	,,
,, ,, two ready-use tanks - - - - - - -	64	,,
Total capacity - - -	1,596	,,

Machinery.

For main engines, *see* Engines, page 79; for main motors, *see* Motors, page 82.

Auxiliary Machinery is fitted as follows:—

(1) In mine room:—

Capstan motor, 14 H.P., and gearing.
Hand pump, fixed, for fresh water.
,, ,, portable, for bilges.

(2) In control room:—

2 periscope motors, each 5 H.P., and control gear.
2 hydroplane motors.
1 steering motor.
2 ventilating fan motors.

In auxiliary machinery room:—

1 trimming and auxiliary bilge pump.*
1 turbo-blower, 3 stage, capacity 706 cubic feet per minute to 22½ lbs. per square inch, for one hour, running at 5,500 r.p.m.

(3) In engine room:—

1 ballast pump and motor; motor 14–16 H.P.; pump delivers 110 gallons per minute against a head of 197 feet of water.
1 oil fuel pump.
1 auxiliary lubricating oil pump.
1 circulating water hand-pump.
1 dirty oil hand-pump.
1 motor transformer for W/T.
1 motor transformer for order transmitters.
1 H.P. air compressor (driven off starboard main engine).
1 auxiliary H.P. air compressor } A motor is fitted to drive either of these
1 circulating water pump - } separately.

Mast.

One hinged W/T mast is fitted aft by engine-room hatch.

Mines.

Eighteen mines are stowed in sets of three, one above the other, in six cylindrical tubes of about 3¼ feet internal diameter. These tubes are placed on the middle line and are inclined at about 26° from the vertical. The mines are kept from rotating in the tubes by guide rails, along which run the guide rollers on the arms of the sinkers.

The mines are released from inside the boat, and can be laid either when on the surface or when submerged.

Separate releasing gear, operated by hand, is fitted for each mine alongside the tube. Elaborate interlocking gear is added to ensure that the mines are not released out of their proper order, &c.

The mine-tubes, which are quite independent of each other, are entirely open at the lower end, but are closed on top by a hinged cover or grating.

* One pump serves both purposes, and is of the same type as the main ballast pump in engine room.

The uppermost mine can be adjusted after leaving harbour, but there is no access to the other two.

The mines carried are of the usual German type, with four firing horns. They are painted black in home waters, and green or blue in the Mediterranean. Similar mines are carried by surface craft, but can be distinguished by a fifth and central horn, which is never fitted to mines laid by submarines.

The mine-sinkers are fitted with a delay-release, usually adjusted so as to allow the mine to ascend to its set depth after about half-an-hour.

Mine-Compensating Tanks.

Are fitted on either side of the mine-room. They are divided by two transverse bulkheads into three sections which can be flooded separately as the mines are released. Flooding is effected by Kingston connections.

Motors (Main).

Each of the two main motors has two armatures on the same shaft.

For details of performance, *see under* Battery, page 78.

Navigation and Signal Lights.

Steaming light is fitted on periscope, side lights on conning tower.

An anchor light is fitted for lighting upper deck, and there is a shaded stern light, which can be used for flashing and can be dimmed when necessary by inserting a resistance. Morse key and resistance are in conning tower. A pilot light is also fitted in conning tower to show whether the stern light is burning or not.

Net Cutter.

These submarines are fitted with the type of net cutter since adopted for "U." boats (*see* page 52).

Periscopes.

Two are fitted, the foremost one passing through forward end of fairwater and being used only from control room, the after one being mounted on conning tower and used from either control room or conning tower. They are 20 feet 4 inches in length; diameter of lower tube, 5·9 inches; diameter of upper tube, 2·16 to 3·15 inches. Worked by motor, with alternative hand gear. When worked by motor, they can be raised 8 inches per second.

Propellers.

Two, of manganese bronze—

	U.C. 34–39.	*U.C. 25–33* and *40–45.*
Diameter - - -	4 feet 2·8 inches.	4 feet 3·2 inches.
Pitch - - - -	3 ,, 6·5 ,,	3 ,, 0·4 ,,
Area - - - -	4·22 square feet.	5·7 square feet.

Safety and Salvage Arrangements.

Generally as in "U." boats (*see* page 53), but, in place of a telephone buoy, a plain buoy, filled with compressed air, is fitted.

Sound-Signalling Apparatus.

Receiving Apparatus.—As in "U." boats (*see* page 54).

No *sending apparatus* is fitted.

Sounding Gear.

As in *U.B. 18–47* (*see* page 72).

Speed.

Maximum speed is about 12 knots on the surface and 7 knots submerged.

Surface cruising speed is about 8 knots.

See also under Engines, p. 79, and Battery, p. 78.

Steering and Hydroplane Gear.

Two vertical rudders, of the balanced type, are fitted. The vertical rudders and foremost hydroplanes are worked electrically, the after hydroplanes by hand.

These submarines are practically always steered from the control room. There is no steering position in the conning tower, but there is a gyro repeater just below conning tower hatch for the convenience of the captain or officer of the watch. The deck steering position, which is on fore end of fairwater, is only used for harbour work.

Tanks.

See also under Ballast Tanks, Fuel Tanks, and Torpedo Tubes and Torpedoes.

Regulator Tanks.

Fitted abreast the conning tower on either side, between the pressure hull and the outer hull. Capacity of each tank, 10 tons.*

Trimming Tanks.

Situated near the extreme ends of the boat, inside the pressure hull :—

Capacity of forward trimming tank - - - 2·7 tons.*
" " after trimming tank - - - - 2·2 "

Forward trimming tank is also used for torpedo compensating.

Water meters are fitted for measuring the amount of water flooding into or being pumped out of the trimming, compensating and regulator tanks. The regulator tanks are each fitted with a gauge to indicate the amount of water in them.

Torpedo Tubes and Torpedoes.

There are three 19·7-inch torpedo tubes, viz., two external tubes right forward, built into the outer hull, with only the breech and the mouth showing, and one internal tube right aft, amidships. The external tubes are provided with bow caps, and their breeches are fitted with heavy iron guards.

The normal stowage only provides for one spare torpedo in addition to the three torpedoes stowed in the tubes, and this spare torpedo has to be taken apart and stowed in three sections in the three after compartments of the boat to allow the doors to be closed. Recently, however, some boats of the class have carried as many as three spare torpedoes, two lashed on deck in rear of the bow tubes, and one stowed below, either right aft in sections or between the two engines.

The two external tubes are loaded from on deck, but fired from conning tower or control room. The safety pins of the pistols are usually removed on leaving harbour, so that there is no need to touch the torpedoes again.

These two torpedo tubes are fitted with gear which enables the gyroscopes of the torpedoes to be angled from inside the boat. This arrangement consists of a system of rods and levers working in watertight casings through glands in the shell of the pressure hull and torpedo tubes. A pointer with a graduated arc inside the boat shows when the desired number of degrees has been set on the torpedo angling gear. All other adjustments to the torpedoes in these tubes must be made when the boat is on the surface.

For details of the stern (internal) tube and types of torpedoes carried, *see under* "U." boats (page 58).

Torpedo Compensating Tanks.

Are fitted inside the pressure hull below the torpedo tubes. The forward torpedo compensating tank is also used as a trimming tank for adjusting the longitudinal trim.

Ventilation.

The general system is the same as in "U." boats (*see* page 59). For details, *see* Plate 21.

Vents.

See Plate 19.

Two large domes, each provided with a master vent valve and operated by hand-wheels in the control room, are fitted for venting the three after ballast tanks (port and starboard); the forward ballast tank has an ordinary vent valve, also operated from control room. Hand wheels are fitted on the control shafting in the mine room forward and torpedo room aft to enable the vents of No. 4 tank and No. 1 tanks respectively to be opened from these compartments as well as from the control room.

* Fresh-water tons.

In detail the venting system is as follows :—

No. 4 (the foremost) ballast tank vents through a vent valve fitted on top of it.
No. 3 ballast tank (starboard) through one pipe into central dome.
No. 3 ,, ,, (port) ,, ,, ,, ,, ,,
No. 2 ,, ,, (starboard) ,, ,, ,, ,, ,,
No. 2 ,, ,, (port) ,, ,, ,, ,, ,,
No. 1 ,, ,, (starboard) through two pipes into after dome.
No. 1 ,, ,, (port) ,, ,, ,, ,,

The venting leads serve also as blowing leads for the L.P. turbo-blower.

Emergency vent valves are fitted on the tanks to enable the venting system to be shut off and the tanks blown with compressed air in the event of damage to the leads.

Wireless Telegraphy.

One hinged mast is fitted aft by engine-room hatch. The aerial leads forward to the bow.

The jumping wires are adapted for use as an auxiliary aerial.

There is a good-sized silent cabinet at after end of mine room on the port side.

A 1-kilowatt commercial quenched spark set is fitted, transmitting wave-length 980 to 2,690 feet, receiving wave-length 650 to 20,000 feet. It is arranged to receive damped waves only. A three-valve amplifier is carried.

Transmitting range, using mast aerial: By day, 100 to 200 miles reliable, 300 to 700 miles extreme; by night, about double the distance. The use of the jumping wires in place of the mast aerial roughly halves the range.

As a rule these submarines make little use of their wireless telegraphy installation except for listening.

U.C. 1–15.

(*See* Plates 10 and 18.)

GENERAL REMARKS.

Properly speaking, this series was composed of two classes, viz., *U.C. 1–10* and *U.C. 11–15*, but the differences are very slight.

These boats were built solely for minelaying. They have only a single hull and no external tanks. They are exceedingly vulnerable to gunfire anywhere, but particularly in the conning tower.

They are easily transportable by rail, and several of them were sent overland to the Mediterranean. "*U.C. 12*," which commissioned at Bremen on 2nd May 1915 and was paid off at Kiel on 4th June 1915, was despatched from Kiel by rail on 22nd June, arrived at Pola on 24th June, and was assembled and ready for sea on 27th June 1915.

Of the whole series, only "*U.C. 4*" and "*U.C. 11*" are now in existence (April 1918), and therefore only the main details are here given.

For minelaying methods, *see under* Mines, *U.C. 16–79* class, page 81.

RECOGNITION.

Shape.

See Silhouettes.

Colour.

As in "U." boats, *see* page 38.

DETAILS.

Accommodation.

Very cramped. For a total complement of 2–3 officers and 13–15 men only six bunks are fitted. Two or three hammocks are supplied by way of further accommodation, but there is not much room to sling them.

Electric heaters and an electric cooking stove are fitted. In addition, there are pipe radiators fed from the engine exhaust.

Air Service.

The compressed air bottles are fitted in the mine room between the tubes.

Anchor Gear.

A mushroom-head anchor, weight 3 cwt., is fitted in the bow ballast tank. The cable consists of about 55 fathoms of ·63-inch chain.

The capstan, fitted right forward in superstructure, is worked by a motor inside the hull.

Ballast Tanks.

See Tanks, page 87.

Battery.

Placed underneath the central compartment, at either end of it, forming the deck and seats in the living spaces.

Compasses.

A complete Anschütz gyro installation is fitted, with two receivers, one in control room and one for the bridge steering position. The master compass, with the gyro-motor generator, is situated at after end of engine room. There is also a portable magnetic compass for the bridge.

Conning Tower.

Placed a little forward of the midship section, and circular in shape, diameter $4\frac{1}{4}$ feet. It is constructed of $\frac{5}{8}$-inch plating, flanged and riveted to hull and fitted with a cast-steel dome. The plating is stiffened inside with one circumferential bulb-angle. The usual watertight hatch is fitted in the dome, but none between conning tower and hull.

Deck and Superstructure.

A light superstructure, about 3 feet above the hull, runs the whole length of the boat, curving down to the waterline forward and thus forming a rounded "whale" bow.

To permit of free flooding, it is not connected directly to the hull plating, but to light brackets riveted to this, and the side plates of the superstructure are about 3 inches off the pressure hull. The fore part of the deck is formed by the steel gratings covering the tops of the mine tubes.

Dimensions.

See tabulated details, page 28.

Endurance.

The usual duration of cruises is two to five days.

Engine (Main).

U.C. 1–10.—One 6-cylinder, 4-cycle, reversible Diesel engine. Diameter of cylinders $6\frac{5}{16}$ inches, stroke 9 inches.

U.C. 11–15.—One 4-cylinder, 4-cycle, non-reversible Diesel engine. Diameter of cylinders $7\frac{7}{8}$ inches, stroke 9·8 inches.

Brake H.P. in both cases, 80 at 500 revolutions. Air injection and air starting. Exhaust through a muffler box in superstructure.

Fuel.

About $2\frac{1}{2}$ tons of heavy oil is carried, specific gravity 0·87.

Fuel consumption at the full speed of 5·7 knots, about four gallons per hour.

Hatches.

Besides the conning tower hatch, there is only one access hatch; this is fitted in hull, over after living quarters, and trunked up to the superstructure deck.

Hull.

The pressure hull is constructed in three sections, of which the foremost and after ones are about 43 feet, and the midship section about 24 feet in length. The foremost and midship sections are approximately cylinders, the after section is roughly a frustum of a cone, tapering considerably towards the stern. Ship-shaped extremities of $\frac{1}{8}$-inch plating are built on to the pressure hull. The pressure hull is of $\frac{7}{16}$-inch plating. The frames are bulb-angles $5\frac{1}{4}$ inches by $2\frac{1}{2}$ inches by $\frac{5}{16}$ inch, and are spaced about 55 inches apart in foremost section and from 26 inches to 34 inches apart in the other two sections. The three sections are secured by bolted flanged joints, the flanges being $1\frac{1}{8}$ inches thick, and bolts $1\frac{1}{8}$ inches in diameter; joining material, thin rubber.

Hydrophones.

Similar to those in *U.B. 18–47*; *see* under Sound-Signalling Apparatus, page 72.

Hydroplanes.

See Steering and Hydroplane Gear, page 87.

Internal Arrangements.

See Plate 18.

The mine room occupies nearly the whole of the foremost section. Air and oxygen bottles are fitted between the mine tubes.

The midship compartment contains living quarters and control room, the latter immediately under the conning tower.

The after compartment contains the engine and motor room.

Jumping Wires.

Similar to those in "U." boats, *see* page 50.

Keel.

About 3 feet 4 inches in width and 13 inches in depth, built of plate, filled in with 18 tons of pig iron. The centre portion forms a detachable safety weight (about 3 tons).

Machinery.

For main engine, *see* Engine, page 85; for main motor, *see* Motor, below.

Auxiliary Machinery.

(*a*) Combined capstan and periscope raising motor.
(*b*) Wireless telegraphy alternator.
(*c*) Gyro motor generator.
(*d*) Ventilating fan motor, fitted for ventilating both boat and battery.
(*e*) Ballast pump and auxiliary compressor motor.

The main air compressor is driven direct off forward end of main engines.

In addition, a bilge pump and water-cooling pump are fitted to the main engines, and a portable hand bilge pump is carried.

Mast.

One telescopic W/T mast is fitted in conning tower.

Mines.

The main armament consists of 12 mines, stowed in pairs, one above the other, in six tubes, $3\frac{1}{4}$ feet in diameter, placed on the middle line and inclined aft from the top to the bottom (the inclination varies from 25° to 28°, increasing from aft forward).

The mines are released from inside the boat, separate releasing gear, operated by hand, being fitted for each mine alongside the tube. Elaborate interlocking gear is added to prevent the upper mines being released before the lower, &c.

Six mine compensating tanks are fitted, capable of being filled separately as the mines are released.

For details of mines, *see* under *U.C. 16–79*, page 81.

Motor (Main).

Two armatures, fitted in tandem, on the same shaft as, and abaft, the main engine. Shunt wound, separate yokes, with eight main and eight inter-pole windings each.

Working voltage 215, giving 175 H.P. for the two armatures, at about 620 revolutions.

A switchboard, installed above the motors, enables the two armatures to be connected in series or in parallel, and the half batteries similarly.

As a generator, the two armatures in parallel can generate 550 amperes at 160 volts, running for prolonged periods at from 400 to 550 revolutions.

Net Cutter.

Probably a net cutter is now fitted, of similar type to that of larger boats, *see* under "U." boats, page 52.

Periscope.

Only one periscope is fitted, which can be used either from conning tower or from the control room below it. The tube is of delta metal, length over all 17 feet, diameter at bottom 5·9 inches, at top 3·4 inches. The periscope is bifocal, magnifications 1·5 and 6. Field of view 40°, with magnification 1·5. There is a range-finding attachment, depending on a known dimension. Electric and hand raising and lowering gear are fitted.

Propeller.

One 3-bladed propeller, diameter 3 feet 3 inches to 3 feet 6 inches; pitch about 1 foot 5 inches.

Safety and Salvage Arrangements.

Similar to those in "U." boats (*see* page 53), but less elaborate.

Sounding Gear.

A hydro-pneumatic sounding machine is fitted, worked electrically. The drum is carried in the superstructure just before the conning tower. The lead is lowered straight over the side of the boat. The line consists of a rubber tube with an outer covering of plaited jute, and the lead is hollowed to form an air chamber. When sounding, a sufficient air pressure is admitted to the tube to expel the water, and, when an equilibrium is reached, a gauge inside the hull registers the depth.

Speed.

On Surface.

With motors for a short period - - - - - - 7·5 knots.
With Diesel engine, maximum - - - - - - 5·7 knots.

Submerged.

Maximum - - - - - - - - - - - 5·0 knots.

Steering and Hydroplane Gear.

The vertical rudder is of the balanced type, cut away in the wake of the diving rudder, which is fitted abaft the propeller. The vertical rudder has a surface area of about 25 square feet, and the diving rudder about 19 square feet. The bow hydroplanes have a surface area of about 12 square feet each. Only hand steering gear and hydroplane gear are fitted. A portable steering wheel is fitted on the bridge; the only other steering position is in control room.

Tanks.

The small bow compartment, separated from the mine room by a pressure-proof bulkhead, serves as a ballast tank, capacity 8 tons. The other ballast tanks, two in number, capacity 13 tons, are under the control room.

Compensating tanks are fitted on either side of the control room, and special mine-compensating tanks on either side of the mine room.

Two trimming tanks are fitted right forward, one on either side inside pressure hull. A single trimming tank is fitted right aft, also inside pressure hull. The trimming tanks can be filled with oil fuel, if required.

The oil-fuel tanks, four in number, are under the main engines.

Ventilation.

The system is similar to that in "U." boats (*see* page 59).

Wireless Telegraphy.

A telescopic W/T mast is fitted in conning tower. The aerial consists of two wires leading both forward and aft; total length, 55 feet. The installation is a ½-kilowatt Telefunken, quenched spark and valve receiving. No silent cabinet is fitted.

These boats very seldom use W/T.

Part III.
Section 5.
Sub-
marines.

APPENDIX I.

TACTICS OF ATTACK.

A.—ATTACK WITH TORPEDO.

1. Procedure of Submarine on her Hunting Ground.

German submarines usually remain on the surface until forced to submerge by the appearance of patrols, or owing to weather conditions. In fog or very heavy weather, *i.e.*, state of sea more than about 6, submarines proceed submerged or lie on the bottom, according to depth of water and other circumstances.

When on the surface, German submarines, as a rule, keep constantly under way, and usually follow a zigzag course, in view of the danger from Allied submarines.

2. Limitations of Weather.

The worst conditions for submarine operations are given by a calm sea, and if, in addition, there is a long swell running, it becomes almost impossible for a submarine to attack unseen.

At the other extreme, if the state of the sea is more than about 6, a submarine will become unmanageable when near the surface. If it is possible for her to make the final approach on a course at right angles to the direction of the sea, she may attack successfully in worse weather; but, unless she is thus favourably situated, she will probably give up any attempt to operate near the surface, and will dive deep until the weather moderates.

3. Sighting the Objective.

By remaining on the surface, the submarine ensures for herself a very much greater range of vision than she can obtain when submerged. As a general rule, she will sight the smoke of her quarry, or the tops of the masts and funnels, on their appearing above the horizon. In order to retain greater freedom of movement, and to prevent the situation from developing too rapidly, she may avoid closing at first, whilst manœuvring for a position right ahead of her quarry. If the ship appears to be zigzagging, and the attack is made in open waters where patrols are not likely to be encountered, she may for a time steer the same course as her quarry. This enables her to make a good estimate of the mean course steered, and possibly to anticipate alterations of course when making her attack, if the cycle of the zigzag is short or its form regular.

4. The Approach.

If, owing to less visibility or the presence of patrols, &c., the circumstances are unfavourable for manœuvring on the surface for position, the submarine will steer either straight towards or straight away from her objective until a definite change of bearing is detected. The direction of this change tells her which bow she is on, and she then brings the target abeam, and steers to intercept it.

As soon as the tops of the funnels or the crow's nest appear above the horizon, the submarine dives. Her speed during the further approach is adjusted in order, as far as possible, to keep the bearing of her quarry unchanged.

If the submarine has been able to reach a position ahead before diving, she usually steers the same course as her quarry, and allows the latter to overtake her, then turning in to attack, if possible, when a fresh leg of the zigzag is commenced.

At distances over 4,500 yards she will freely show as much as 6 feet of periscope, as there is no danger of being sighted, and it is important for her frequently to check her estimation of the quarry's course. Inside this distance, she will use her periscope more and more sparingly, showing only a few inches of it, and that only for a few seconds at a time.

5. Choice of Position.

If the ship is zigzagging, the choice of position for attack will be limited. If, however, the submarine's initial position is such as to enable her to attack from either side, she will usually choose the weather side in order to approach down wind. If there are any white horses, the feather made by the periscope will then be moving in the *same* direction as the sea and consequently will be very difficult to detect. If there are no white horses, and when the sun is low and its bearing corresponds roughly to the direction of approach, the submarine's best side for attack is that on which the sun is behind her.

In very heavy weather, as mentioned in paragraph 2, the only direction in which an attack is possible may be that at right angles to the sea.

Twilight.

In twilight, when choice of position is possible, a submarine prefers to attack towards the direction of the light, against which the target will be sharply silhouetted, whilst, on the other hand, her own periscope is then very difficult to see.

6. The Attack.

If approaching on a course at right angles to the bearing of her quarry, the submarine, on getting close in, will reduce speed to allow the ship to draw ahead of her ; at the same time she will turn on to the desired firing course.

If originally almost ahead of her quarry, she must judge correctly the time to turn on to the firing course, towards or away from the target, according as she intends to use a bow or a stern tube.

If the submarine has misjudged her attack, or if her intentions are upset by the ship zigzagging, she can fire an angled shot (*see* paragraph 8).

7. Firing Course.

Theoretically the best chances of the torpedo hitting are obtained when the course of the torpedo is at right angles to that of the target ; but, as can be proved by constructing a diagram, errors in estimating the course and speed of the enemy, and hence the deflection for firing, have a much smaller effect on the result of the shot when the torpedo approaches from abaft the beam than when it approaches from before it. This holds good in all cases, though the *best* position for firing varies with the speed of the target.

German submarine commanders are well aware of the advantages of the overtaking shot, but records of attacks show that, either owing to mistakes or force of circumstances, shots are not infrequently fired from other directions.

8. Choice of Tubes and Angled Shots.

German submarines (with the exception of the converted mercantile and new cruiser types) are not fitted with beam tubes. Most of them have more bow than stern tubes, and the most frequent shot attempted is a direct shot from a bow tube. This not only allows of a second tube always being kept ready for emergencies, or, in the case of a valuable target or convoy, of two or more torpedoes being fired, but is also the easiest form of attack, provided it develops according to plan. If, owing to the manœuvring of the target or to errors of judgment, the submarine finds the bow shot impossible, she may still use her stern tube for a direct shot or may fire an angled shot from either a bow or a stern tube.

The torpedo tubes of German submarines are all fitted to enable the gyroscopes to be angled up to the last moment before firing. In the Submarine School detailed instruction is given in the use of angled shots, and practice carried out with them ; but on service they do not appear to be favoured by submarine commanders, probably on account of the difficulty of judging the deflection, which varies with the range. Also, although the gyro can be set for every 15° from 0° to 90°, on service only angles of 90° are used.

9. Firing Range.

German submarines, when attacking single ships, endeavour, if possible, to close to a firing range of about 300 yards. At this range a 38-knot torpedo takes about 15 seconds only to reach the target, and a hit may be obtained on a 400-foot target even should her speed have been estimated as much as 8 knots in error.

When attacking a convoy or other formation, the submarine will endeavour to fire at a range of not less than 500 yards or more than 1,000 yards, in the case of a single shot.

10. Procedure after Firing.

After firing, the submarine will probably turn away in the direction of her quarry's stern, in order to avoid ramming or being rammed, and also in order to gain time for her escape by forcing her quarry or any escorting vessels to turn 16 points to attack her.

If the danger is not too great, the submarine will endeavour to observe through her periscope the result of the shot, and, in the event of a miss, or if no serious damage results, she may fire again. She will then immediately dive to a depth of about 150 feet, if possible, and proceed for at least 15 minutes at this depth. If no

N 2

pursuit is then heard, she will very probably come to the surface cautiously for another look round.

11. Night Attack.

Night attacks appear to be increasingly favoured by German submarine commanders.

In very bright moonlight a night attack with torpedoes may be made by the submarine when submerged. Under ordinary conditions she will attack on the surface, with more or less buoyancy according to the degree of visibility and state of the sea. She will, when possible, choose a *dark* background with reference to clouds and moon. Under these conditions it may be difficult to see a submarine on the surface, even at 400 yards.

The last stage of the attack will probably be made using the electric motors, in order to prevent the submarine's presence being betrayed by the noise and smell of the Diesel engines.

B.—ATTACK WITH GUNFIRE.

1. Methods of Firing.

Controlled fire is invariably used.

Range-finders are not generally carried in German submarines except those of the converted mercantile and cruiser types. The range, when opening fire, is therefore usually estimated by the control officer, who "spots" the gunlayer on to the target.

(*a*) *Modified Bracket System.*

Submarines frequently use a "bracket" system of firing. With this system great importance is attached to the necessity of obtaining a short shot. Having bracketed over the ship, a system of "barrage" firing is used, by which, with the range decreasing, a short shot is first registered and the range on the sights then left unaltered until an "over" is obtained. The range on the sights is then brought down the requisite amount to cause a short shot to be registered and enable the process to be repeated. With an increasing range, this procedure is, of course, reversed.

(*b*) *System of Maximum Fire.*

When it is desired to fire with the maximum effect—for example, to smother the fire of an armed adversary—the same general system is used, but in this case the gunlayer fires as rapidly as possible without waiting to spot the fall of shot, whilst an observer makes the necessary alterations to the sight to cause him to fire a succession of "barrages" (as described above).

2. Type of Ammunition.

Only high-explosive shell are supplied to submarines, but these are fitted with three different types of fuze.

(i) *Nose-fuzed Shell.*

Used against cargo steamers and when ranging, as, besides their making a large hole in the hull, the splash is large and therefore easy to observe.

(ii) *Internal-fuzed Shell.*

Used against troop transports and vessels carrying ammunition. The explosion takes place in the interior of the ship, resulting in fires, &c.

(iii) *Time and Percussion-fuzed Shell.*

A small proportion of these shells are supplied, primarily for use as time-fuzed shell against the *guns' crews* of armed merchantmen who offer resistance, in order to drive them away from their guns.

3. Conditions of Firing.

The guns' crews of submarines are in a very exposed position. In anything but fair or moderate weather, therefore, they will have difficulty in obtaining any decisive results with their gunfire, if the ship attacked is handled with courage and skill.

The foremost gun is in a more exposed position than the after gun, and the gun's crew will have the greatest difficulty when firing right ahead in bad weather.

APPENDIX II.

PROCEDURE WHEN HUNTED WITH HYDROPHONES.

1. General.

A submarine's best natural means of protection against pursuit is a rough sea, in which any sounds she makes will be drowned by water noises. In really heavy weather she will dive deep, *i.e.*, to over 100 feet, to avoid the "pumping" which occurs anywhere nearer the surface. In a moderate sea, however, it is possible for a submarine to keep under way submerged, watching the movements of her pursuers through her periscope, and only diving deep to avoid being rammed or fouling the pursuing vessels' sweeps, when they are very close to her. The submarine will avoid proceeding *against* the sea when using her periscope, and consequently is not likely to be sighted.

Should the pursuit commence when the submarine is in deep water, at no great distance from which there is water of moderate depth, *i.e.*, 20–30 fathoms or less, she will probably make for a position inside the 30-fathom line, so as to be able to lie on the bottom. She is almost certain to do so if she has been hunted during the previous night or earlier on the same day, as her batteries will be running low.

If the submarine knows she is being hunted, she will from time to time stop to listen on her own hydrophones, in order to form an estimate of the positions and number of her pursuers. It is not necessary for her to stop in order to listen, but she will obtain a much more accurate estimate by doing so.

2. In Deep Water.

Except in case of the most serious emergency, the submarine will not lie on the bottom in depths exceeding 30 fathoms. If pursued when in greater depths, with no shallow water in the neighbourhood, she will keep under way submerged, running her main motors at the lowest possible speed consistent with good depth-keeping. To enable her to do so, she must be accurately trimmed, and any sudden disturbance of her trim, such as may be occasioned by the explosion of depth charges, may cause her either to rise to the surface or make a sudden dive and get out of control. Some German submarines, when pursued with hydrophones, run their two electric motors each at a different number of revolutions in order to break the sound-waves produced by the two propellers ; this they call "proceeding silently."

In order further to reduce sounds to a minimum, all auxiliary machinery in the submarine will, if possible, be stopped. The hydroplanes and vertical rudder can be operated by hand instead of by power, and compressed air can be used for trimming, in place of the usual pumps, or the boat can be trimmed by sending men forward or aft.

In deep water German submarines will usually proceed at a depth of about 150 feet, this being considered to provide entire security against depth charges, and at the same time to leave a reasonable margin of safety for regaining control before a dangerous depth has been reached should the boat make an involuntary dive. For short periods German submarines can dive to well over 200 feet without injury.

3. In Shallow Water.

In shallow water a submarine, if possible, will usually take to the bottom and lie there with all machinery stopped. She may either anchor herself there with very heavy negative buoyancy (obtained by filling her regulator tanks), or she may lie just touching the bottom, with only a small amount of negative buoyancy, allowing the tide to carry her away from her original position. But if there is a heavy ground swell, or the bottom is rocky and injury is likely to result from bumping, she may keep under way. Lying on the bottom is the only measure at present adopted by German submarines which renders them entirely inaudible to their pursuers ; it also gives the submarine the best conditions for ascertaining the type and number and movements of the hunting craft by means of her own hydrophones. In a tideway the tendency will be for the pursuers to drift away from the submarine, especially if they have to stop their own engines in order to listen.

The one great risk a submarine runs when lying on the bottom is that her tanks may not be absolutely tight. If they are not, a leakage of air or oil may occur through vent valves or seams, and the bubbles rising to the surface may betray her. She will, of course, take every precaution to prevent this.

In narrow and shallow waters inshore, the submarine, if she keeps under way, will be compelled to use her periscope occasionally for navigational purposes and also in order to avoid collision with patrol craft, if the latter are numerous and suitably placed.

Part III.
Section 5.

Sub-
marines.

APPENDIX III.

NAVIGATION.

1. General.

The navigation of submarines, like that of destroyers and other smaller vessels of the German Navy, is entrusted to a warrant officer, who is a specialist in the subject, and is known as a *Steuermann* or *Obersteuermann*.

There is a considerable amount of evidence to show that, as a rule, these navigating warrant officers are well up in their work and reliable in taking and working star and sun sights. In addition, many boats carry a second *Steuermann*, or a "war pilot," as he is sometimes called, whose special qualification is knowledge of the locality in which the submarine is to operate. These "war pilots" are usually ex-mercantile officers, who obtained an intimate knowledge of Allied ports by years of trading to them in time of peace.

The commanding officer does not, as a rule, interfere in the navigation or concern himself greatly with it. He gives the navigating warrant officer directions to follow a certain route or work out the course and speed to reach a certain point at a given time. These warrant officers are the right hands of the submarine commanders, and assist in working out firing course, deflection, &c., during the progress of an attack.

2. In German Waters.

Submarines based on German North Sea ports proceed on their cruises along swept channels in the German Bight or *viâ* the Kiel Canal and Little Belt (occasionally the Sound) into the Kattegat. They are escorted by minesweeping or patrol vessels, and also by destroyers, through the areas in which danger from mines is apprehended.

On their return from their cruises submarines are said to report by W/T beforehand the expected hour of their arrival at a pre-arranged rendezvous, where they are then met by surface escorting craft. The latter apparently accompany them to the vicinity of their base.

Submarines return either through the German Bight or *viâ* the Sound.* In the latter case they are apparently met near the north entrance to the Sound and escorted to Kiel.

3. In the North Sea.

The submarines whose operations are confined to the North Sea are coastal (U.B.) and minelaying (U.C.) boats, working either from Flanders or from a North Sea port.

They depend more on light vessels and buoys than on sights for their position. A line of soundings may occasionally be used. Navigation is complicated by fog and tidal streams, the latter varying at different depths, and by the frequent alterations of course and speed necessary to avoid patrols or to reach a position for attack.

4. In the Channel.

The same general considerations apply as in the North Sea, but the main navigational difficulty is in the passage through the Straits of Dover. In view of the strength of the tides, the position must be accurately determined before attempting the passage, and submarines sometimes lie on the bottom for several hours to await an opportunity for getting a fix without being sighted by patrols.

In spite of these difficulties, the route *viâ* the Straits of Dover was till recently used by all classes of boats as a frequent alternative to the route northabout, the advantages being the saving of time and better weather. But it appears that the larger submarines are now (April 1918) disinclined to attempt the passage, and any "U." boats encountered in the Channel will probably have come, and will intend to return, northabout.

The submarines of the Flanders Flotilla ("U.B." and "U.C." types) still always use the Dover route when entering and leaving the Channel, but the increased activity of this flotilla off the east coast of England now observable (April 1918) clearly indicates that they too are finding the passage of the Straits more difficult than it used to be.

* More rarely *viâ* the Little Belt.

5. In the Atlantic.

The Dover route having fallen into disfavour (*see* above), the larger boats operating in the Atlantic now usually proceed northabout round Scotland. In doing so during the winter months they all appear to prefer to pass through the Fair Isle Channel, except the converted mercantile type, which probably always pass round the north of the Shetlands, keeping for the greater part of the way to the deep channel outside the 100-fathom line, and making a landfall at Utsire or some other point further north on the Norwegian coast. During the summer months all "U." boats, both outward and homeward bound, apparently prefer to pass round Muckle Flugga.

In proceeding southward they sometimes pass through the Minches, but more usually proceed west of the Hebrides. Their subsequent navigation depends upon the area of operations assigned to them, but in general it may be stated that, failing satisfactory sights, they will not hesitate to close the land in order to get a satisfactory fix. On such occasions they will be ready for instant diving, as it is vital to them to avoid being sighted and reported. A keen look-out is kept, as a rule, by an officer, a petty officer, and two seamen, all provided with excellent glasses.

6. In the Mediterranean.

Conditions of navigation vary greatly in different parts of the Mediterranean. The main difficulty, which has to be negotiated by every submarine leaving and entering the Adriatic, is the passage of the Straits of Otranto. This usually entails proceeding submerged for a considerable period (20 hours has been mentioned).

New submarines joining the Mediterranean flotilla are no longer sent overland, but proceed northabout round Scotland and through the Straits of Gibraltar, which they probably pass during the night.

SUBMARINE U. 157.

(CONVERTED MERCANTILE CLASS).

(With 2 - 5·9" guns).

(Stern View).

Ordnance Survey, May, 1918

SUBMARINES.

U. 110.

(U. 105-114).

(With 4·1" gun forward and 22-pr. gun aft.)

U. 103.

(U. 99-104).

(With 4·1" gun forward and 22-pr. gun aft.)

Ordnance Survey, May, 1918.

SUBMARINE.

U. 51-56
(With 2 - 22-pr. guns).

U. 51-56.
(With 4·1-inch gun).

SUBMARINE.

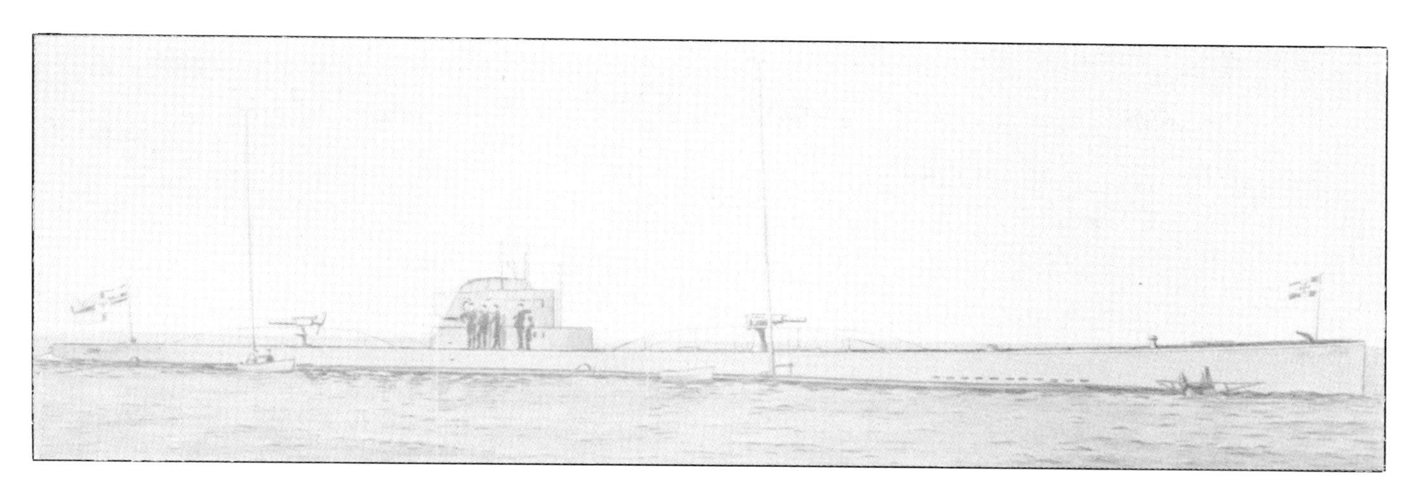

U. 53.

(U. 51-56).

(With 2—22-pr. guns).

U. 53.

View of Conning Tower.

Ordnance Survey, May, 1918.

SUBMARINES.

U. 38-41.

(With 22-pr. guns; present armament is 1 - 4·1" and 1 - 22-pr. gun).

U. 34.

(With 1 - 4·1" gun; in addition, it is believed, a 22-pr. gun has now been mounted aft.)

U. 34.

Ordnance Survey, May, 1918.

SUBMARINE.

U. 35.
(Alongside Spanish Cruiser).

U. 35.
(Conning Tower).

U. 35.
(The foremost 22-pr. has now been replaced by 1 - 4·1 inch gun).

SUBMARINES.

U. 32.

(Since date of Photograph a 4·1 inch gun has been mounted on fore side of Conning Tower, and a permanent Bridge Screen has been added).

U. 24-25.

(Now armed with 2-22-pr. guns, one before and one abaft Conning Tower).

U. 16.

(1-4-pr. gun now mounted on fore side of Conning Tower).

SUBMARINES.

U.B. 105.
(U.B. 102-110).

U.B. 49.
(U.B. 48-53).

CONNING TOWER.
U.B. 49.
(Gun is a 22-pr.; many boats of this class carry a 4·1" gun).

SUBMARINES.

U.B. 30.
(U.B. 30-41).

U.B. 1-8.
(U.B. 9-17 very similar).

Ordnance Survey, May, 1918.

SUBMARINES.

U.C. 52.

(U.C. 46-54).

(Other submarines of U.C. 16-79 class generally similar, but many of them have a rounded bow).

U.C. 1-10.

(U.C. 11-15 very similar).

Note:—The painting in waves, as above, is now obsolete.

Ordnance Survey, May, 1918.

SUBMARINE.

U.C. 48.

View taken in Dock.

Ordnance Survey, May, 1918.

U. 139-142.

(ROUGH SKETCH DRAWN FROM MEMORY).

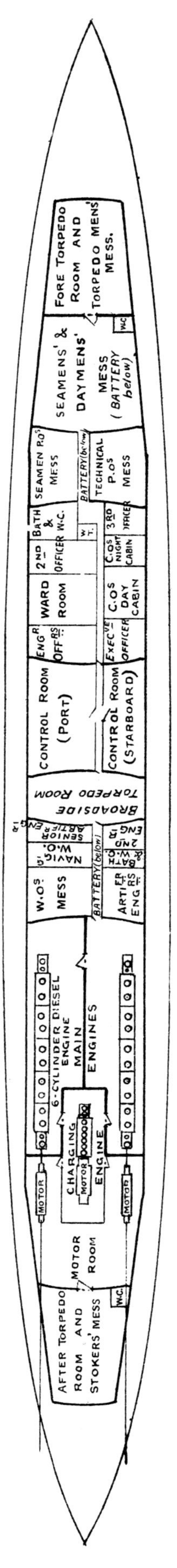

U. 151-154.

(ROUGH SKETCH DRAWN FROM MEMORY).

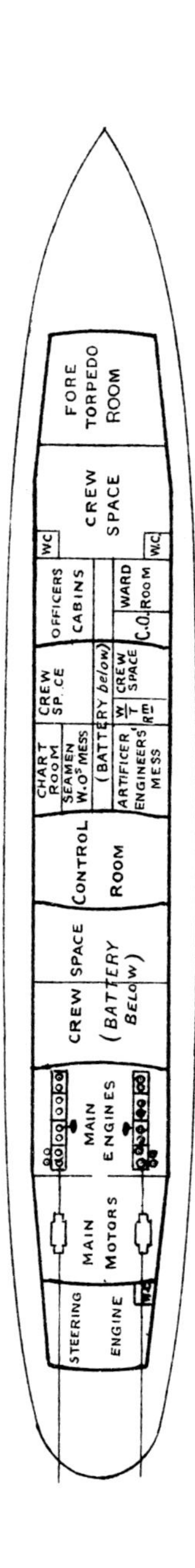

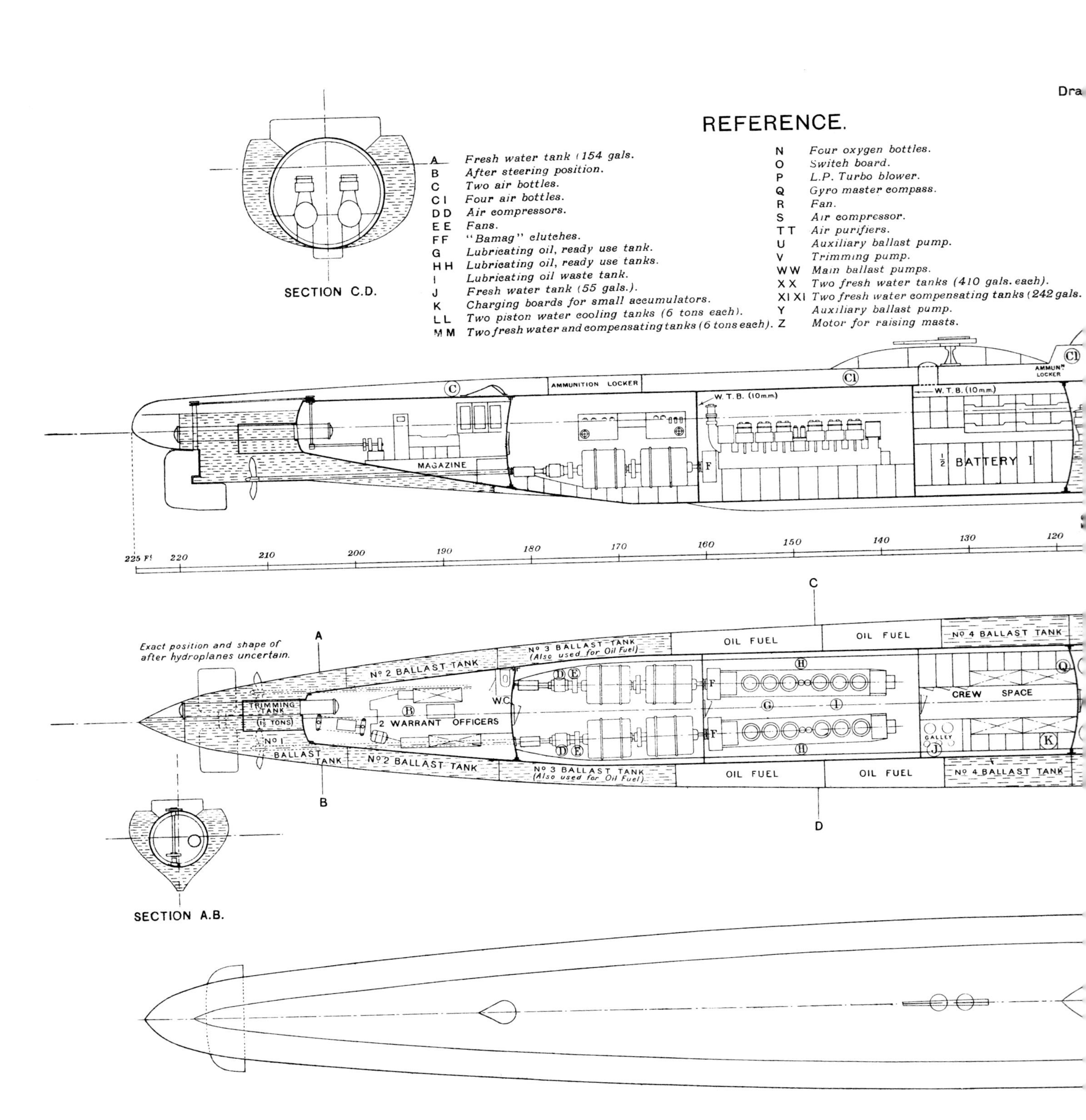
Dra
REFERENCE.
A Fresh water tank (154 gals.
B After steering position.
C Two air bottles.
C I Four air bottles.
D D Air compressors.
E E Fans.
F F "Bamag" clutches.
G Lubricating oil, ready use tank.
H H Lubricating oil, ready use tanks.
I Lubricating oil waste tank.
J Fresh water tank (55 gals.).
K Charging boards for small accumulators.
L L Two piston water cooling tanks (6 tons each).
M M Two fresh water and compensating tanks (6 tons each).
N Four oxygen bottles.
O Switch board.
P L.P. Turbo blower.
Q Gyro master compass.
R Fan.
S Air compressor.
T T Air purifiers.
U Auxiliary ballast pump.
V Trimming pump.
W W Main ballast pumps.
X X Two fresh water tanks (410 gals. each).
XI XI Two fresh water compensating tanks (242 gals.
Y Auxiliary ballast pump.
Z Motor for raising masts.
SECTION C.D.
AMMUNITION LOCKER
AMMUNN. LOCKER
W. T. B. (10mm.)
MAGAZINE
½ BATTERY I
225 Ft 220 210 200 190 180 170 160 150 140 130 120
Exact position and shape of after hydroplanes uncertain.
No 2 BALLAST TANK
No 3 BALLAST TANK (Also used for Oil Fuel)
OIL FUEL
No 4 BALLAST TANK
TRIMMING TANK (1½ TONS)
2 WARRANT OFFICERS
WC
CREW SPACE
GALLEY
No 1 BALLAST TANK
SECTION A.B.

Plate 12.
C.B. 1182 S.
April 1918.

-70.

...oximate only.

Total length, about - - - 225′ 0″
Beam at midship section, about 21′ 0″
Draught, including keel, about 12′ 0″
Displacement on the surface, about 850 Tons.
" submerged, about 1200 "
Armament: One 4·1″ gun forward, and one 22 pdr. gun aft. (Drawing shows two 22 pdr. guns.)
Five Torpedo tubes, 17·7″ (four forward and one aft).

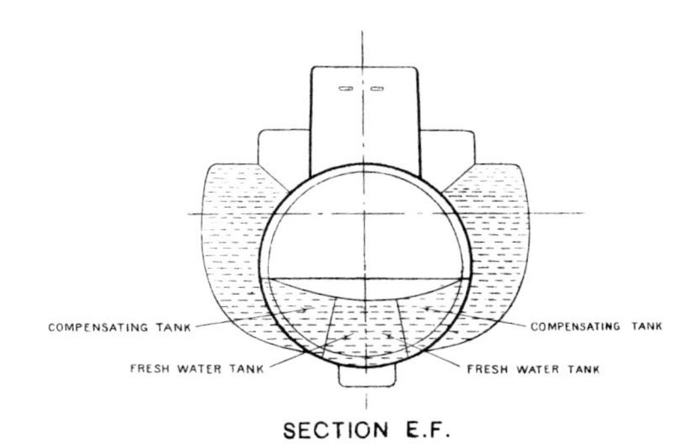

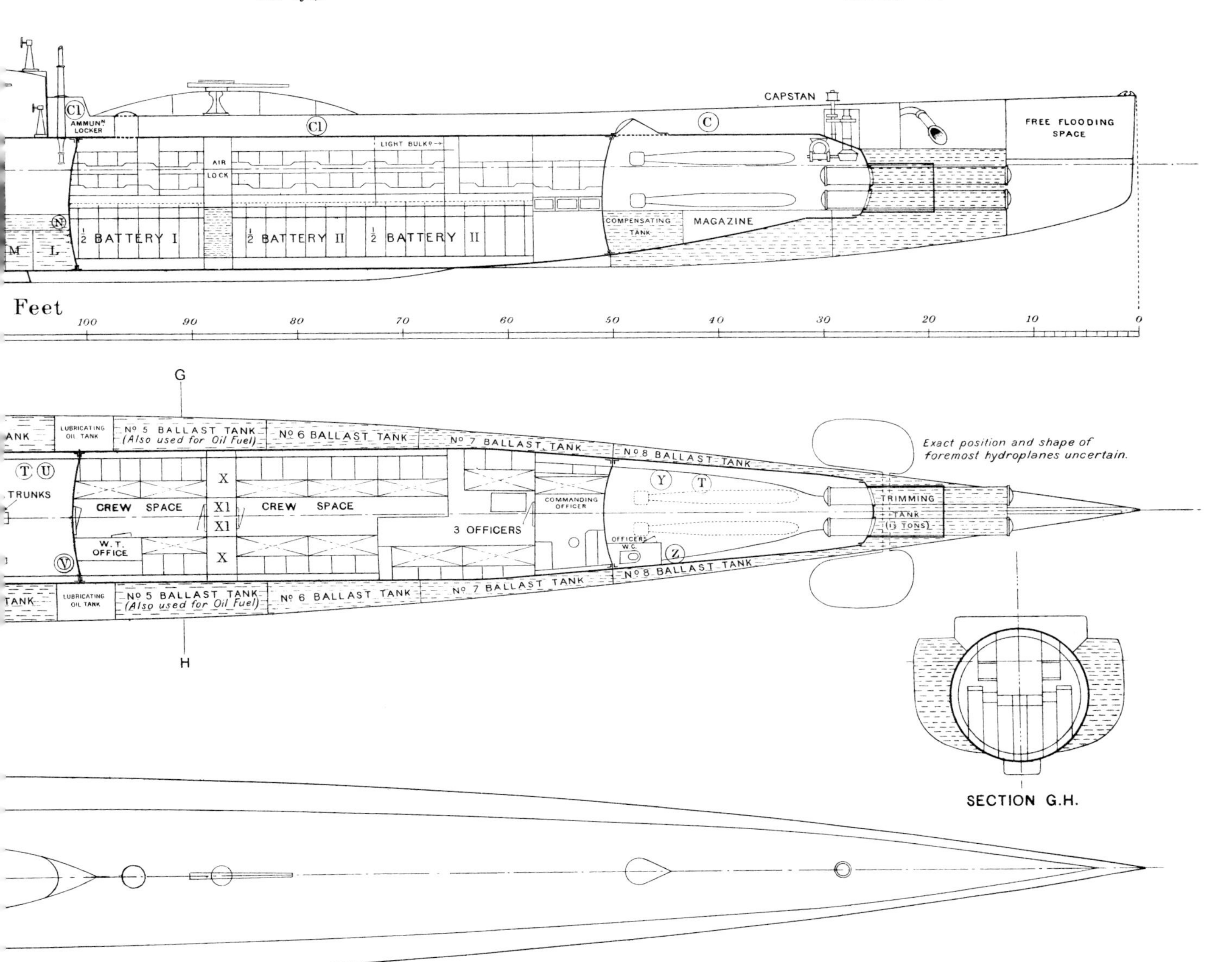

Ordnance Survey, May, 1918.

U 5

Length (over-
Breadth (abo
Draught (on

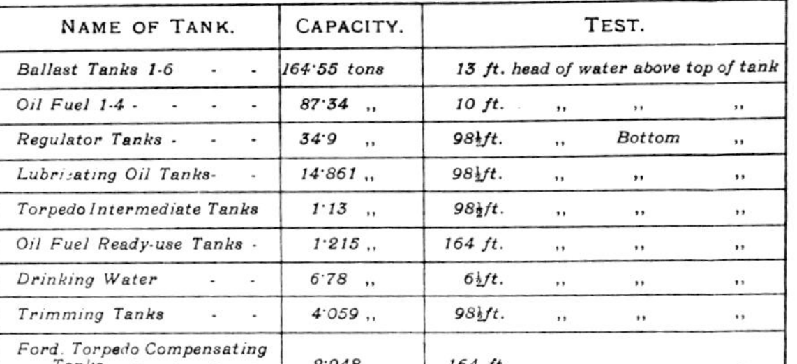

Name of Tank.	Capacity.	Test.
Ballast Tanks 1-6	164·55 tons	13 ft. head of water above top of tank
Oil Fuel 1-4	87·34 ,,	10 ft. ,, ,, ,,
Regulator Tanks	34·9 ,,	98½ft. ,, Bottom ,,
Lubricating Oil Tanks	14·861 ,,	98½ft. ,, ,, ,,
Torpedo Intermediate Tanks	1·13 ,,	98½ft. ,, ,, ,,
Oil Fuel Ready-use Tanks	1·215 ,,	164 ft. ,, ,, ,,
Drinking Water	6·78 ,,	6½ft. ,, ,, ,,
Trimming Tanks	4·059 ,,	98½ft. ,, ,, ,,
Ford. Torpedo Compensating Tanks	2·948 ,,	164 ft. ,, ,, ,,

CAPACITIES IN TONS OF FRESH WATER

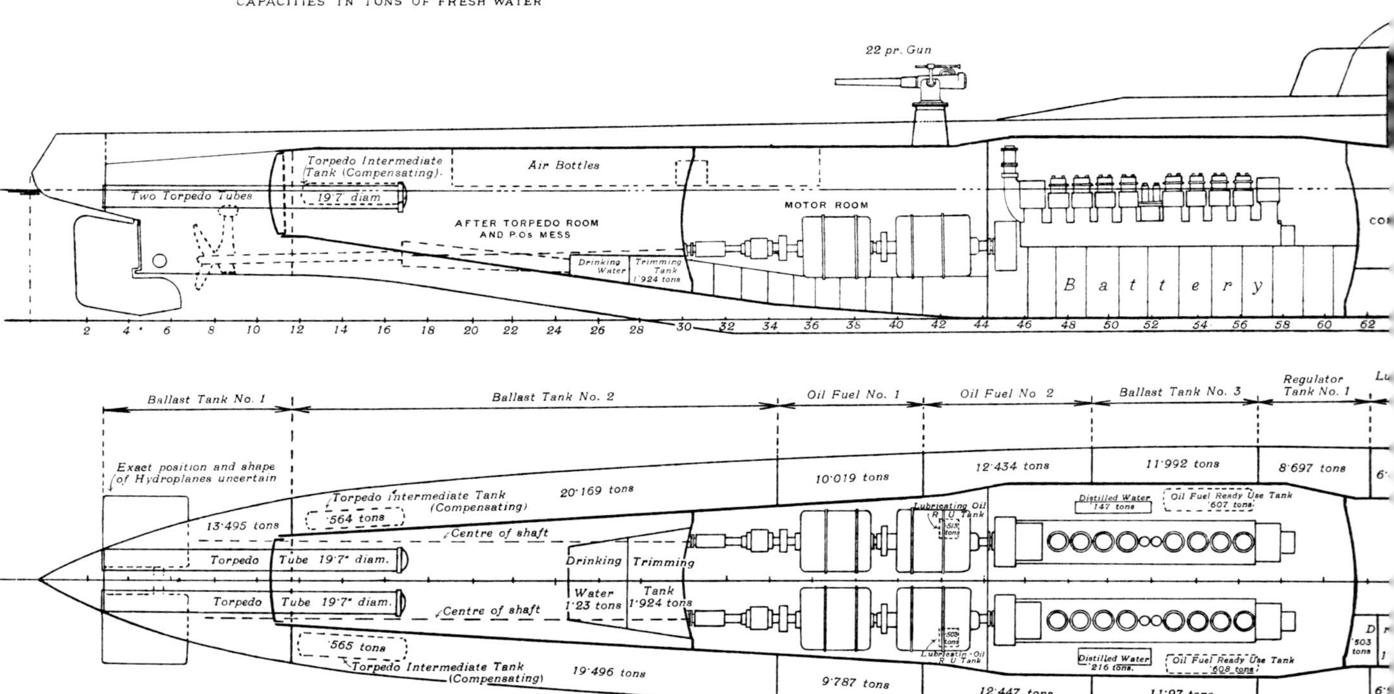

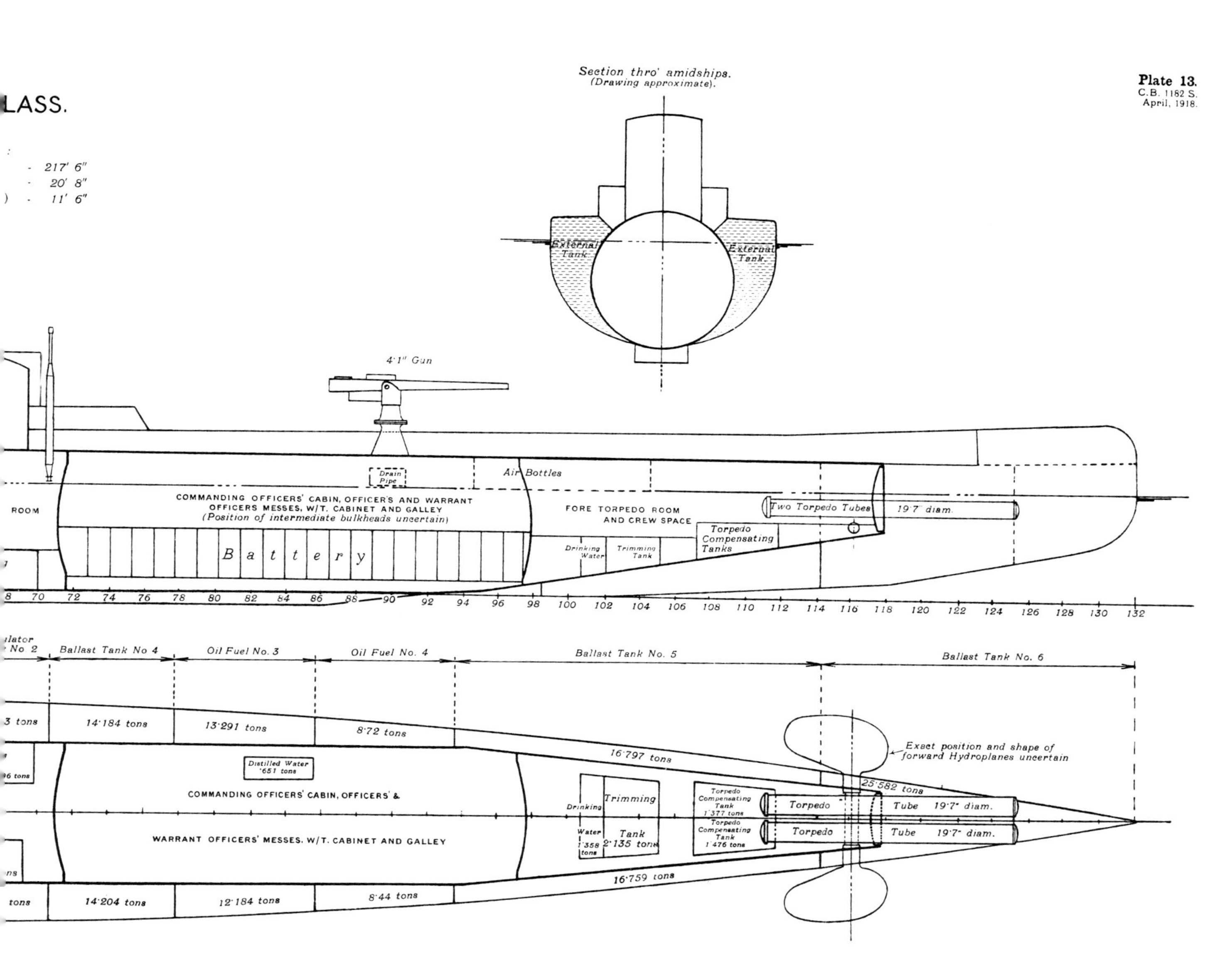

Ordnance Survey, May. 1918.

LOHMANN CLUTCH WITH CENTERING DEVICE

REFERENCE.

2, 3.	Casing bolted to Diesel shaft.
B. B.	Movable friction blocks.
15.	Clutch operating sleeve.
7, 8.	Adjusting gear for 10.
10.	Adjustable friction plates.
X.	Movable cone-shaped centering ring.

BALL THRUST BLOCK

TYPE I.

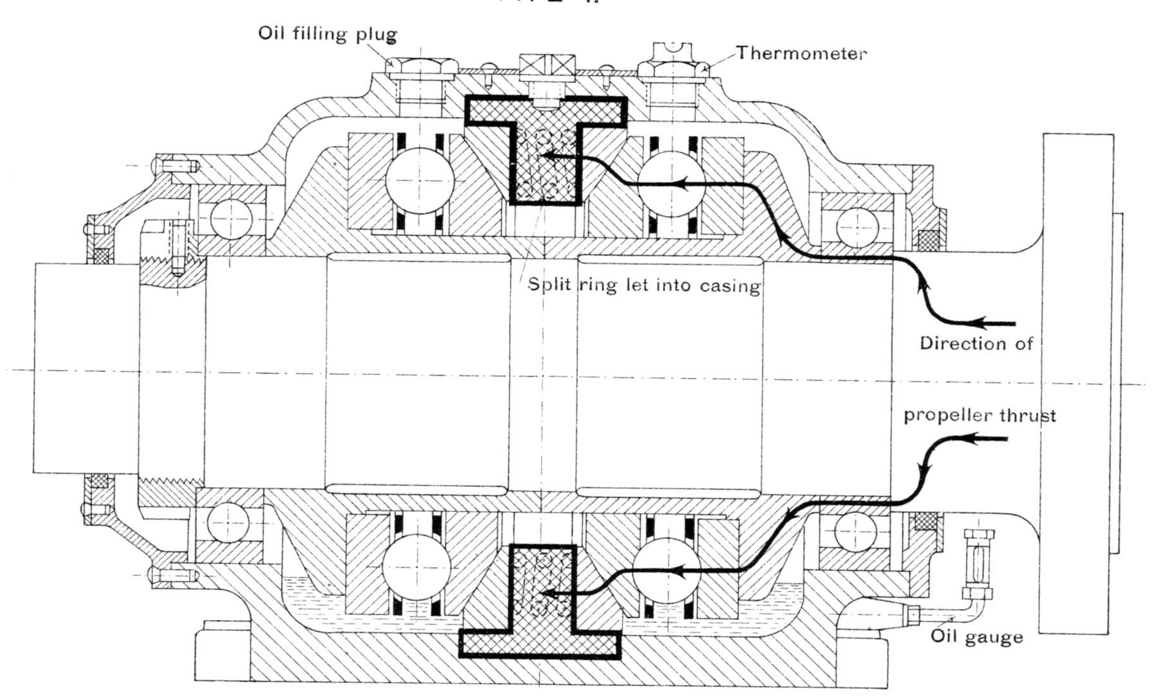

BALL THRUST BLOCK

TYPE II.

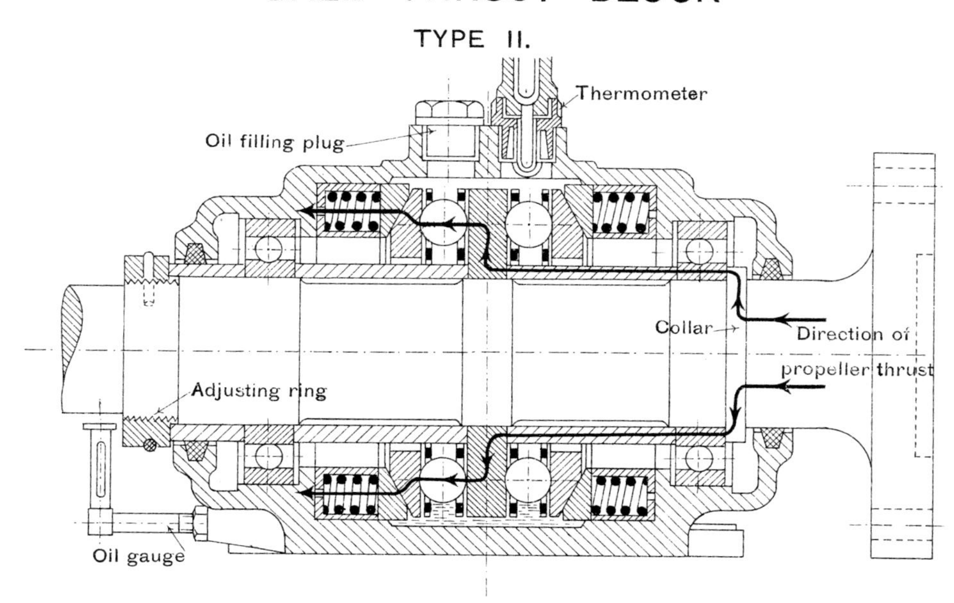

Ordnance Survey, May, 1918.

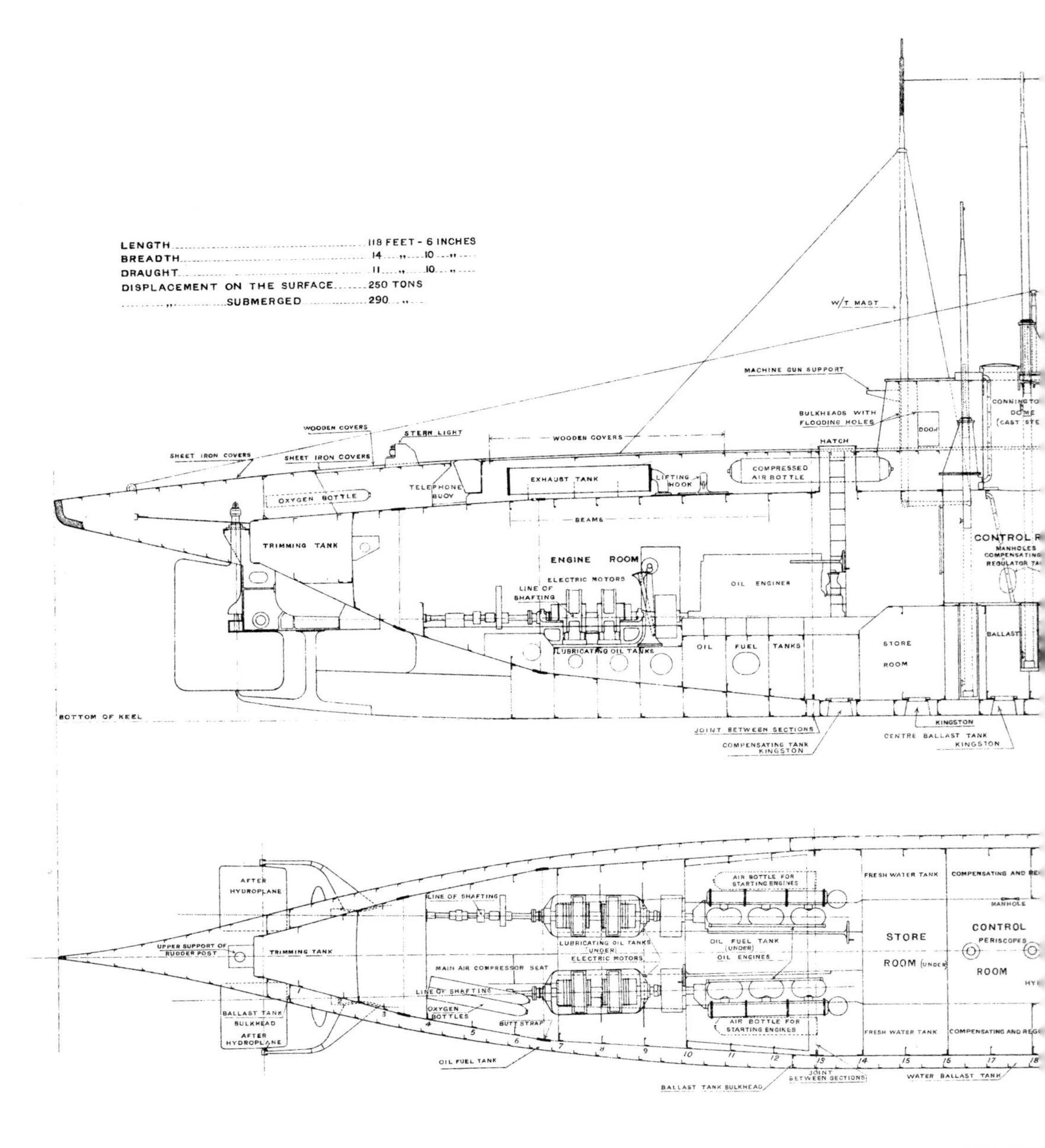
LENGTH 118 FEET - 6 INCHES
BREADTH 14 „ 10 „
DRAUGHT 11 „ 10 „
DISPLACEMENT ON THE SURFACE 250 TONS
„ SUBMERGED 290 „
W/T MAST
MACHINE GUN SUPPORT
BULKHEADS WITH FLOODING HOLES
HATCH
WOODEN COVERS
SHEET IRON COVERS
STERN LIGHT
TELEPHONE BUOY
OXYGEN BOTTLE
EXHAUST TANK
LIFTING HOOK
COMPRESSED AIR BOTTLE
BEAMS
TRIMMING TANK
ENGINE ROOM
ELECTRIC MOTORS
LINE OF SHAFTING
OIL ENGINES
LUBRICATING OIL TANKS
OIL FUEL TANKS
STORE ROOM
CONTROL R
MANHOLES
BALLAST
BOTTOM OF KEEL
JOINT BETWEEN SECTIONS
COMPENSATING TANK KINGSTON
KINGSTON
CENTRE BALLAST TANK KINGSTON
AFTER HYDROPLANE
UPPER SUPPORT OF RUDDER POST
AIR BOTTLE FOR STARTING ENGINES
OIL FUEL TANK (UNDER)
MAIN AIR COMPRESSOR SEAT
OXYGEN BOTTLES
BUTT STRAP
BALLAST TANK BULKHEAD
OIL FUEL TANK
FRESH WATER TANK
COMPENSATING AND REG
STORE ROOM (UNDER)
CONTROL ROOM
PERISCOPES
WATER BALLAST TANK

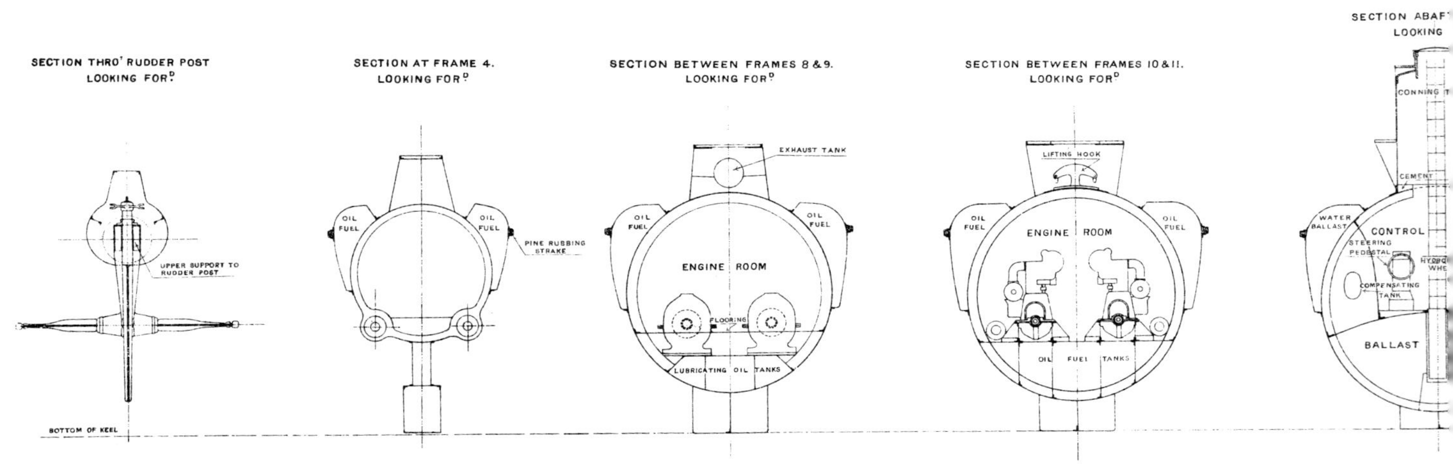
SECTION THRO' RUDDER POST LOOKING FOR^{D}
UPPER SUPPORT TO RUDDER POST
BOTTOM OF KEEL
SECTION AT FRAME 4. LOOKING FOR^{D}
OIL FUEL
PINE RUBBING STRAKE
SECTION BETWEEN FRAMES 8 & 9. LOOKING FOR^{D}
EXHAUST TANK
ENGINE ROOM
FLOORING
LUBRICATING OIL TANKS
SECTION BETWEEN FRAMES 10 & 11. LOOKING FOR^{D}
LIFTING HOOK
ENGINE ROOM
OIL FUEL TANKS
SECTION ABAF
LOOKING
CONNING T
CEMENT
WATER BALLAST
CONTROL
STEERING PEDESTAL
COMPENSATING TANK
BALLAST

Plate 16.
C.B. 1182 S.
April, 1918.

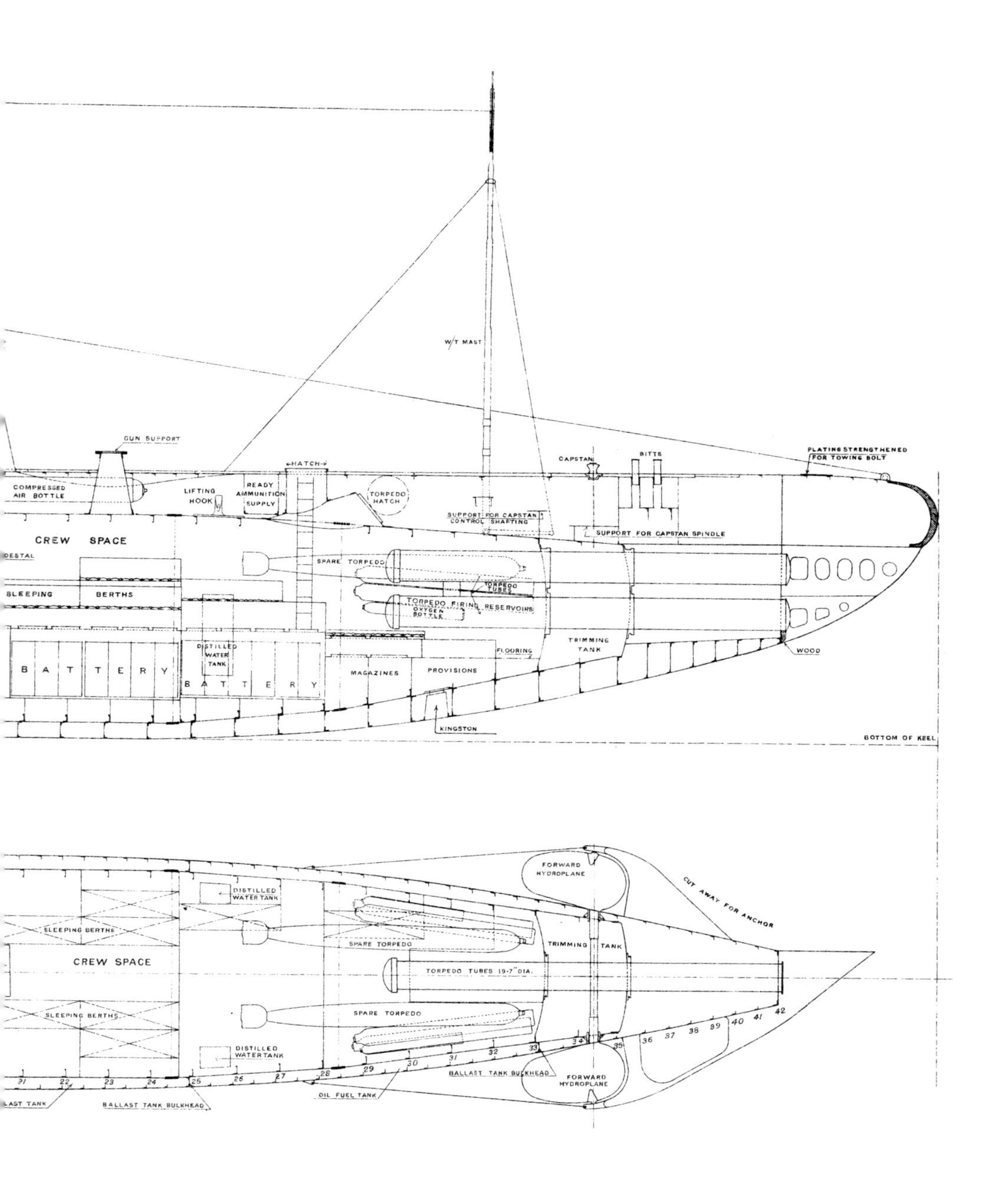

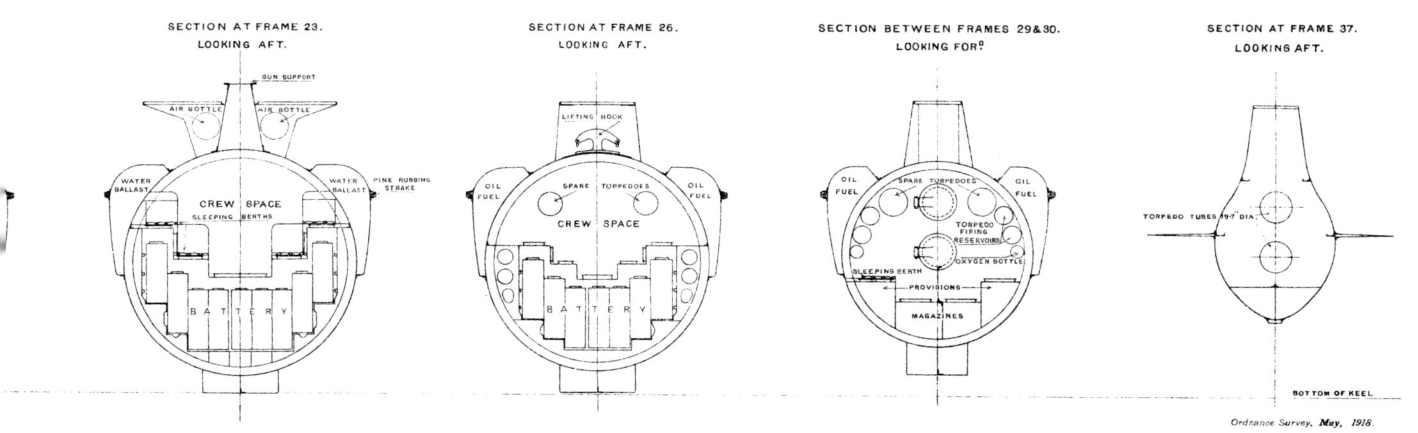

Ordnance Survey, May, 1918.

U.C

[*There are slight differences betw good many of these boats have bow here shown.*]

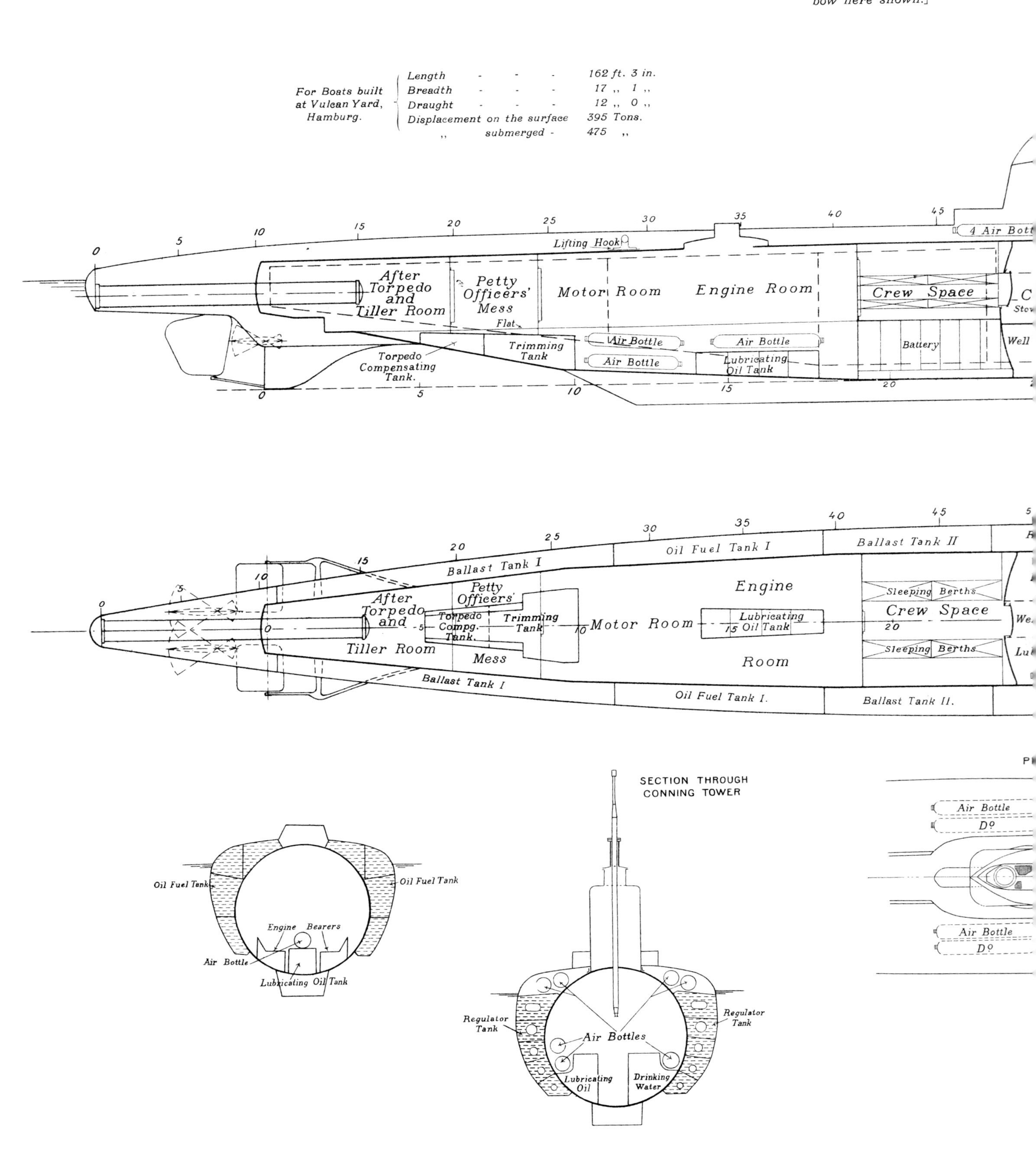

…oats built in different Yards, and a …anging stem instead of the rounded

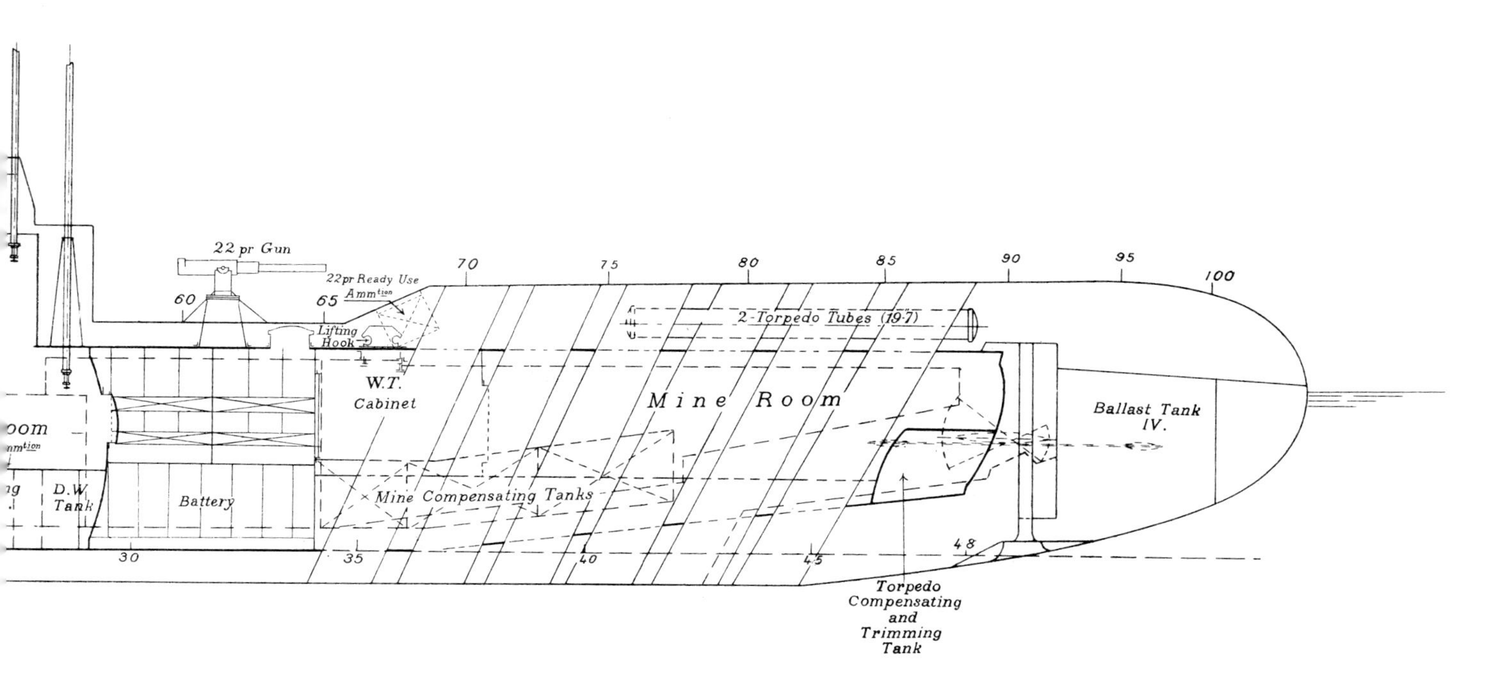

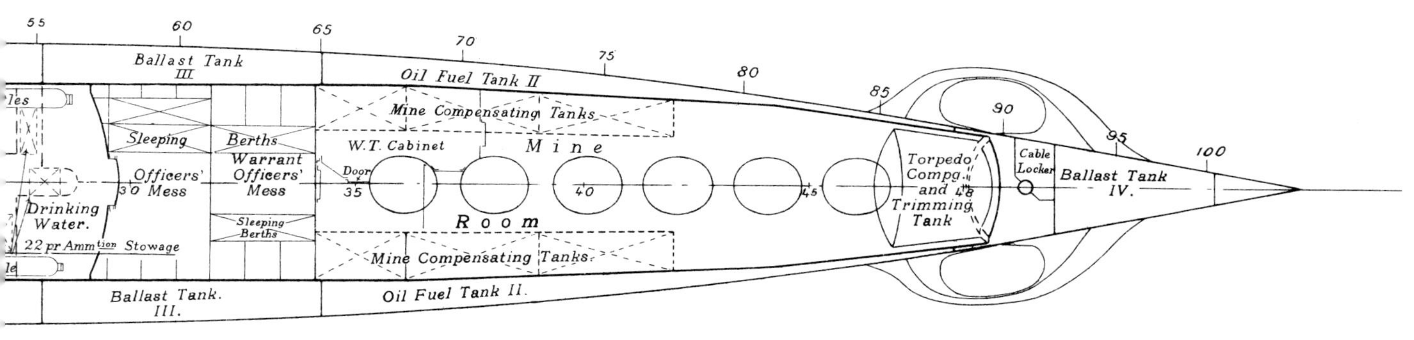

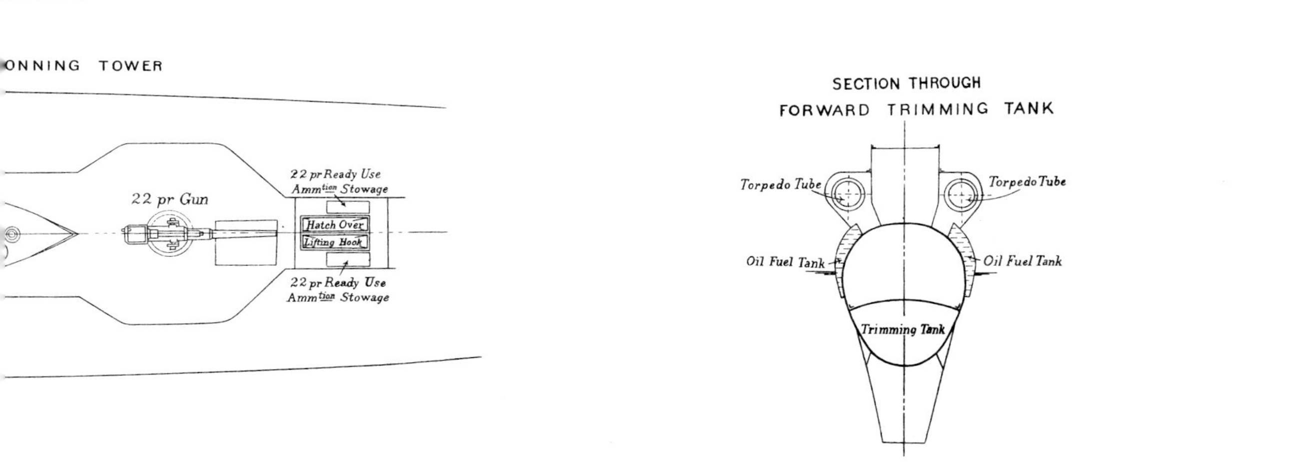

Ordnance Survey, May, 1918.

U.C

Section

Length (
Breadth (
Draught (

Displacement { S
S

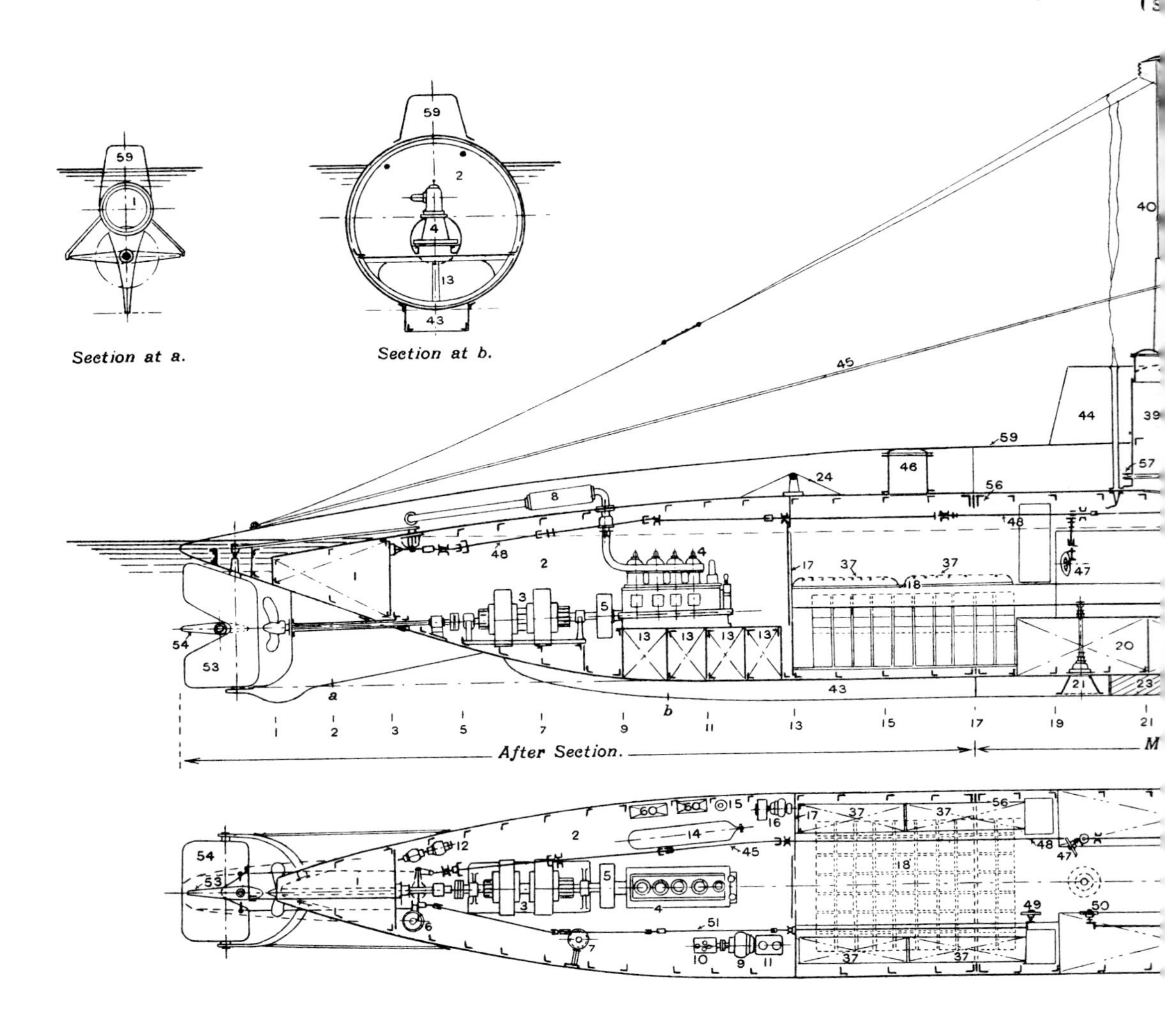

Section at a.

Section at b.

REF

1. After Trimming Tank, 1·37 tons
2. Engine Room.
3. Main Motors, each 59-kw.
4. Diesel-Benz Engine, 80 H.P., 4-cycle, non-reversible.
5. Friction Clutch.
6. Anschütz Gyroscopic Compass (master compass).
7. W.C.
8. Silencer.
9. Motor for auxiliary air compressor and ballast pumps, 8·8—10·3-kw., 110-160 volts.
10. Auxiliary Air Compressor.
11. Ballast Pumps.
12. Gyro. Motor Generator.
13. Oil Fuel Tanks, 706 gallons.
14. Starting Air Bottle.
15. Air Vessel for engine fuel spraying.
16. Fan, ·85 – 1·3-kw., 110-160 volts.
17. Ordinary Bulkhead.
18. Accumulators, 70 in number } arranged in 2 batterie
19. Ditto 42 ditto } 56 cells each.
20. Ballast and Fresh Water Tanks, 13 tons.
21. Kingston, diameter 8·46".
22. Compensating Tanks. 2·7 tons.
23. Safety Weight.
24. Lifting Shackles.
25. Telephone and Light Buoy.
26. Sounding Machine.
27. Pressure Bulkhead.
28. Mine Tubes (open to sea).
29. Foremost Trimming Tanks, 2·2 tons.
30. Mine Compensating Tanks, 3·6 tons.
31. Air Bottles, 8 in number, 185 gallons.
32. Foremost Ballast Tank, 8 tons.

Plate 18.
C.B. 1182.S.
April, 1918.

15,

Views.

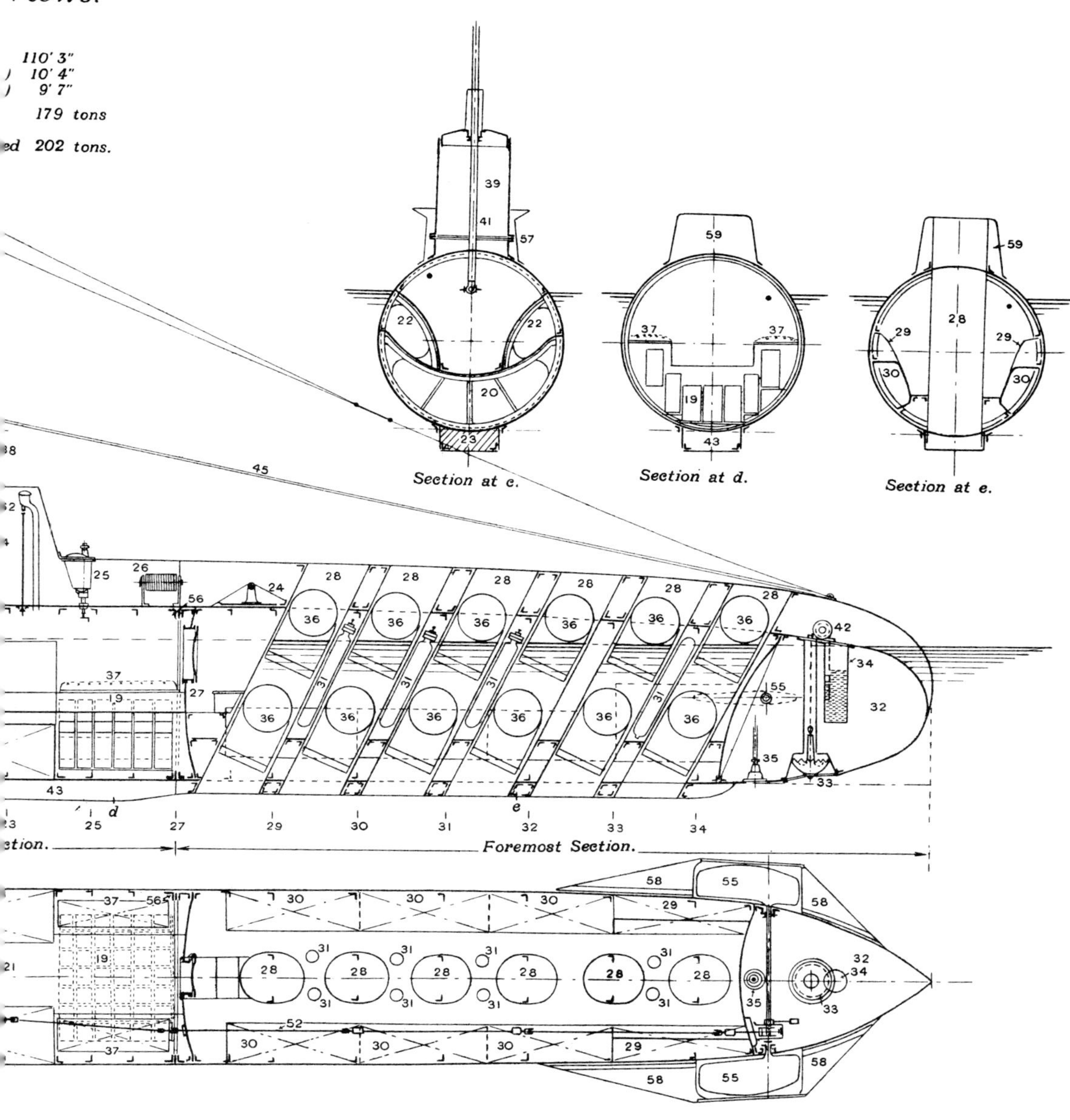

NCE.

3. Mushroom-head Anchor, 3 cwt.
4. Cable Locker, 55 fms., ·63".
5. Kingston.
6. Mines, 12 in number, charge 264lbs. T.N.T., negative buoyancy 660 lbs.
7. Bunks, 6 in number.
8. Steering Wheel, fitted to unship.
9. Conning Tower, diameter 4' 3", open to pressure hull.
0. Telescopic Mast, top 26 ft. above water-line.
1. Periscope, diameter 3·5"—5·9", raised by electric motor of 4·5 H.P. which also works capstan.
2. Capstan.
3. Ballast Keel, 18·3 tons.
4. Fairwater surrounding Conning Tower.
5. Jumping Wires, 1·2".
6. After Hatchway.
47. Hand Wheel for Vertical Rudder.
48. Gearing of Vertical Rudder.
49. Hand Wheel for Diving Rudder.
50. Ditto Foremost Hydroplanes.
51. Gearing of Diving Rudder.
52. Ditto Foremost Hydroplanes.
53. Vertical Rudder, area 25 sq. ft.
54. Diving Rudder, area 19·4 sq. ft.
55. Foremost Hydroplanes, area 11·8 sq. ft. each.
56. Connecting Flanges, fitted with bolts and rubber packing.
57. Connecting Flanges for Conning Tower.
58. Hydroplane Guard.
59. Superstructure (open to sea).
60. Lubricating Oil Tanks, 189 gallons.
61. Aerial, 2 wires, length 50 ft.
62. Main vent from tanks.

Ordnance Survey, *May, 1918.*

SUBM

FLOODING, VENTING, AND

NOTE.—The Emergency V
with compresse

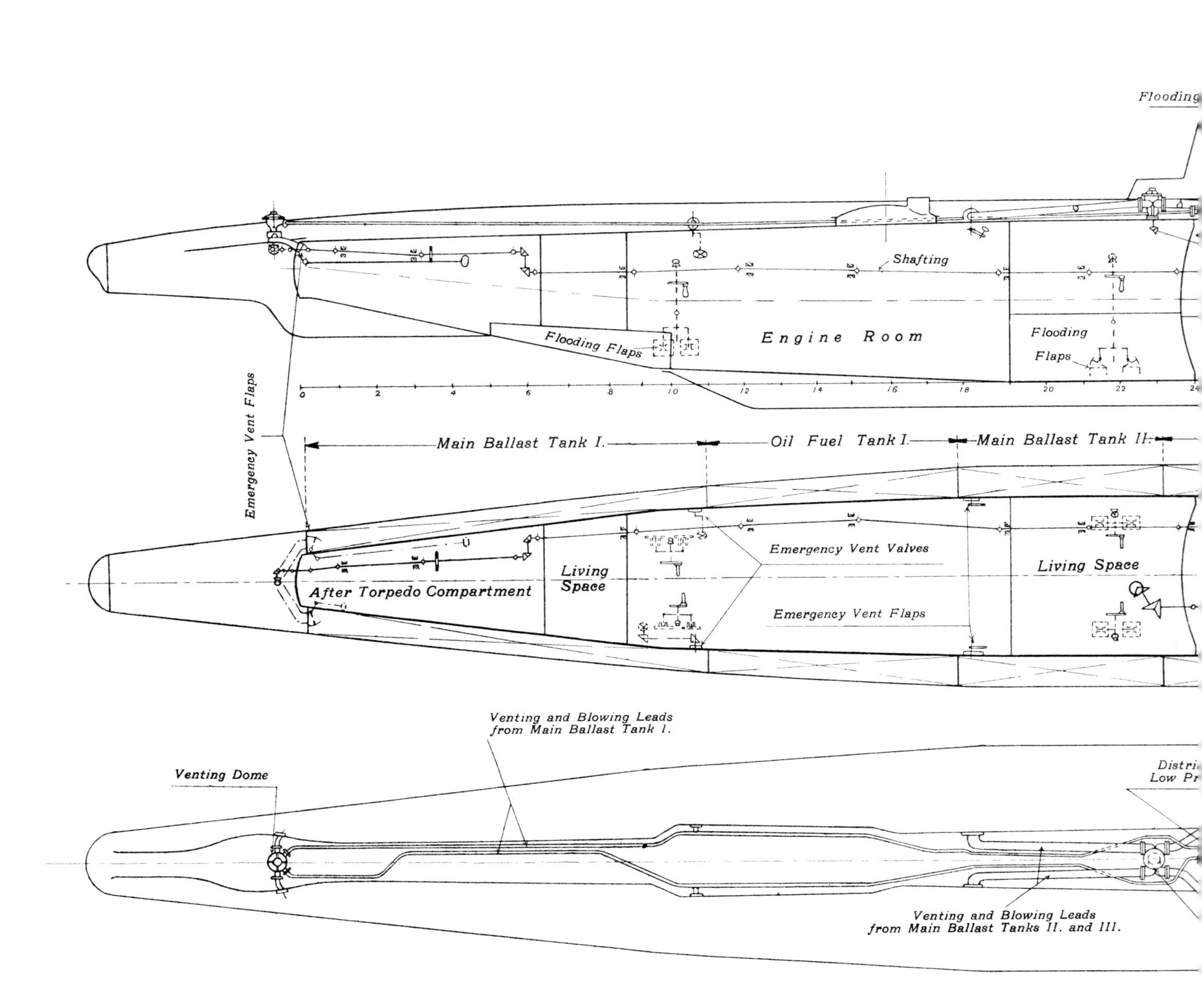

S—U.C. 16-79 CLASS.

NG ARRANGEMENTS OF THE BALLAST TANKS.

s, Valves, and Slides are only closed when blowing en the L.P. Blowing Leads are damaged.

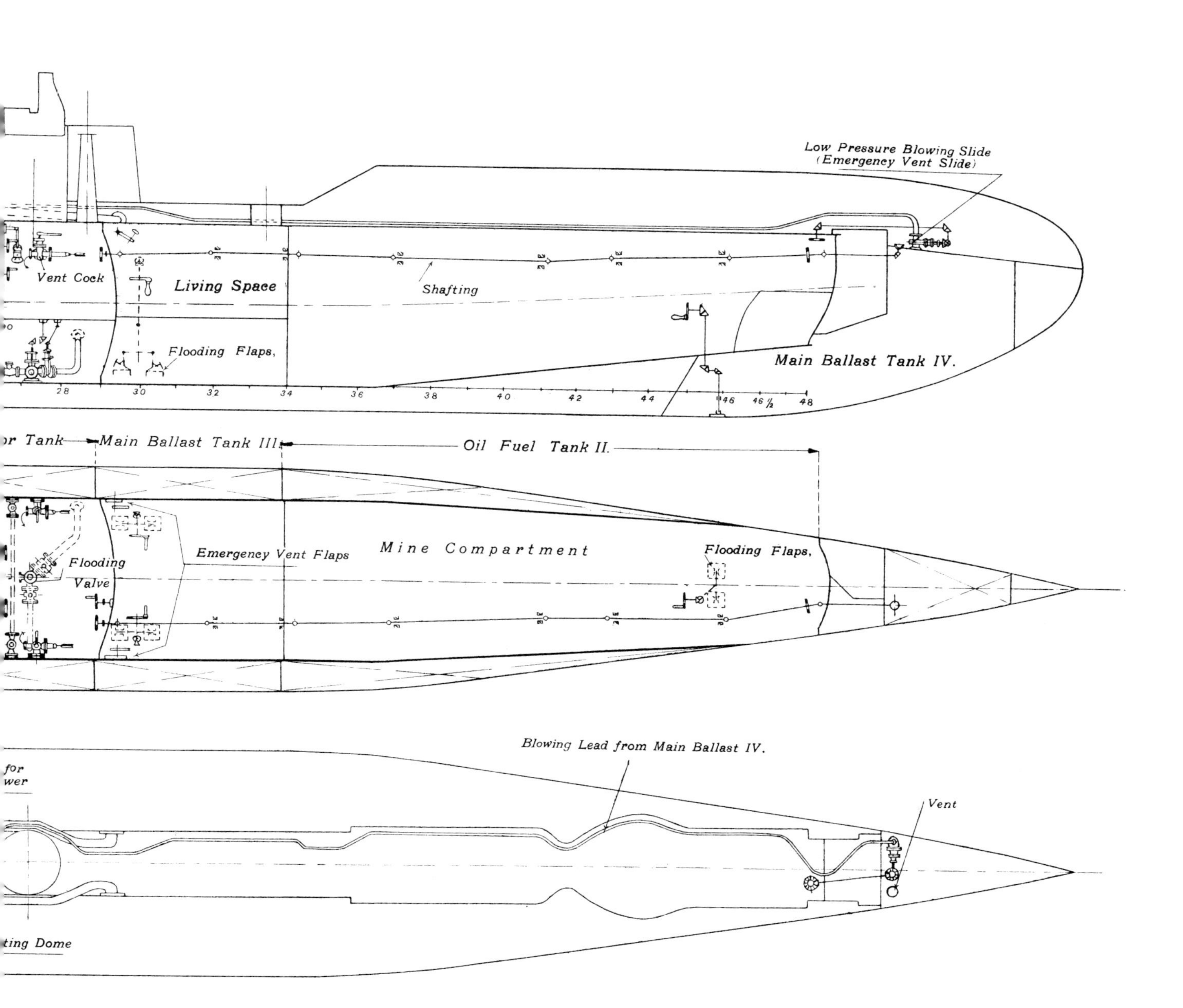

Ordnance Survey, May, 1918.

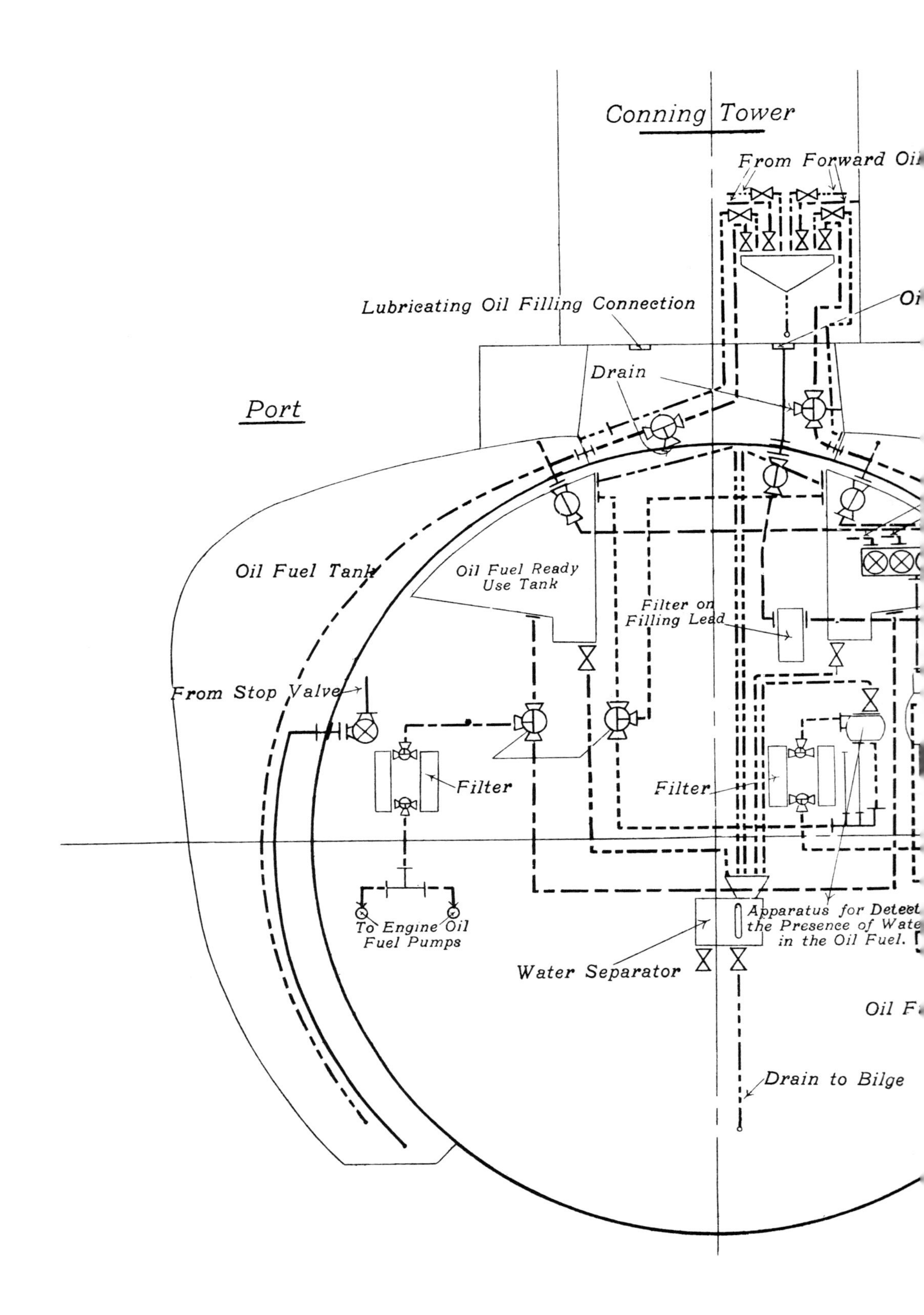
Conning Tower
From Forward Oi
Oi
Lubricating Oil Filling Connection
Drain
Port
Oil Fuel Tank
Oil Fuel Ready
Use Tank
Filter on
Filling Lead
From Stop Valve
Filter
Filter
Apparatus for Detect
the Presence of Wate
in the Oil Fuel.
To Engine Oil
Fuel Pumps
Water Separator
Oil F
Drain to Bilge

BMARINES.
16-79 CLASS.
OF OIL FUEL LEADS.

Plate 20.
C.B. 1182. S.
April, 1918.

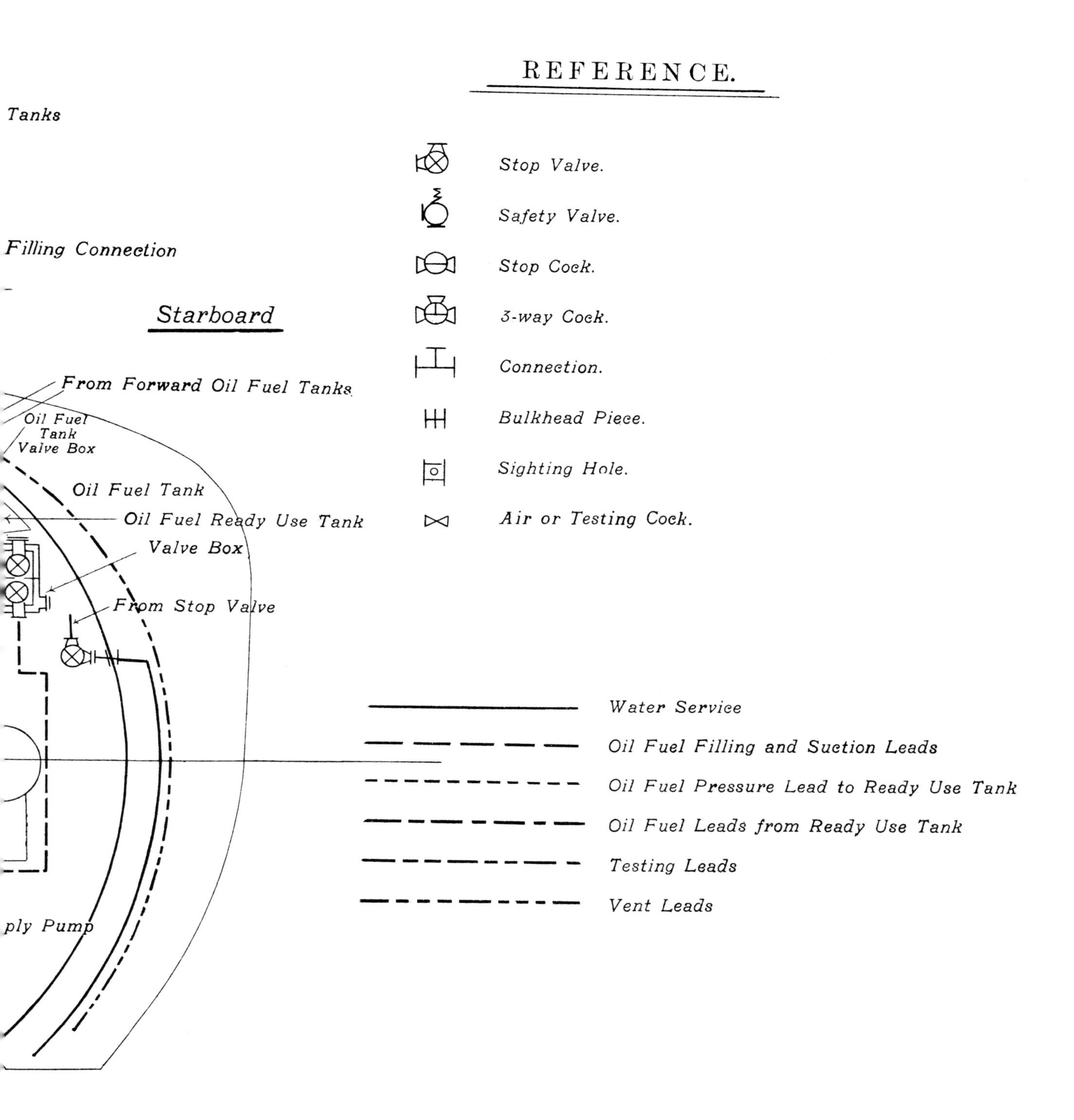

Ordnance Survey, May, 1918.

SUBMARINES—

VENTILATIO

UNDER NORMAL

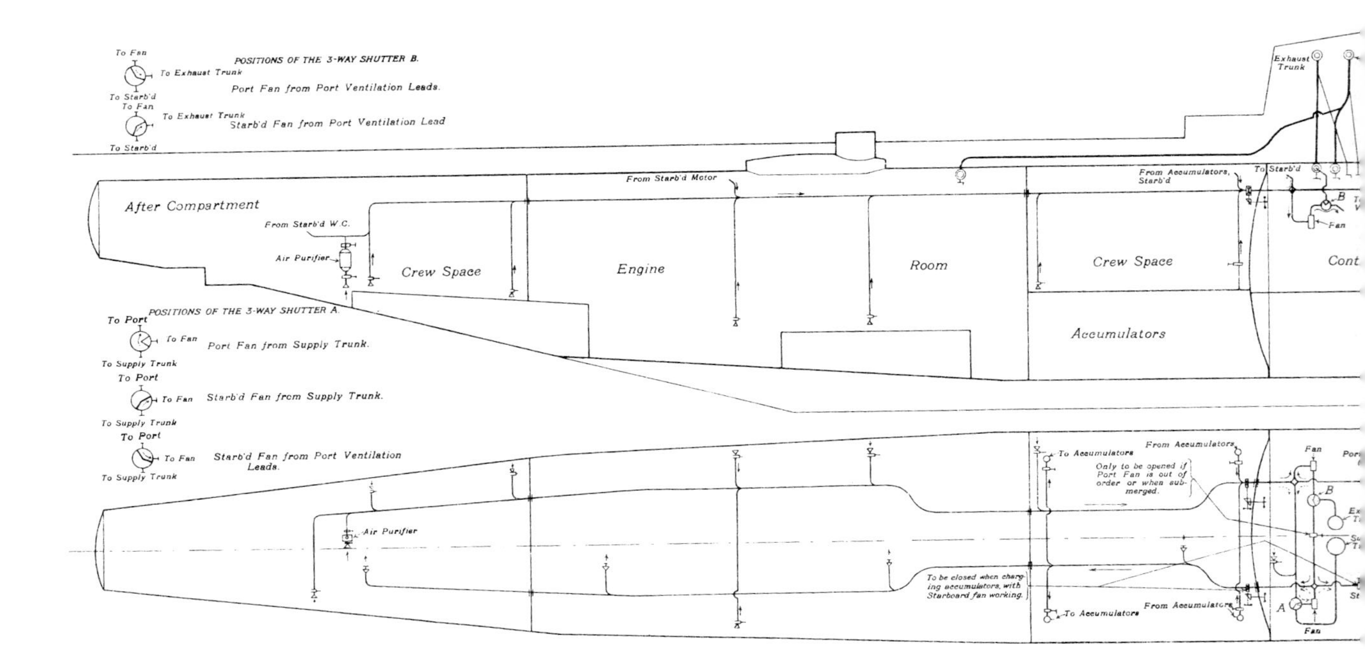

16-79 CLASS.

NGEMENTS.

ED CONDITIONS.

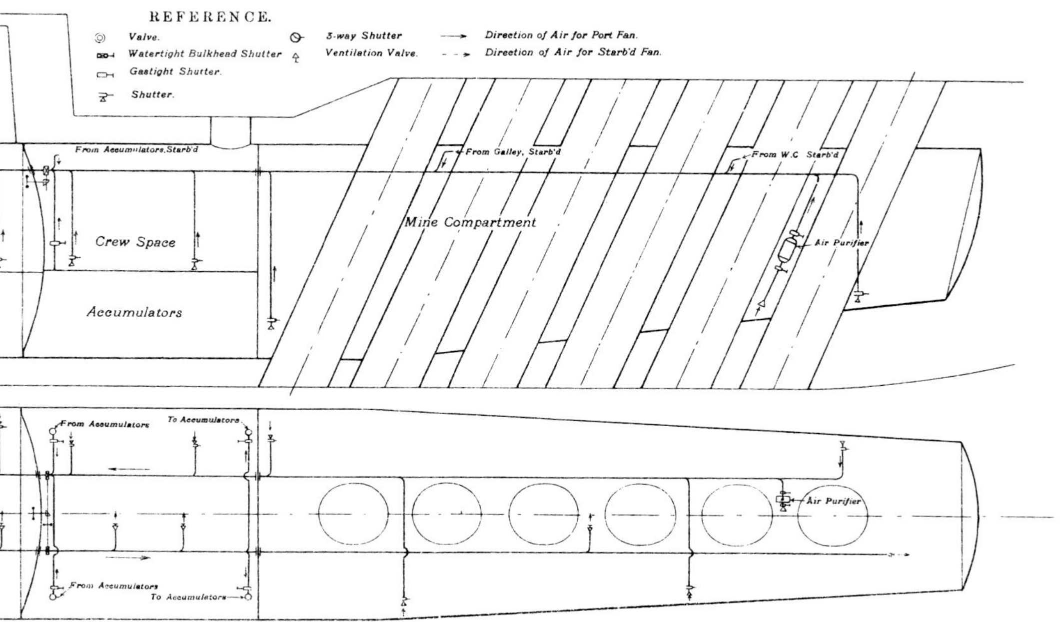

SUBMARINE TELEPHONE AND LIGHT BUOYS.

Fig. 1.

Fig. 3.

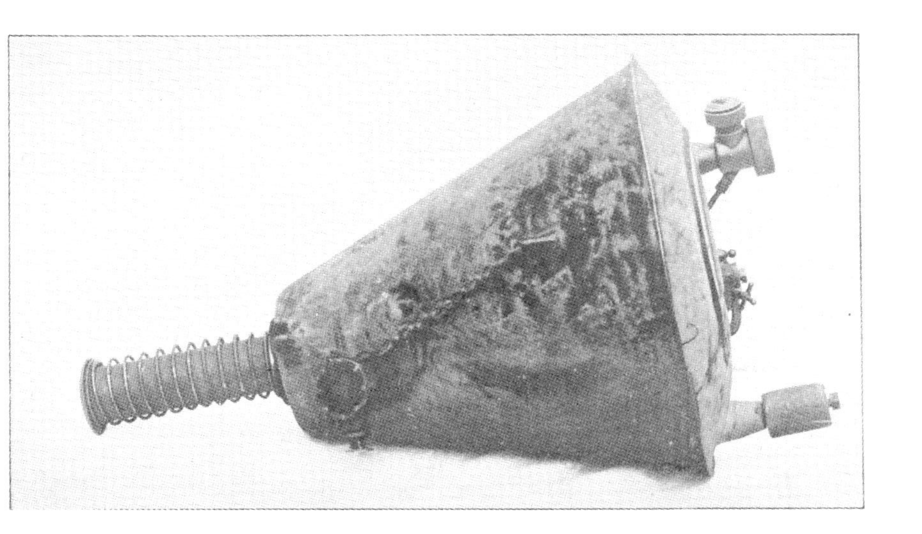

Fig. 2.

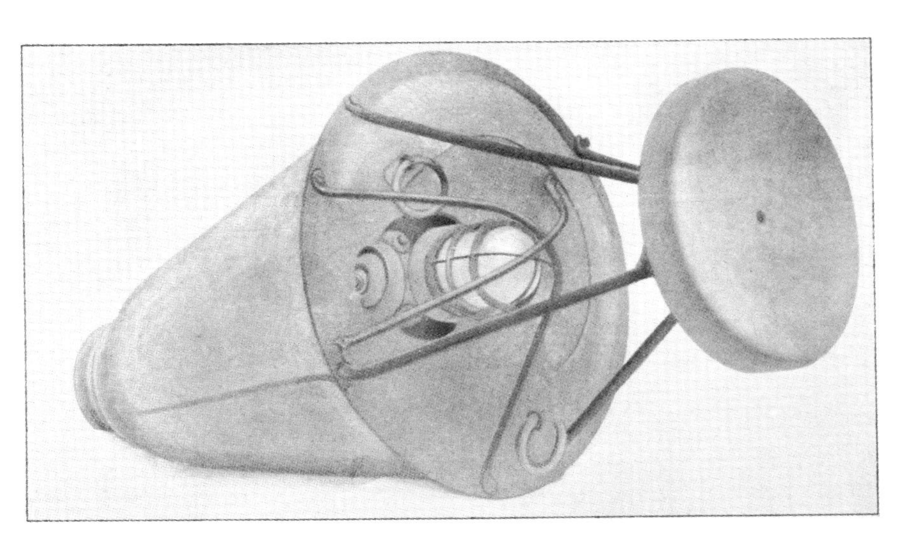

Fig. 4.

SECTION VI
MISCELLANEOUS VESSELS

GERMAN NAVY.

PART III.
SECTION 6.

MISCELLANEOUS VESSELS.

Name.	Classification.	Page.	Plate No.	Name.	Classification.	Page.	Plate No.
A. 1 & 2 - -	Tugs	24	—	1 & 2, B.H. - -	Tugs	24	—
A. 5 & 6 - -	Tugs	24	—	I. & II., B.H. - -	Tugs	25	—
A. 14 & 15 - -	Tugs	24	—	Biene - - -	Hulk	20	—
Acheron - - -	T.S.	20	—	Blitz - - -	G.B.	18	—
Adeline Hugo Stinnes III.	Collier	30	—	Bombe - - -	Tender	22	—
Adler - - -	Hosp. Ship	31	—	Bosnia - - -	Repair Sh.	18	108
Aeolus - - -	Tug	25	—	Bremen - - -	Transp.	29	108
Alarm - - -	Fire Float	22	—	Bürgermeister -	Transp.	29	108
Albatross - -	M.V.	8	101	Bussard - - -	Tug	24	—
Alexandrine - -	Hulk	21	—	C. - - - -	G.B.	5	—
Alster - - -	Aux.	27	—	C. 1.—9 - - -	M. Lighters	22	—
Ammon - - -	Aux. Transp.	27 29	—	Camäleon - -	Hulk	20	—
Answald - -	Aux.	27	—	Cap Finisterre -	Transp.	29	108
Anvers - - -	Aux.	27	—	Carlota - - -	Yacht	26	—
Arndgast - -	Tug	25	—	Cassel - - -	Transp.	29	—
Artemisia - -	Transp.	29	—	Caurus - - -	Tug	24	—
Asta - - -	Yacht	26	—	Charlotte - -	Hulk	20	—
Baden - - -	Hulk	20	—	Chemnitz - -	Transp.	29	—
Badenia - - -	Transp.	29	—	City of Bradford -	Aux.	27	108
Basilisk - - -	Hulk	20	—	Clara Blumenfeld -	Collier	30	—
Batavia - - -	Transp.	29	—	Comet - - -	Hulk	21	—
Bayern - - -	Hulk	20	—	Comet - - -	Yacht	26	—
Berlin - - -	Aux.	27	108	Condor - - -	G.B.	6	—
Bermuda - -	Transp.	29	—	Cordoba - - -	Aux.	27	—

o (33) AS 2016—8 Pk 2109 6 7/17 E & S

A

Name.	Classification.	Page.	Plate No.
Crokodill	Hulk	20	—
Dania	Aux.	27	—
Daunsfeld	Tug	24	—
Delphin	G.B.	18	106
Deutschland	Aux. M.V.	28	—
D. I. 1 & 2	Tugs	24	—
D. I. 4	Tug	24	—
Drache	G.B.	18	106
Duisburg	Transp.	29	—
Eber	G.B.	5	—
Eider	Store Ship	22	—
Eisvogel	Tug	24	—
Ellerbek	Yacht	26	—
Erlangen	Aux.	27	—
Ersatz Hohenzollern	Yacht	14	—
Eskimo	Aux.	27	—
F. 1—4	M. Lighters	22	—
Falke	Hulk	21	—
Farewell	S.V.	23	—
Fleiss	Tug	24	—
Flink	Tug	24	—
Föhn	Tug	24	—
Fortifikation	Tender	22	—
Frankfurt	Transp.	29	—
Friedrich der Grosse	Hulk	20	—
Friedrichsort	Tender	23	—
Fritz Hugo Stinnes V.	Collier	30	—
Fuchs	G.B.	18	106
Fürst Bülow	Aux.	27	—
G. 1—4	M. Lighters	22	—
Gauss	Aux.	27	—
General	Aux.	27	—
Giessen	Transp.	29	—
Glenearn	Aux.	27	108
Glückauf	Collier	30	—
Graf Waldersee	Transp.	29	108
Graf Zeppelin	Aux.	27	—
Greif	Hulk	21	—
Greiffenfels	Transp.	29	—
Grete Hugo Stinnes VIII.	Collier	30	—
Grille	T.S.	18	—
Grossherzog von Oldenburg.	Cable Ship	31	108
Habsburg	Transp.	29	—
Hansa	Hosp. Ship	31	—
Hay	G.B.	18	106
Helene Blumenfeld	Collier	30	—
Helga	Tug	24	—
Helgoland	S.V.	23	—
Heppens	Moor. V.	23	—
Hermann	Tug	24	—
Hertha	Yacht	26	—
Hessen	Aux.	27	—
Hilde Hugo Stinnes X.	Collier	30	—
Hilfe	Fire Float	22	—
Hohenzollern	Yacht	15	103
Hulda	Yacht	26	—
Hummel	Hulk	20	—
Hyäne	S.V.	18	103
Imperator	Hosp. Ship	31	—
Indianola	Aux.	27	108
Inkula	Aux.	27	108
Irene	Hulk	20	—
Jade	Pilot Vessel	23	—
Jagd	Hulk	20	—
Jutta	Yacht	26	—
K. 1—2	Oilers	23	—
K. 1—5	Water Tanks	25	—
Kehrwieder	Hosp. Ship	31	—
Kigoma	Aux.	27	—
König Friedrich August.	Aux. / Transp.	27 / 29	—
König Wilhelm	T.S.	20	—
Königin Luise	Trans.	29	108
Kraft	Tug	25	—
Kronprinz	T.S.	20	—
Langlütgen	Tender	23	—

Name.	Classification.	Page.	Plate No.
Legde	S.V.	24	—
Leipzig	T.S.	20	—
Lensahn	Hosp. Ship	31	—
Loreley	Yacht	18	—
M. 1—60	M.V.	10	102
M.A. 1 & 2	Tugs	24	—
Marie	Tug	24	—
Mars	Hulk	20	—
Mellum	Moor. V.	23	—
Mentor	T.S.	23	—
Meteor	Hulk	20	—
Minos	Aux.	27	—
Minseroog	S.V.	24	—
Mosquito	Tender	23	—
Mottlau	Tug	25	—
Möwe (*ex* **Pungo**)	Raider	16	104
Möwe	Aux.	27	—
Mücke	Hulk	20	—
Mürwik	T.S.	23	—
Natter	Hulk	20	—
Nautilus	M.V.	8	101
Niederwald	Aux.	27	—
Niobe	Hulk	20	—
Nixe	T.S.	20	—
Norder	Tug	25	—
Nordsee	Tender	18	—
Odin	Aux. M.V.	28	—
Oldenburg	Hulk	20	—
Orion	Yacht	26	—
Panther	G.B.	5	100
Passat	Tug	25	—
Patricia	Transp.	29	108
Pawnee	Aux.	27	—
Pelikan	M.V.	10	101
Pfeil	G.B.	18	—
Phœnix	Aux.	27	—
Portia	Hosp. Ship	31	—
Pretoria	Transp.	29	108
Prinz Adalbert	Aux. M.V.	28	—
Prinz Ludwig	Transp.	29	—
Prinz Sigismund	Aux.	27	—
Prinz Waldemar	Aux. M.V.	28	—
Prinzess Wilhelm	Hulk	21	—
Prinzessin	Transp.	29	108
Radaune	Yacht	26	—
Reiher	Tug	25	—
Rhein	Hulk	20	—
Rio Negro	Aux.	27	—
Rival	Tug	25	—
Rossall	Hosp. Ship	31	108
Rugia	Aux.	27	—
Rüstringen	Pilot Vessel	23	—
Sachsen	Hulk	20	—
Salamander	Hulk	20	—
Santa Elena	Aux.	27	—
Scheibenhof	Tug	25	—
Schillig	Desp. V.	23	—
Schneewittchen	Guard Boat	22	—
Schwaben	Transp.	29	—
Schwalbe	G.B.	18	—
Schwan	Launch	23	—
Seeadler	G.B.	6	100
Seeadler (*ex* **Pass of Balmaha**).	Raider	17	105
Seestern	Torp. Workshop.	23	—
Sierra Ventana	Hosp. Ship	31	—
Silvana	Aux.	27	—
Sirius	Guard Boat	22	—
Skorpion	Hulk	20	—
Sonderburg	Tug	25	—
Sophie	Hulk	20	—
Sperber	Hulk	21	—
Stark	Tug	25	—
Steigerwald	Aux.	27	—
Stein	Hulk	21	—

Name.	Classification.	Page.	Plate No.	Name.	Classification.	Page.	Plate No.
Stephan - - -	Cable Ship	31	108	W. 81 - - -	Oiler	23	—
Strande - - -	Torp. Workshop.	23	—	W. 83 - - -	Oiler	23	—
Sturm - - -	Tug	25	—	Waltraute - -	Hulk	20	—
Syria - - -	Transp.	29	—	Wangeroog - -	Pilot Vessel	23	—
T. 1 & 2 - -	Tugs	24	—	Wega - - -	Guard Boat	22	—
Thalatta - -	Yacht	26	—	Weichsel - -	Tug	25	—
Titania - - -	Hosp. Ship	31	—	Weih - - -	Tug	25	—
Ulan - - -	Hulk	20	—	Wespe - - -	Hulk	20	—
Uranus - - -	Hulk	20	—	Wik - - -	Moor. V.	23	—
Vesuv (?) - -	Subm. Depôt	11	—	Wilhelmshaven -	Pilot Vessel	23	—
Victoria Luise -	Transp.	29	108	Wolf - - -	Raider	16	—
Viola - - -	Hosp. Ship	31	—	Worms - - -	Transp.	29	—
Viper - - -	Hulk	20	—	Wotan - - -	Aux. M.V.	28	108
Voslapp - - -	Tug	25	—	Württemberg - -	T.S.	18	—
Vulkan - - -	Subm. Depôt.	12	102	Yorck - - -	Aux.	27	108
W. 1—4 - - -	M. Lighters	22	—	Ypiranga - -	Transp.	29	—
W. 1—4 - - -	Water Tanks	25	—	Zieten - - -	F.P.V.	13	103

List of Plates.

(At end of Part III.)

No. 100.—Gunboats *Panther* and *Seeadler*, photographs.

No. 101.—Mining Vessels *Albatross*, *Nautilus*, and *Pelikan*, photographs.

No. 102.—Mining Vessels M 1.—60 and Submarine Salvage Ship *Vulkan*, photographs.

No. 103.—Fishery Protection Vessel *Zieten*, Imperial Yacht *Hohenzollern* and Surveying Vessel *Hyäne*, photographs.

No. 104.—Raider *Möwe*, photographs and plan.

No. 105.—Raider *Seeadler*, sketch and plans.

No. 106.—Gunboats *Drache*, *Hay*, *Delphin*, and *Fuchs*, photographs.

No. 107.—Armed Trawlers, photographs.

No. 108.—Auxiliary Vessels, silhouettes.

Abbreviations.

Aux. = Auxiliary.
Desp. V. = Despatch Vessel.
F.P.V. = Fishery Protection Vessel.
G.B. = Gunboat.
Hosp. Ship = Hospital Ship.
M. Lighter = Mining Lighter.
M.V. = Mining Vessel.
Moor. V. = Mooring Vessel.
S.V. = Surveying Vessel.
Subm. Depôt = Submarine Depôt Ship.
T.S. = Training Ship.
Transp. = Transport.

GUNBOATS.

C.

General Remarks.—Provision was made for this vessel in the 1913 and 1914 estimates, and the order for her was reported to have been given to the Imperial Dockyard, Danzig, in December 1913. There is no evidence, however, that she was ever laid down.

Cost.—

	£
Hull, machinery, &c. - - - - - - - - -	73,385
Gun armament - - - - - - - - -	25,440
Torpedo armament - - - - - - -	Nil.
Total - - - - - - - - -	£98,825

General Dimensions.—

Length, L.W.L. - - - - - - - -	219 ft. 10 ins.
Breadth, extreme - - - - - - - -	33 „ 6 „
Draught, designed load - - - - - - -	10 „ 6 „
Displacement, designed load - - - - - -	1,132 tons.

Armament.—

4—4·1-in. (10·5 c.m.) Q.F. guns.
2—1-pr. (37 mm.).
2 machine guns.

Machinery and Boilers.—

Main Engines - - - - - -	Two sets, reciprocating.
Boilers - - - - - - -	Four, Schulz, Navy type.
Propellers - - - - - - -	Two.

Speed, Horse Power, and Fuel.—

Speed (designed).	Horse-Power (designed).	Fuel. (*a*) Coal. (*b*) Oil.
Knots. 14	1,550	—

Panther *and* Eber.

(*See* Plate 100.)

Name.	Designation before Launch.	Programme Year.	Where Built.	Laid Down.	Launched.	First Commissioned.
Panther -	A.	1900–01	Imperial Dockyard, Danzig.	16.7.00	1.4.01	15.3.02
Eber - -	B.	1902–03	Vulcan Works, Stettin	1.8.02	6.6.03	15.9.03

General Remarks.—The *Eber* was interned at Bahia on 5th September 1914, and her officers and men commissioned the auxiliary cruiser "*Cap Trafalgar.*"

Cost.—*Panther*—£89,040 ; *Eber*—£93,933.

Complement.—130.

3

Panther *and* Eber—*cont.*

General Appearance.—*See* Plate.

General Dimensions.—

Length between perpendiculars - - - - - - -	203 ft. 5 ins.
" L.W.L. - - - - - - - - - -	210 " 4 "
Breadth, extreme - - - - - - - - -	31 " 10 "
Draught, designed load - - - - - - - -	10 " 6 "
Displacement, designed load - - - - - - -	984 tons.

Hull.—

Material.—Composite. Sheathed, and covered with Muntz metal.

Conning Tower.—Special steel, ·3″ thick.

Armament.—

Guns.—

2—4·1-in. (10·5 cm.) Q.F. L/40. One on forecastle, one aft.
6—1-pr. (37 mm.) Maxim automatic; two on each side of bridge, and two aft.
2—·31-in. (8 mm.) Maxim automatic.

Searchlight.—

1—On top of chart house.

Boats.—*See* Section 1.

Machinery and Boilers.—

Main Engines.—

Two sets, in separate compartments.
Type—Horizontal, triple expansion.

Boilers.—

Four, Schulz, Navy type.

Propellers.—Two.

Speed, Horse Power, and Fuel.—

Speed (designed).	Horse-Power (designed).	Fuel. (*a*) Coal. (*b*) Oil.
Knots. 14	1,300	(*a*)275

Seeadler *and* Condor.

(*See* Plate 100.)

Name.	Programme Year.	Where Built.	Launched.	First Commissioned
Seeadler - -	—	Imperial Dockyard, Danzig - - - -	1892	1893
Condor - -	—	Blohm and Voss, Hamburg - - - -	1892	1893

General Remarks.—These vessels, which were formerly reckoned as light cruisers, were re-classified as gunboats, *Condor* in 1913, *Seeadler* in 1914. Of the other two vessels of the class, the *Geier* was interned at Honolulu on 9th November 1914, and seized by the U.S. on 6th April 1917, and the *Cormoran* was sunk by the Germans at Tsingtau shortly before the capitulation on 7th November 1914.

The *Seeadler* is now (June 1917) employed as Mining Depôt Ship.

General Appearance.—*See* Plate.

Seeadler *and* Condor—*cont.*

Complement.—162.

General Dimensions, &c.—

Length between perpendiculars	249 ft. 4 ins.
„ L.W.L.	261 „ 2 „
Breadth, extreme	34 „ 5 „
Draught, designed load	14 „ 6 „
Displacement, designed load	1,602 tons.

Hull.—

Material.—Steel, sheathed with wood and coppered.

Constructive Details.—Eight main water-tight compartments. Double-bottom under engine and boiler rooms.

Armament.—

Guns and Ammunition Supply.—

8—4·1-in. (10·5 cm.) Q.F. L/35; two on forecastle, two on poop, and four in sponsons on the broadside, all protected by shields.
5—1-pr. (37 mm.); two on forecastle, one either side of bridge, and one right aft on poop.
2—·31-in. (8 mm.) Maxim automatic.

Torpedo Tubes—

2—13·8-in. (35 cm.) above water; one on each side on upper deck amidships.

Searchlight.—

1—On platform above the bridge.

Boats.—*See* Section 1.

Steering Gear.—Steam.

Machinery and Boilers.—

Main Engines.—

Two sets, in separate water-tight compartments divided by an athwartship bulkhead.
Type—Horizontal, triple expansion.
Diameter of cylinders, 16·4 in., 27·5 in., and 43 in.
Length of stroke, 24·6 in.

Boilers.—

Four in number.
Type—Cylindrical, direct tube, three furnaces in each.

Propellers.—Two.

Speed, Horse-Power, and Fuel.—

Speed (designed).	Horse-Power (designed).	Fuel. (*a*) Coal. (*b*) Oil.
Knots. 16	2,800	(*a*) 305

Endurance.—3,380 miles at 10 knots with a coal expenditure of 20·5 tons per day for all purposes.

MINING VESSELS.

Albatross.

(*See* Plate 101.)

General Remarks.—1906–07 programme. Built at the Weser Yard, Bremen, launched 13th October 1907 and first commissioned on 19th May 1908. Driven ashore on the island of Gothland by Russian naval forces on 2nd July 1915; subsequently salved and interned by Sweden.

Complement.—198.

Cost.—Same as *Nautilus* (*see* p. 9).

General Appearance.—*See* Plate.

General Dimensions.—

Length, between perpendiculars - - - - - - 295 ft. 0 ins.
Breadth, extreme - - - - - - - - - 42 ,, 8 ,,
Draught, designed load - - - - - - - 13 ,, 2 ,,
Displacement, designed load - - - - - - - 2,165 tons.

Armament.—

Guns.—**8**—15-pr. (8·8 cm.), Q.F. L/35.

Mines.—It is reported about 600 can be carried. They are launched over the stern from the quarter-deck.

Searchlights.—

2—one on each mast.

Machinery and Boilers.—

Engines.—

Two sets.
Type—Vertical, triple expansion.

Boilers.—Eight, Schulz.

Propellers.—Two.

Speed, Horse-Power, and Fuel.—

Speed (designed).	Horse-Power (designed).	Fuel. (*a*) Coal. (*b*) Oil.
Knots. 20	6,500	(*a*) 443

Nautilus.

(*See* Plate 101.)

General Remarks.—1905–06 programme. Built at the Weser Yard, Bremen, launched on 18th August 1906 and first commissioned on 19th March 1907. Modernised in 1912–13.

Complement.—198.

Nautilus—*cont.*

Cost.—

	£
Hull and machinery	107,632
Gun armament	24,462
Mining armament	31,311
Total	£163,405

General Appearance.—*See* Plate. Two raking masts and funnels; swan bow, topgallant forecastle, superstructure deck extending from bridge nearly right aft, and a deeply overhanging stern.

Accommodation.—The officers' accommodation, and that of part of the crew also, is on the upper deck under the forecastle; the remainder of the crew live forward on the main deck.

General Dimensions, &c.—

Length, between perpendiculars	295 ft. 0 ins.
,, extreme	305 ,, 2 ,,
Breadth, extreme	42 ,, 8 ,,
Draught, designed load	13 ,, 2 ,,
Displacement, designed load	1,945 tons.

Armament.—

Guns.—

8—15-pr. (8·8 cm.), Q.F. L/35, on superstructure deck.

Mines.—About 200 mines are carried. There are two large two-storied store rooms on the lower deck, one forward and one aft, each about 60 ft. long, and extending across the entire width of the ship. 50 mines can be accommodated on each storey.

The main deck aft contains all the mines ready for dropping, stowed on special trollies on rails, which run along the main deck to two ports, about 6 ft. by 3 ft., cut in the stern, through which the mines are dropped some 20 ft. abaft the rudder and screws.

The mines are hoisted in by two cranes, one on each side forward.

It is stated that this vessel is intended particularly for service in coastal waters, and has therefore been given a light draught. In order that she may be employed in winter she has been specially strengthened to resist ice.

Searchlights.—

2—one on each mast.

Machinery and Boilers.—

Engines.—

Two sets.

Type.—Vertical, triple expansion.

Boilers.—

Eight, Schulz.

Working pressure, 225 lbs.

Total heating surface, 17,000 sq. ft.

Propellers.—Two.

Speed, Horse-Power and Fuel.—

Speed (designed).	Horse-Power (designed).	Fuel. (*a*) Coal. (*b*) Oil.
Knots. 20	6,000	(*a*) 443

o AS 2016—8 B

Pelikan.

(*See* Plate 101.)

General Remarks.—Mining ship. Built at the Imperial Dockyard, Wilhelmshaven, launched on 29th July 1890, and completed in 1892. Underwent a large refit in 1908.

Complement.—195.

General Appearance.—*See* Plate.

General Dimensions, &c.—

Length between perpendiculars - - - - - -	259 ft. 2 ins.
Breadth, extreme - - - - - - - - -	38 „ 1 „
Draught, designed load - - - - - - -	14 „ 9 „
Displacement, designed load - - - - - -	2,320 tons.

Armament.—

Guns and Ammunition Supply.—

4—15-pr. (8·8-cm.) Q.F., L/30.
4—·31-in. (8 mm.) Maxim automatic.

Mines.—Four sets of rails run along main deck to stern, where the mines are dropped clear of the screws and rudder.

Searchlights.—**2.**

Steering Gear.—Electrical.

Machinery and Boilers.—

Engines.—Two; vertical, triple expansion.

Boilers.—Two double-ended, cylindrical, originally fitted; probably replaced in 1913 by Schulz water tube.

Propellers.—Two.

Speed, Horse-Power, and Fuel.—

Speed (designed).	Horse-Power (designed).	Fuel. (*a*) Coal. (*b*) Oil.
Knots. 15	3,000	(*a*) 404

M. 1 to M. 60.

(*See* Plate 102.)

General Remarks.—These boats represent a compromise between a trawler and a small torpedo boat. They have all been built during the war, the first of them having been met at sea in January 1916. They are fitted for both minelaying and minesweeping, but appear to have been intended primarily for the latter purpose. They are also used as patrol craft and for escorting merchant vessels.

When used as minesweepers, these boats work in flotillas of 11 or 12; they are reported as manœuvring well in very close formation. As in the case of minesweeping T.B.'s, each flotilla or division has a destroyer attached as leader.

Complement.—About 40.

Colour.—Light grey.

M. 1 to M. 60—*cont.*

General Appearance.—*See* Plate. All these boats have a tall funnel and foremast; some of them have a stump mainmast. Those without mainmast have a derrick with samson post amidships.

General Dimensions.—

Length - - - - - - - -	about 164 ft. 0 ins.
Breadth - - - - - - - -	„ 20 „ 0 „
Draught - - - - - - - -	„ 9 „ 0 „

Armament.—

Guns.—

2—22-pr. (8·8 cm.), Q.F., one on forecastle, one in after part of vessel.
2—machine guns, on bridge.

Torpedo Tubes.—
Probably none are fitted.

Mines.—
Rails on either side of the deck for a total of about 70 mines.

Searchlights.—
1—on foremast.

Machinery and Boilers.—

Engines.—

Boilers.—Probably burn coal and oil.

Auxiliary Machinery.—A steam winch is fitted amidships.

Propellers.—Two.

Speed, Horse-Power, and Fuel.—

Speed (designed).	Horse-Power (designed).	Fuel.
Knots. *About* 17 - -	—	Coal and oil.

DEPÔT AND SALVAGE SHIPS FOR SUBMARINES.

* Vesuv.

General Remarks.—This vessel, supposed to be an improved *Vulkan* (*see* p. 12), formed part of the 1912–13 programme, and was reported to be building at Danzig Dockyard in 1914. It is, however, not certain that she was ever laid down. In any case, nothing has been heard of her during the war.

General Dimensions.—

Length, L.W.L. - - - - - - - - -	295 ft. 3 ins.
Breadth, extreme - - - - - - - - -	64 „ 0 „
Draught, designed load - - - - - -	13 „ 9 „
Displacement, designed load - - - - - -	2,755 tons.

Engines.—Two steam engines, one in each hull.

Speed.—9 knots.

* Name doubtful.

Vulkan.

(*See* Plate 102.)

Programme year - - - - - - - - 1907–8.
Launched - - - - - - - - - 28th September 1907.
First commissioned - - - - - - - 4th March 1908.

Cost.—Uncertain, but according to Press, about £97,847.

General Remarks.—This was the first salvage and docking ship for submarines built by any Power. Designed and patented by the German engineer, von Klitzing, and built at Howaldt's Works, Kiel.

Complement.—124.

General Description.—This vessel consists of two parallel hulls, each displacing 900 tons, and each 225 ft. long, separated amidships by a well or space 42 ft. 8 ins. wide. The total breadth is 64 ft. 0 ins., the maximum draught 13 ft. 9 ins. The hulls unite together above water at bow and stern; they are further connected amidships by two transverse semi-circular double gantries which rise across the central well, and the legs of which rest on either hull The gantries are provided with powerful tackles to which a sunken submarine can be shackled, and by means of which it can be hoisted up between the two hulls clear of the water. Hinged bars are provided on the inner side of one hull and are swung across and fixed in sockets on the inner side of the second hull, for the raised submarine to rest on. The lifting power is about 500 tons.

Machinery and Boilers.—

Engines.—Electric motors driving two propeller shafts, the current for which is supplied by two independent turbo-generators, either of which is run alone when the vessel is to steam slowly. All switches and controlling mechanism for the entire machinery are operated from the bridge (situated on a deck built across the tops of the gantries) so that the ship can be controlled directly from there without communication with the engine-room. The turbo-dynamos also supply power for lifting submarines, and are utilised for recharging the accumulators of submarines.

Boilers.—Four Melhorn, water-tube.

Speed, Horse-Power, and Fuel.—

Speed (designed).	Horse-Power (designed).	Fuel. (*a*) Coal. (*b*) Oil.
Knots. 12	1,200	—

FISHERY PROTECTION VESSEL.

Zieten.

(*See* Plate 103.)

General Remarks.—Iron gunboat, built at the Thames Iron Works, Blackwall; launched on 9th March 1876, and completed at Kiel in the same year. Before the war was used only for fishery protection work in the North Sea.

Complement.—106.

General Dimensions, &c.—

Length, L.W.L.	226 ft. 5 ins.
Breadth, extreme	27 „ 10 „
Draught, designed load	11 „ 6 „
Displacement, designed load	994 tons.

Constructive Details.—Two rifle-proof conning towers, with umbrella shields, one before foremast and one abaft mainmast.

Armament.—

Guns.—

6—4-pr. (5 cm.) Q.F., L/40.

Torpedo Tubes.—

2—13·8-in. (35 cm.) submerged; one in bow and one in stern.

Searchlights.—

2—One before foremast and one right aft.

Machinery and Boilers.—

Main Engines.—Two horizontal simple engines made by Penn.

Boilers.—Six, cylindrical.

Propellers.—Two.

Speed, Horse-Power, and Fuel.—

	Speed (designed).	Horse-Power (designed).	Fuel. (*a*) Coal. (*b*) Oil.	
	Knots. 13	2,350	(*a*) 138	

B 3

Miscellaneous Vessels.

IMPERIAL YACHTS.

Ersatz Hohenzollern.

Name.	Designation before Launch.	Programme Year.	Where Building.	Ordered.	Laid Down.	Launched.	Commissioned for Trials.	Completed Trials.
—	Ersatz *Hohenzollern.*	1913–14	Vulcan Works, Stettin.	24.4.13	14.6.13	*1.4.15	—	—

* Approximate date.

General Remarks.—It does not appear (June 1917) that this vessel has been completed.

Cost.—£489,236.

Complement.—455.

General Dimensions, &c.—

Length, extreme	527 ft. 8 ins.
Length, L.W.L.	450 „ 9 „
Breadth, extreme	62 „ 4 „
Draught, designed load	19 „ 3 „
Displacement, designed load	7,185 tons.
Freeboard (approximate)	21 ft. 0 ins.

Constructive Details.—Special attention is paid to safety in the event of collision. The double-bottom extends almost over the whole length of the ship, and is carried up at the sides to about 8 ft. above the water-line. In addition a longitudinal bulkhead, extending the greater part of the ship's length, is fitted on either side. There is a large number of transverse bulkheads which extend above the water-line, and in places are carried right up to the upper deck. The main bulkheads are of specially strong construction and contain no doors below the water-line.

Machinery and Boilers.—

Main Engines.—Three sets of turbines on three shafts, the power being transmitted to the respective propeller shafts in each case by means of a Föttinger-transformer.

Boilers.—10, Schulz, Navy type—eight fitted to burn coal and oil and two oil only.

Propellers.—Three.

Speed, Horse-Power, and Fuel.—

Speed (designed).	Horse-Power (designed).	Fuel. (*a*) Coal. (*b*) Oil.
Knots. 24 (?)	25,000 (?)	(*a*) 984 (*b*) 492

Hohenzollern.

(*See* Plate 103.)

General Remarks.—1891-92 programme. Built by Vulcan Works, Stettin, launched 27th June 1892, and completed in 1893. Completed an extensive refit in 1907, during which the vessel was re-boilered with Schulz boilers.

Complement.—344.

General Dimensions.—

Length between perpendiculars - - - - -	382 ft. 7 ins.
,, extreme - - - - - - - - -	400 ,, 0 ,,
Breadth, extreme - - - - - - - -	45 ,, 11 ,,
Draught, designed load - - - - - - -	22 ,, 10 ,,
Displacement, designed load - - - - -	4,210 tons.

Constructive Details.—Framing very light; great height between decks; many watertight compartments, and a double bottom.

Armament.*—Doubtful.

Guns.—

3—4·1-in. (10·5 cm.) Q.F., L/35; one aft and two in sponsons (one each side just abaft topgallant forecastle).
12—4-pr. (5 cm.) Q.F., L/40.
6—·31-in. (8 mm.) Maxim automatic.

Searchlights.—

2—One on either side of bow on the upper deck in small sponsons.

Machinery and Boilers.—

Engines.—Two; vertical, triple expansion.

Boilers.—Fitted with new Schulz boilers in 1906. No forced draught, but fans are fitted for ventilation.

Propellers.—Two; four-bladed.

Speed, Horse-Power, and Fuel.—

Speed (designed).	Horse-Power (designed).	Fuel. (*a*) Coal. (*b*) Oil.
Knots. 21·5	9,500	(*a*) 492

Steam Trials.—

Nature of Trial.	Date.	Speed.	Horse-Power.	Revolutions.	Remarks.
Full power	—	21·53	—	110	—

* When employed as an Imperial yacht the *Hohenzollern* carried only 2—4-pr. guns.

RAIDERS.

Möwe.

(*See* Plate 104.)

General Remarks.—Auxiliary cruiser, ex-merchantman. Launched, May 1914, at Tecklenborg Yard, Geestemünde, as the fruit and passenger steamer "*Pungo.*"

Complement.—About 200.

General Appearance.—*See* Plate 104. A second (dummy) funnel can be fitted.

General Dimensions, &c.—

Length between perpendiculars - - - - - 385 ft. 0 ins.
Length of boat deck - - - - - - - - 155 ,, 0 ,,
Length of forecastle - - - - - - - 53 ,, 0 ,,
Breadth - - - - - - - - - 49 ,, 0 ,,
Depth of hold - - - - - - - - 32 ,, 7 ,,
Gross register tonnage - - - - - - About 4,500 tons.

Masts.—Two raking pole masts, the height of which can be varied. A crow's nest may be fitted on each mast.

Constructive Details.—Four decks.

Armament.—

Guns.—

4—5·9-inch. Two under forecastle, firing through dropping ports. One either side between foremast and bridge, concealed by dropping bulwarks.

1—4·1-inch mounted on poop, disguised as hand-steering gear.

2—22-pr. (probably) one each side of the after portion of boat deck.

? **2**—machine guns.

Torpedo Tubes.—**4**—17·7-inch (probably) above water training. Two mounted between break of forecastle and foremast, two between mainmast and break of poop. All concealed by dropping bulwarks.

Mines.—A supply of mines is carried.

Depth Charges.—A number are carried.

Searchlights.—Fitted on collapsible platform abaft bridge.

Machinery and Boilers.—

Main Engines.—Reciprocating, triple expansion.
Cylinders—28-in., 45-in., and 74-in. Stroke—53·2-in.

Boilers.—

Propellers.—One.

Speed, Horse-Power, and Fuel.

Speed (designed).	Horse-Power (designed).	Fuel. (*a*) Coal. (*b*) Oil.
Knots. 14	3,200	—

Estimated Coal Consumption.—At 14 knots, 69 tons per day; at 12 knots, 47 tons; at 10 knots, 32 tons; at 8 knots, 20 tons.

Wolf.

General Remarks.—Auxiliary cruiser, *ex* merchantman.

Complement.—Originally 340, depleted by prize crews.

General Appearance.—Screw steamer. Slightly sloping bow, counter stern. Well decks forward and aft. Two masts, no rake. One moderately thick funnel midway between, and nearly as high as, the masts. Whistle on fore side, exhaust pipe on after side of funnel. Hurricane deck and boat deck amidships, with two bridges and monkey's island.

General Dimensions, &c.—

Length - - - - - - - - - About 420 feet.
Gross register tonnage - - - - - - About 5,000 tons.

Masts.—Two masts, without rake.

Derricks.—Derricks on foremast not noticeable. Derrick on fore side of mainmast for seaplane. Two short derrick masts at break of poop, one either side, for mines.

Armament.—

Guns.—

4—5·9-in. (probably) mounted one on either side, just forward of foremast and between mainmast and hurricane deck. Normally concealed by bulwarks.

2—4·1-in. (probably) mounted on forecastle, firing through dropping ports.

Two light Q.F. guns, mounted on fore end of boat deck.

Torpedo Tubes.—Four training deck tubes, mounted one on either side, just before bridge and just abaft mainmast. Normally concealed by bulwarks.

Mines.—A supply of 75 mines can be stowed in No. 4 hold, abaft mainmast.

Rangefinder Position.—In observation top, three-quarters of the way up foremast.

Wireless Telegraphy.—Three wires between spreaders, slung from mast to mast. A small installation in seaplane.

Boats.—Three are carried at davits on either side of boat deck.

Steering Gear.—An emergency hand wheel is fitted on poop.

Seaplane.—A seaplane is housed on deck, just before mainmast.

Seeadler.

(*See* Plate 105.)

General Remarks.—Sailing commerce destroyer with auxiliary power, *ex* British ship *Pass of Balmaha.* Built on the Clyde, 1888.

General Appearance.—*See* Plate 105.

General Dimensions, &c.—

Length, L.W.L. - - - - - - -	235 ft. 6 ins.
Beam - - - - - - - - - - -	39 „ 0 „
Depth of hold - - - - - - -	22 „ 6 „
Gross register tonnage - - - - - -	1,571 tons.

Masts.—Three masts, with six yards on each. The foremast and mainmast are of practically equally height, the mizen about 10 ft. shorter. As a rule, no spanker is set, but a spare gaff and boom are carried, so that the vessel may be disguised as a barque.

The *original* heights of the masts were as follows :—

Fore truck to L.W.L. - - - - - - -	146½ ft.
Main truck to L.W.L. - - - - - - -	147½ „
Mizen truck to L.W.L. - - - - - - -	130 „

Armament.—

Guns.—

2—4·1-in. mounted one on each side, just abaft break of forecastle.

? **2**—machine guns.

Magazines and Stowage.—The magazine is under forecastle.

Torpedo Tube.—A submerged tube is reported to be fitted on port side of engine room.

Mines.—A supply of mines is carried.

Smoke Boxes.—Apparatus for producing a smoke screen is also carried.

Rangefinder Positions.—Rangefinder normally mounted on forecastle, but can be moved to top of forward deckhouse.

Wireless Telegraphy.—An aerial is slung from fore truck to main truck, masked by a very thick triatic stay. The wireless room is on the main deck the port side, abreast the galley.

Boats.—Two large motor boats are carried, one on either side just before mizen, on heavy skid beams. Their propellers are hidden by painted canvas covers. Three other boats are stowed on forward deck house.

Machinery.—An auxiliary 4-cylinder Diesel engine is fitted, maximum speed 120 r.p.m. The engine room is just before mizen mast. The speed is reported to be 11 knots, and radius of action between 20,000 and 30,000 miles. The exhaust is through a short thick funnel 2 ft. high at after end of fore deck house.

SPECIAL

Classification and Name.	Date of Launch.	Complement.	Hull. Length. (a) P.P. (b) L.W.L. (c) Extreme.	Hull. Breadth (extreme).	Hull. Designed Load. Draught (maximum).	Hull. Designed Load. Displacement.	Armour. Maximum Thickness. W.L.	Armour. Maximum Thickness. Deck.	Armour. Maximum Thickness. Gun Protection.	Armour. Maximum Thickness. Conning tower.	Armament.
			Ft. ins.	Ft. ins.	Ft. ins.	Tons (Eng.)	Ins.	Ins.	Ins.	Ins.	
REPAIR SHIP.											
Bosnia (*ex* Merchant Vessel.)	–/99.	—	(*c*)483 7	57 4	—	*9,683	—	—	—	—	—
FLEET TENDER.											
Nordsee	7/14	—	(*b*)175 10	30 10	—	778	—	—	—	—	2–22-pr.
TRAINING SHIPS.											
Württemberg	11/78	518	(*a*)298 7 (*b*)308 9 (*c*)321 6	60 4	19 8	7,250	16	2¼	10	—	8–1-pr.; 4–m.; 2–S.T. (17·7″); 2–S.T. (13·8″); 1–T. (13·8″).
Schwalbe	8/87	116	(*a*)203 5	30 9	13 2	1,100	—	—	—	7·9	8–4·1-in.; 5–1-pr.; 2–m.; 2–T.
Grille	9/57	69	(*a*)170 7 (*b*)172 3	24 3	8 2	344	—	—	—	—	2–1-pr.
TENDERS TO GUNNERY SCHOOL.											
Drache	6/08	64	(*a*)177 2	29 6	9 10	758	—	—	—	—	4–15 or 22-pr.; 4–4-pr.
Hay	8/07	56	(*a*)134 6 (*b*)140 9	28 6	9 10	630	—	—	—	—	8–4-pr.
Delphin	1/06	44	(*a*)124 8 (*b*)133 10	27 .3	9 10	440	—	—	—	—	4–22-pr.; 2–4-pr.; 4–m.
Fuchs	7/05	62	(*a*)134 6 (*b*)140 9	28 6	9 10	630	—	—	—	—	2–4·1-in.; 2–15-pr.
MISCELLANEOUS.											
Blitz	8/82	} 134	(*a*)245 1	32 6	13 2	1,366	—	—	—	—	6–15-pr.; 4–m.; 1–S.T.; 2–T.
Pfeil	9/82	}									
Loreley	–/84	—	(*a*)200 2	27 7	13 9	909	—	—	—	—	2–4-pr. 1–m.
SURVEYING VESSEL.											
Hyäne	6/78	104	(*a*)137 10	25 3	10 6	485	—	—	—	—	1–15-pr.; 1–4-pr; 3–1-pr.

* Gross Register tons.

VESSELS.

Machinery.			Fuel.					Remarks.	Classification and Name.
No. of Propellers.	Speed. (a) Designed. (b) Mean on trial.	Horse Power. (a) Designed. (b) Mean on trial.	Maximum Coal Stowage.	Daily Expenditure. At Sea-going full speed.		Endurance. At 10 Knots.			
	Knots.		Tons.	Tons.	Tons.	Miles.	Miles.		
									REPAIR SHIP.
1	(a) 11·5	(a) 3,900	—	—	—	—	—	*See* silhouette, Plate 108.	Bosnia. (*ex* Merchant Vessel).
									FLEET TENDER.
—	—	(a) 1,700	—	—	—	—	—	Used as Leader Ship of North Sea Auxiliary Patrol.	Nordsee.
									TRAINING SHIPS.
2	(a) 15·0	(a) 6,000	650	101·5	38·6	—	3,825	Torpedo School Ship.	Württemberg.
2	(a) 14·0	(a) 1,500	(a) 300	—	—	—	—	Wireless School Ship	Schwalbe.
1	(a) 13·0	(a) 700	59	—	—	—	—	Used for Pilotage instruction and for Admiralty Staff tours.	Grille.
									TENDERS TO GUNNERY SCHOOL.
—	(a) 14·0 (b) 15·3	(a) 1,600 (b) 1,670	146	—	—	—	—	Steel Gunboat (*See* Plate 106.)	Drache.
1	(a) 12·5	(a) 1,100	75	—	—	—	—	Do.	Hay.
1	(a) 9·5	(a) 450	55	—	—	—	—	Do.	Delphin.
1	(a) 12·5	(a) 1,100	75	—	—	—	—	Do.	Fuchs.
									MISCELLANEOUS.
2	(a) 15·0	(a) 2,700	(a) 250	19	—	3,000	—	Steel Gunboats, used as tenders to High Sea Fleet.	Blitz. Pfeil.
1	(a) 12·0	(a) 600	(a) 162	—	—	—	—	*Stationnaire at Constantinople.*	Loreley.
									SURVEYING VESSEL.
1	(a) 8·0	(a) 340	(a) 108	—	—	—	—	(*See* Plate 103)	Hyäne.

HARBOUR SHIPS.

(Ships removed from the Effective List, not yet definitely appropriated or disposed of, are included.)

Ship.	Former Classification.	Launched.	Displacement.	Stationed at	Present Employment.
			Tons.		
Bayern	Battleship, 2nd Class.	1878	7,250	Baltic	Hulks.—Used as targets in armour experiments.
Oldenburg	Battleship, 3rd Class.	1884	5,140	„	
Sachsen	Battleship, 2nd Class.	1877	7,250	„	
Mücke	Armoured Coast Defence Gunboats.	1877	1,091	Kiel	Probably converted into repair ships, attached to 40,000-ton floating dock.
Natter		1880	1,091	„	
Skorpion		1877	1,091	Mürwik	Torpedo range.
Basilisk		1878	1,091	One at Wilhelmshaven.	Used for supplying steam for heating purposes to ships in the dockyard.
Biene		1876	1,091		
Camäleon		1878	1,091		
Crokodill		1879	1,091		
Hummel		1881	1,091	7 (?) at Danzig.	The number at Danzig is uncertain, as some have been reported sold or utilised as targets.
Salamander		1880	1,091		
Viper		1876	1,091		
Wespe		1876	1,091		
Baden	Battleship, 2nd Class.	1880	7,250	Cuxhaven	Hulk.—Used for accommodating persons attending the Mining School.
Rhein	Mining Ship	1866	394	„	Hulk.—Repair ship for torpedo boats of the Minesweeping Divisions.
Ulan	Special Vessel.	1875	370	„	Hulk.—Mine store.
Acheron (*late* Moltke).	Training Ship.	1877	2,800	Kiel	Hulks.—Used for accommodating the crews of submarines.
Irene	Light Cruiser.	1887	4,223	Wilhelmshaven.	
Sophie	Cruiser, 3rd Class.	1881	2,124	Heligoland	
Meteor	Torpedo Vessel.	1890	931	Kiel	
Waltraute	—	—	—	Eckernförde	
Niobe	Sailing Frigate.	*1853	1,269	Wilhelmshaven.	Hulk.—For harbour accommodation of Staff of Vice-Admiral Commanding Scouting Forces.
Jagd	Cruiser, 3rd Class.	1888	1,233	Eckernförde	Hulk.—For testing torpedoes on range.
König Wilhelm (*late* Turkish *Fatikh*).	Battleship, 3rd Class.	1868	9,600	Mürwik	Training Ship for boys. Engines still in place, but never goes to sea.
Charlotte	Cruiser, 2nd Class.	1885	3,233	„	Hulk.—Tender to *König Wilhelm*—overflow ship.
Uranus (*ex* Kaiser)	Battleship, 3rd Class.	1874	7,555	„	Hulk.—Used for accommodation of officers and men going through torpedo and wireless courses.
Kronprinz	„ „	1879	5,707	Kiel-Wik	Hulks.—Machinery schools, stokers' instructional ships, and instructional ships in electro-technical science.
Leipzig	Unarmoured Corvette.	1875	3,925	Wilhelmshaven.	
Friedrich der Grosse	Battleship, 3rd Class.	1874	6,711	Ellerbek	Coal hulk.
Mars	Gunnery School Ship.	1877	3,260	Kiel-Wik	Hulk.
Nixe	Cruiser, 3rd Class.	1885	1,722	Sonderburg	Accommodation Ship attached to Gunnery School.

* Purchased from England in 1861.

Harbour Ships—*cont.*

Ship.	Former Classification.	Launched.	Displacement.	Stationed at	Present Employment.
			Tons.		
Falke - - -	Cruiser, 3rd Class.	1891	1,549	—	Hulk.
Comet - - -	Torpedo Vessel.	1892	971	Emden	Hulk.—Attached to light cruiser "Arcona."
Greif - - -	Cruiser, 3rd Class.	1886	2,027	—	——
Sperber - - -	Sloop	1888	1,100	Sonderburg (?).	Hulk.—For targets (Scheibenhulk).
Alexandrine - -	Light Cruiser.	1885	2,373	Danzig	Hulk.—Workshop.
Prinzess Wilhelm -	„ „	1887	4,223	„	Mining Depôt Ship.
Stein - - -	Training Ship.	1879	2,856	Wilhelmshaven.	Hulk.—Offices.

VESSELS OF MINOR IMPORTANCE.

Classification and Name or Description.	Built.	Displacement.	I.H.P.	Appropriation.		Remarks.
CRANES (FLOATING).		Tons.				
Schwimmkran No. 1 -	1891–2	630	250	Imperial Dockyards.	Kiel -	To lift 100 tons.
„ „ 1 -	1902	1,132	80		Wilhelmshaven.	To lift 100 tons. I.H.P. of crane engine 460.
„ „ 2 -	1869	431	70		Kiel -	To lift 40 tons.
„ „ 2 -	1875–6	375	48		Wilhelmshaven.	To lift 40 tons.
„ „ 3 -	1908–9	1,800	303		Kiel -	To lift 150 tons.
„ „ — -	1904	748*	60		Danzig -	To lift 100 tons. I.H.P. of crane engine 120.
DREDGERS AND HOPPER BARGES.			I.H.P. of Dredging Engine.			
Bagger I., B.H. - -	1912	541	200–230	Imperial Dockyard, Kiel		Small rock dredger.
„ II., B.H. - -	1912	590	230–300	„ „ „		Large rock dredger.
„ III., B.H. -	1912	886	400	„ „ „		Suction and bucket dredger.
	Propelling machinery		700			
„ IV., B.H. -	—	—	12	„ „ „		Mud dredger.
„ V. - - -	1881–2	261	35	Imperial Dockyard, Wilhelmshaven.		Suction dredger.
„ VI. - - -	1893	941	260	„ „ „		Hopper dredger.
„ VII. - -	1903–5	4,566	1,615	„ „ „		Twin-screw dredger.
„ Danzig - -	1876–7	295	110	Imperial Dockyard, Danzig.		

* Register tons.

C 3

Vessels of Minor Importance—*cont.*

Classification and Name or Description.	Built.	Displacement.	I.H.P.	Appropriation.	Remarks.
DREDGERS AND HOPPER BARGES —*cont.*		Tons.			
Dampfklappenprahm No. 1.	1893	557	135	Imperial Dockyard, Wilhelmshaven.	Hopper barges.
Dampfklappenprahm No. 2.	1893	557	134	” ” ”	
Dampfklappenprahm No. 3.	1893	557	138	” ” ”	
Dampfklappenprahm No. 4.	1897–8	534	131	” ” ”	
Dampfklappenprahm No. 5.	1897–8	534	139	” ” ”	
FIRE ENGINES (FLOATING).					
Alarm - - - -	1901	31	108	Imperial Dockyard, Kiel.	
Hilfe - - - -	1905–6	54	78	Imperial Dockyard, Danzig.	
Spritzendampfer - -	1897	39	130	Imperial Dockyard, Wilhelmshaven.	
GUARD BOATS.					
Schneewittchen - -	1888	70	591	C.-in-C., Baltic Station.	
Sirius - - - -	1890	51	200	C.-in-C., North Sea Station.	
Wega - - - -	1890	51	200	” ”	
MINING LIGHTERS (STEAM).					
C. 1 to C. 9 - - -	1904–6	—	—	Mining Division, Cuxhaven.	
F. 1 to F. 4 - - -	1871–81	—	—	Defences, Friedrichsort.	
G. 1 to G. 4 - - -	1888–90	—	50–80	Defences, Geestemünde.	
Minendepot - - -	1899–1913.	28	60	Mine Depôt, Friedrichsort.	Tender.
Sprengprahm - -	1887	125	450	Inspection of Torpedoes	For experimental purposes.
W. 1 to W. 4 - -	1901	—	120	Defences, Wilhelmshaven.	
MISCELLANEOUS VESSELS.					
Bombe - - -	1881	21	80	Artillery Depôt, Geestemünde.	Tender.
Dampfbeiboot Kl. A. -	1892	18	170	Artillery Depôt, Friedrichsort.	”
” Kl. I. -	1889	5	20	Mining Depôt, Wilhelmshaven.	”
Dampffähre - - -	1902	46	50	Imperial Dockyard, Danzig.	Ferry boat to Holm.
Eider - - - -	1888	395	255	Imperial Dockyard, Kiel.	Store ship.
Fortifikation - -	1908	148	250	Fortifications, Geestemünde.	For transport of troops in mouth of Weser River.

Vessels of Minor Importance—*cont.*

Classification and Name or Description.	Built.	Displacement.	I.H.P.	Appropriation.	Remarks.
MISCELLANEOUS VESSELS—*cont.*		Tons.			
Friedrichsort	1880	66	100	Ammunition Depôt, Dietrichsdorf.	Tender.
Langlütgen	1884	15	(motor) 15	Fortifications, Geestemünde.	Communication Vessel.
Mentor	1909–10	88	640	Submarine Service	Machinery—2 Diesel motors.
Mosquito	1900	17	170	Imperial Dockyard, Kiel.	Launch of Secretary of State of Imperial Navy Office.
Motorboot	1911	3	15	Ammunition Depôt, Dietrichsdorf.	Tender.
Motortaucherfahrzeug	1910	83	32	Imperial Dockyard, Kiel.	For diving purposes.
Mürwik	1870 ?	50	70	Torpedo School Ship, Mürwik.	Tender.
†Schillig	1878	31	—	Coast District Office V., Wilhelmshaven.	Despatch vessel.
Schwan	1911–12	12	100	Imperial Dockyard, Kiel.	Admiral Superintendent's launch.
Seestern	1893	61	80	Torpedo Workshop, Friedrichsort.	Tender.
Strande	1907	76	156	„ „	„
MOORING VESSELS.					
†Heppens	1877	291	—	Coast District Office V., Wilhelmshaven.	
Mellum	1892	363	345	„ „	
Wik	1903	364	385	Imperial Dockyard, Kiel.	
OILERS.					
Öldampfer K 1	1904	466	90	Imperial Dockyard, Kiel.	For lubricating oil.
„ K 2	1906	616	226	„ „	For oil fuel.
„ W 81	1903	214	136	Imperial Dockyard, Wilhelmshaven.	For lubricating oil.
„ W 83	1906	445	220	„ „	For oil fuel.
PILOT VESSELS.					
Jade	1903	352	500	Coast District Office V., Wilhelmshaven.	
Rüstringen	1911	658	950	„ „	
†Wangeroog	1906	163	—	„ „	
Wilhelmshaven	1875	145	200	„ „	
SURVEYING LAUNCHES.					
Farewell	1887	66	200	Imperial Dockyard, Wilhelmshaven.	Harbour Works.
Helgoland	1909	15	120	Works Department, Heligoland.	

† Sailing Vessels.

C 4

Miscellaneous Vessels.

Vessels of Minor Importance—*cont.*

Classification and Name or Description.	Built.	Displacement.	I.H.P.	Appropriation.	Remarks.
SURVEYING LAUNCHES—*cont.*		Tons.			
Legde - - - -	1913	54	80	Imperial Dockyard, Wilhelmshaven.	
Minseroog - - -	1908	118	160	,, ,, ,,	Harbour Works.
Peilboot I. - - -	1912	88	131	North Sea Station -	Coast Surveys.
,, **II.** - -	1912	89	132	,, ,, ,, -	
,, **V.** - -	1912	84	128	,, ,, ,, -	
,, **VI.** - -	1902	66	130	Baltic Station - -	
,, **VII.** - -	1903	65	130	,, ,, - -	
TUGS.					
Bussard - - -	1890	34	60	Imperial Dockyard, Kiel.	
Caurus - - -	1901	322	500	Imperial Dockyard Wilhelmshaven.	
Dampfbeiboot A 1 -	1909	66	100	,, ,, ,,	
,, **A 2** -	1909	66	100	,, ,, ,,	
,, **A 5** -	1904	22	50	,, ,, ,,	
,, **A 6** -	1906	26	75	,, ,, ,,	
,, **A 14** -	1910	59	102	,, ,, ,,	
,, **A 15** -	1911	51	100	,, ,, ,,	
,, **DI 1** -	1902	27	90	Imperial Dockyard, Danzig.	
,, **DI 2** -	1903	30	100	,, ,, ,,	
,, **DI 4** -	1907	43	130	,, ,, ,,	
Dampfboot T 2 - -	1904	22	40	Imperial Dockyard, Wilhelmshaven.	
Daunsfeld - - -	1907	266	513	,, ,, ,,	
Eisvogel - - -	1890	34	60	Imperial Dockyard, Kiel.	
Fleiss - - - -	1889	96	175	Imperial Dockyard, Wilhelmshaven.	
Flink - - - -	1905	97	164	,, ,, ,,	
Föhn - - - -	1890	96	234	Imperial Dockyard, Kiel.	
Helga - - - -	1913	500	750	Heligoland.	
Hermann - - -	1898	57	60	Imperial Dockyard, Kiel.	
Marie - - - -	1882	39	60	,, ,, ,,	
Motorboot 1—B.H. -	1911	6	20	Imperial Dockyard, Wilhelmshaven.	
Motorboot 2—B.H. -	1912	6	22	,, ,, ,,	
Motorschlepper—M.A. 1	1914	—	530	,, ,, ,,	
Motorschlepper—M.A. 2	1914	—	530	,, ,, ,,	
Motorschlepper T 1 -	1909	25	90	,, ,, ,,	

Vessels of Minor Importance—*cont.*

Classification and Name or Description.	Built.	Displacement.	I.H.P.	Appropriation.	Remarks.
TUGS—*cont.*		Tons.			
Mottlau - - -	1875	144	300	Imperial Dockyard, Danzig.	Paddle steamer.
Pinnass 5 - - -	1902	36	84	Imperial Dockyard, Kiel.	
„ **6** - - -	1908	77	100	„ „ „	
Reiher - - - -	1903	98	200	„ „ „	
Rival - - - -	1874	127	250	Imperial Dockyard, Wilhelmshaven.	Paddle steamer.
Scheibenhof - -	1908	35	79	Imperial Dockyard, Kiel	For towing targets.
Schlepper I.—B.H. -	1912	93	200	Works Department, Heligoland.	
„ **II.—B.H.** -	1912	93	200	„ „ „	
Sonderburg - - -	1909	443	500	Naval Gunnery School, Sonderburg.	Also water tank vessel.
Stark - - - -	1905	138	257	Imperial Dockyard, Wilhelmshaven.	
Sturm - - - -	1904	342	556	„ „ „	
Weichsel - - -	1904	342	560	Imperial Dockyard, Danzig.	
Weih - - - -	1891	34	60	Imperial Dockyard, Kiel.	
TUGS AND PUMPING VESSELS.					
Æolus - - - -	1897	109	270	Imperial Dockyard, Kiel.	
Arngast - - -	1909	266	600	Imperial Dockyard, Wilhelmshaven.	
Kraft - - - -	1890	620	1,250	„ „ „	
Norder - - - -	1883	557	1,000	Imperial Dockyard, Kiel.	
Passat - - - -	1908	241	417	„ „ „	
Voslapp - - -	1912	384	600	Imperial Dockyard, Wilhelmshaven.	
WATER TANK VESSELS.					
Wasserfahrzeug - -	1899	346	245	Imperial Dockyard, Danzig.	Drinking water.
„ **K 1** -	1890	334	240	Imperial Dockyard, Kiel	„ „
„ **K 2** -	1877	128	57	„ „ „	„ „
„ **K 3** -	1898	561	86	„ „ „	„ „
„ **K 4** -	1906	616	226	„ „ „	„ „
„ **K 5** -	1884	59	55	„ „ „	„ „
„ **W 1** -	1874	59	60	Imperial Dockyard, Wilhelmshaven.	„ „
„ **W 2** -	1902	403	180	„ „ „	„ „
„ **W 3** -	1898	424	180	„ „ „	„ „
„ **W 4** -	1912	958	600	„ „ „	„ „
Sonderburg - - -	(*See above* under Tugs.)				

Miscel-laneous Vessels.

Vessels of Minor Importance—*cont.*

CLASSIFICATION and NAME or DESCRIPTION.	Built.	Dis-place-ment.	I.H.P.	Appropriation.	Remarks.
YACHTS (SAILING).		Tons.			
Asta - - - -	1894	51	—	C.-in-C., North Sea Station.	
Carlota - - -	1906	16	—	Naval School, Mürwik.	
Comet - - - -	1887	155	—	C.-in-C., North Sea Station.	
Hertha - - -	1895	22	—	Inspector of Training.	
Jutta - - - -	1900	16	—	Naval School, Mürwik.	
Orion - - - -	1896	165	—	C.-in-C., Baltic Station.	
Thalatta - - -	1888	—	—	" " "	
YACHTS (STEAM).					
Ellerbek - - -	1885	24	100	Imperial Navy Office and Commissioner of Kaiser Wilhelm Canal.	
Hulda - - - -	1889	14	180	C.-in-C., Baltic Station	At disposal of the Emperor.
Radaune - - -	1886	24	120	Imperial Dockyard, Danzig.	Dockyard Yacht.
Schneewittchen - -	(*See above* under Guard Boats.)				

ARMED MERCHANT CRUISERS AND OTHER AUXILIARIES.

D.A.P.=German-American Petroleum Co., Hamburg.
D.O.A.=German East Africa Line.
H.S.A.=Hamburg South America Line.
K.H.=A. Kirsten, Hamburg.
N.=Neptun Line, Bremen.
H.A.=Hamburg America Line.
H.B.A.=Hamburg-Bremen Africa Line.
Kos.=Kosmos Line.
N.D.L.=North German Lloyd.
S. & B.=Sartori & Berger.

Ship.	Line.	Built.	Gross Tonnage.	Speed.	Dimensions.		Remarks. (June 1917.)
					Length.	Breadth.	
				Knots.	Feet.	Feet.	
Alster	K.H.	1914	1,500	—	354·6	48·1	In home waters.
Ammon	Kos.	1914	7,000	—	—	—	Depôt ship for mine-sweepers.
Answald	H.B.A.	1909	5,401	11	419·6	54·5	Seaplane carrier.
Anvers	*ex* Norwegian	1883	860	—	215·0	31·7	Depôt ship at Borkum.
***Berlin**	N.D.L.	1908	17,324	18	590·2	69·7	Interned at Trondhjem. 6—*4·1-in.*, 6 *m.*, 2 mine ports in stern.
***City of Bradford**	*ex* British	1903	1,349	13½	256·5	34·5	Repair ship for seaplanes.
Cordoba	H.S.A.	1895	4,889	10	376·5	46·3	In home waters.
Dania	H.A.	1904	3,898	11½	355·3	44·0	,, ,,
Erlangen	N.D.L.	1902	5,285	12	378·8	47·5	,, ,,
Eskimo	*ex* British	1910	3,326	17	331·2	45·0	,, ,,
Fürst Bülow	H.A.	1911	7,614	11½	468·8	58·2	,, ,,
Gauss	N.	1882	442	—	147·6	22·5	,, ,,
General	D.O.A.	1910	8,063	15	449·1	54·3	Accommodation ship at Constantinople.
***Glenearn**	*ex* British	1914	4,828	9½	390·0	50·5	In home waters.
Graf Zeppelin	—	—	—	—	—	—	,, ,,
Hessen	—	—	—	—	—	—	,, ,,
***Indianola**	*ex* British	1912	4,566	10½	390·0	52·5	Seaplane carrier.
***Inkula**	*ex* British	1904	5,137	12	401·0	52·1	Depôt ship for mine-sweepers, Swinemünde. 4 *or* 6—*15-pr.*
Kigoma	D.O.A.	1914	8,100	15½	449·3	55·3	In home waters.
König Friedrich August	H.A.	1912	9,462	15½	475·7	55·3	,, ,,
Minos	N.	1913	718	—	185·9	28·3	Eider barrage.
Möwe, *ex* Pungo	—	1914	4,500	14½	385·0	49·0	*See* "Raiders," p. 16.
Möwe	Argo	1912	1,251	12	241·4	36·0	Jade barrage, 2—*22-pr. A.A.*
Niederwald	H.A.	1904	3,490	12	354·8	45·2	In home waters.
Pawnee	D.A.P.	1913	4,972	—	380·1	52·8	,, ,,
Phœnix	Argo.	1913	1,166	—	236·0	32·5	Mine store at Wilhelmshaven.
Prinz Sigismund	S. & B.	1899	697	15½	205·4	28·6	Kiel harbour barrage.
Rio Negro	H.S.A.	1905	4,556	12	361·2	46·7	In home waters.
Rugia	H.A.	1905	6,598	13	409·4	52·7	Submarine depôt ship at Emden.
Santa Elena	H.S.A.	1907	7,415	12	430·9	54·8	Seaplane carrier.
Seeadler	*ex* British	1888	1,571	11	245·0	39·0	*See* "Raiders," p. 17.
Silvana	H.A.	1897	804	14	205·6	29·5	Depôt ship for armed trawlers, Wilhelmshaven.
Steigerwald	H.A.	1911	4,836	13½	354·0	48·8	In home waters.
Wolf	—	—	About 5,000	—	About 420	—	*See* "Raiders," p. 16.
***Yorck**	N.D.L.	1906	8,909	14½	463·5	57·4	Sheltering at Valparaiso.

* *See* silhouette, Plate 108.

D 2

AUXILIARY MINELAYERS.

Vessel.	Former Service.	Built.	Gross Tonnage.	Dimensions.		Speed.	Armament and Remarks
				Length.	Breadth.		
Deutschland -	Ferry steamer, Sassnitz-Trelleborg.	1909	2,882	354·2	50·8	14	Reported to carry 800 mines and to be armed with 12 Q.F. guns and several machine guns.
Kaiser - -	Hamburg-America tourist steamer.	1905	1,916	303·1	38·4	20	
Odin - - -	Tourist steamer in the Baltic.	1902	1,176	336·5	33·9	16	
Prinz Adalbert -	Mail steamer, Kiel-Korsör.	1895	699	205·7	28·7	15	Reported to carry about 80 mines.
Prinz Waldemar	Mail steamer, Kiel-Korsör.	1893	660	204·1	27·8	14	Reported to carry about 80 mines.
Wotan - -	Oiler - -	1913	5,703	405·0	52·5	10	For Silhouette, *see* Plate 108.

AUXILIARY PATROL VESSELS.

At least 250 trawlers and other small steam craft are known to have been taken over by the German Navy.

For names of trawlers known to be armed, *see* I.D. 1151, "Foreign Fishing Vessels in Home Waters."

Examples of armed trawlers are given on Plate 107.

TRANSPORTS.

This list includes all vessels in home waters (in June 1917) which have been employed as transports or have been reported to be specially equipped for that purpose. It is not a complete list of the ships suitable and available.

It is not believed that any of these vessels have been armed, with the exception of those whose names appear also in the list of Armed Merchant Cruisers and Other Auxiliaries (p. 27), but arrangements have probably been made for arming others rapidly if required.

D.A. = German-Australia Line.
H.A. = Hamburg-America Line.
Kos. = Kosmos Line.
D.O.A. = German East Africa Line.
H.S.A. = Hamburg-South America Line.
N.D.L. = North German Lloyd.

Name of Ship.	Line.	Built.	Tonnage.		Speed.	Reported Appropriation and Remarks.
			Gross.	Net.		
					Knots.	
Ammon	Kos.	1914	7,000	—	--	Heavy Siege Artillery.
Artemisia	H.A.	1901	5,739	3,676	12½	" " "
Badenia	H.A.	1902	6,930	4,309	11	" " "
Batavia	H.A.	1899	11,464	7,300	—	Siege Artillery.
Bermuda	H.A.	1899	7,038	4,468	11½	Heavy Siege Artillery.
*Bremen	N.D.L.	1897	11,540	7,078	15	Field Artillery.
*Bürgermeister	D.O.A.	1902	5,945	3,660	13	Heavy Siege Artillery.
*Cap Finisterre	H.S.A.	1911	14,503	8,749	17	Cavalry and Engineers.
Cassel	N.D.L.	1901	7,642	4,623	12½	Field Artillery.
Chemnitz	N.D.L.	1901	7,542	4,784	12½	Infantry.
Duisburg	D.A.	1900	4,496	2,854	12	Army Service Corps.
Frankfurt	N.D.L.	1899	7,431	4,739	12½	
Giessen	N.D.L.	1907	6,964	4,376	12	Infantry.
*Graf Waldersee	H.A.	1898	13,193	8,375	13	Cavalry and Engineers.
Greiffenfels	Hansa	1914	5,852	3,661	11½	Field Artillery.
Habsburg	H.A.	1906	6,437	4,076	13	Siege Artillery.
König Friedrich August	H.A.	1912	9,462	5,590	15½	" "
*Königin Luise	N.D.L.	1896	10,785	6,565	15	Carried drafts of 1,500 men in peace time.
*Patricia	H.A.	1899	14,466	9,072	13	Carried drafts of 1,700 men in peace time.
*Pretoria	H.A.	1897	13,234	8,415	13	Cavalry and engineers.
Prinz Ludwig	N.D.L.	1894	9,687	5,688	15½	Infantry.
*Prinzessin	D.O.A.	1905	6,387	3,697	14	Army Service Corps.
Schwaben	N.D.L.	1906	5,098	3,225	11½	Field Artillery.
Syria	H.A.	1898	3,597	2,240	11	
*Victoria Luise	H.A.	1900	16,703	8,127	18	Cavalry and Engineers.
Worms	D.A.	1907	4,428	2,778	12	Army Service Corps.
Ypiranga	H.A.	1908	8,103	4,907	13½	Cavalry and Engineers.

* For silhouette, *see* Plate 108.

COLLIERS.

Merchant Vessels known to have been employed as colliers for the German High Sea Fleet.

Bl = B. Blumenfeld, Hamburg. Sti = Hugo Stinnes, Mulheim a. Ruhr.
Kohl = Kohlenhandelsgesellschaft, Flensburg.

Name of Ship.	Line.	Built.	Gross Tonnage.	Net Tonnage.	Speed.
					Knots.
Adeline Hugo Stinnes III. -	Sti	1909	2,709	1,623	10·5
Clara Blumenfeld - - -	Bl	1908	2,331	1,224	12
Fritz Hugo Stinnes V. - -	Sti	1910	1,398	809	10·5
Glückauf - - - - - -	Kohl	1909	2,204	1,213	—
Grete Hugo Stinnes VIII. - -	Sti	1911	1,500	850	10
Helene Blumenfeld (Naval Collier No. 8).	Bl	1905	2,114	967	12
Hilde Hugo Stinnes X. - -	Sti	1911	1,500	880	10

HOSPITAL SHIPS.

Military Hospital Ships, in conformity with Article 1 of the Hague Convention, *i.e.*, equipped by the Government:—

Distinguishing Letter.	Name.	Built.	Gross Tonnage.	Speed.	No. of Masts.	No. of Funnels.	Distance from: Bow to Foremast.	Foremast to Funnel.	Funnel to Funnel.	Funnel to Mainmast.	Mainmast to After Edge of Stern.	Accommodation and Remarks.
				Knots.			Ft.	Ft.	Ft.	Ft.	Ft.	
	Adler	1903	594	—	Two	One	48·9	64·0	—	35·4	57·7	
	Hansa	1881	594	—	One	One	33·1	61·0	—	65·0 to after edge of stern.	—	53 serious cases, 250 minor cases. Separate accommodation for enteric cases.
E	**Sierra Ventana.**	1912	8,262	13½	Two	One	83·7	133·5	—	141·1	96·8	

Auxiliary Hospital Ships, in conformity with Article 2 of the Hague Convention, *i.e.*, privately equipped but recognised by the Government:—

Distinguishing Letter.	Name.	Built.	Gross Tonnage.	Speed.	No. of Masts.	No. of Funnels.	Distance from: Bow to Foremast.	Foremast to Funnel.	Funnel to Funnel.	Funnel to Mainmast.	Mainmast to After Edge of Stern.	Accommodation and Remarks.
				Knots.			Ft.	Ft.	Ft.	Ft.	Ft.	
D	**Imperator**	1897	1,079	15	Two	Two	52·5	41·0	27·9	44·3	63·0	
	Kehrwieder	1900	56	Under 12	One	Two	42·7	36·7	19·0	78·1	—	
	Lensahn	1901	513	—	Two	Two	47·2	44·3	14·4	37·4 to after edge of stern.	49·2	
	Portia	1911	1,189	10	Two	One	55·8	67·3	—	44·3	67·9	
	Titania	1912	1,187	10	Two	One	55·8	67·3	—	44·3	67·9	
	Viola	1912	1,156	10	Two	One	57·4	67·6	—	42·0	68·2	
	Rossall*	1895	2,739	9	Two	One	78·0	76·0	—	82·0	86·0	For silhouette, *see* Plate 108.

* *Ex* British steamer. Is employed as an auxiliary hospital ship, but her name has not been notified by the German Government.

CABLE SHIPS.

(For silhouettes, *see* Plate 108.)

Owners and Name.	Built.	Gross Tonnage.	Net Tonnage.	Speed.
NORDDEUTSCHE SEEKABELWERKE.				Knots.
Grossherzog von Oldenburg	1905	2,691	1,113	12·0
Stephan	1902	4,630	2,467	11·5

GUNBOATS.

PANTHER.

SEEADLER.

Ordnance Survey, July, 1917.

MINING VESSELS.

ALBATROSS.

NAUTILUS.

PELIKAN.

MINING VESSELS.

M.1. TO M.60.

SALVAGE SHIP FOR SUBMARINES.

VULKAN.

VULKAN *(Bow View).*

VULKAN *(Stern View).*

FISHERY PROTECTION VESSEL.

ZIETEN.

IMPERIAL YACHT.

HOHENZOLLERN.

SURVEYING VESSEL.

HYÄNE.

RAIDERS.

MÖWE.
Photograph taken during 1st cruise (January–March, 1916).

MÖWE.
Photograph taken during 2nd cruise (December, 1916—March, 1917).

MÖWE.
Photograph taken during 2nd cruise.

PLAN.

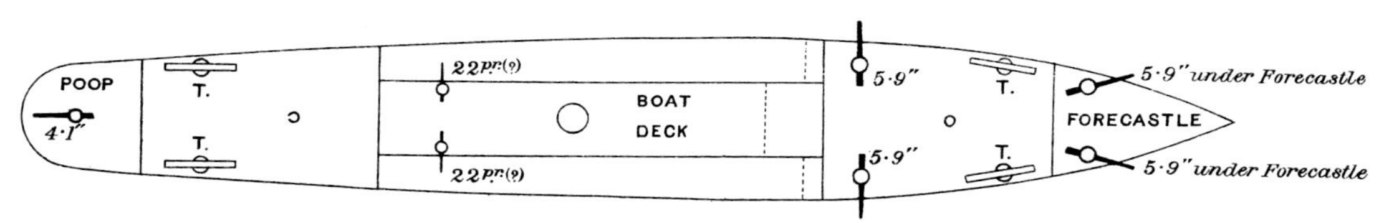

RAIDER.

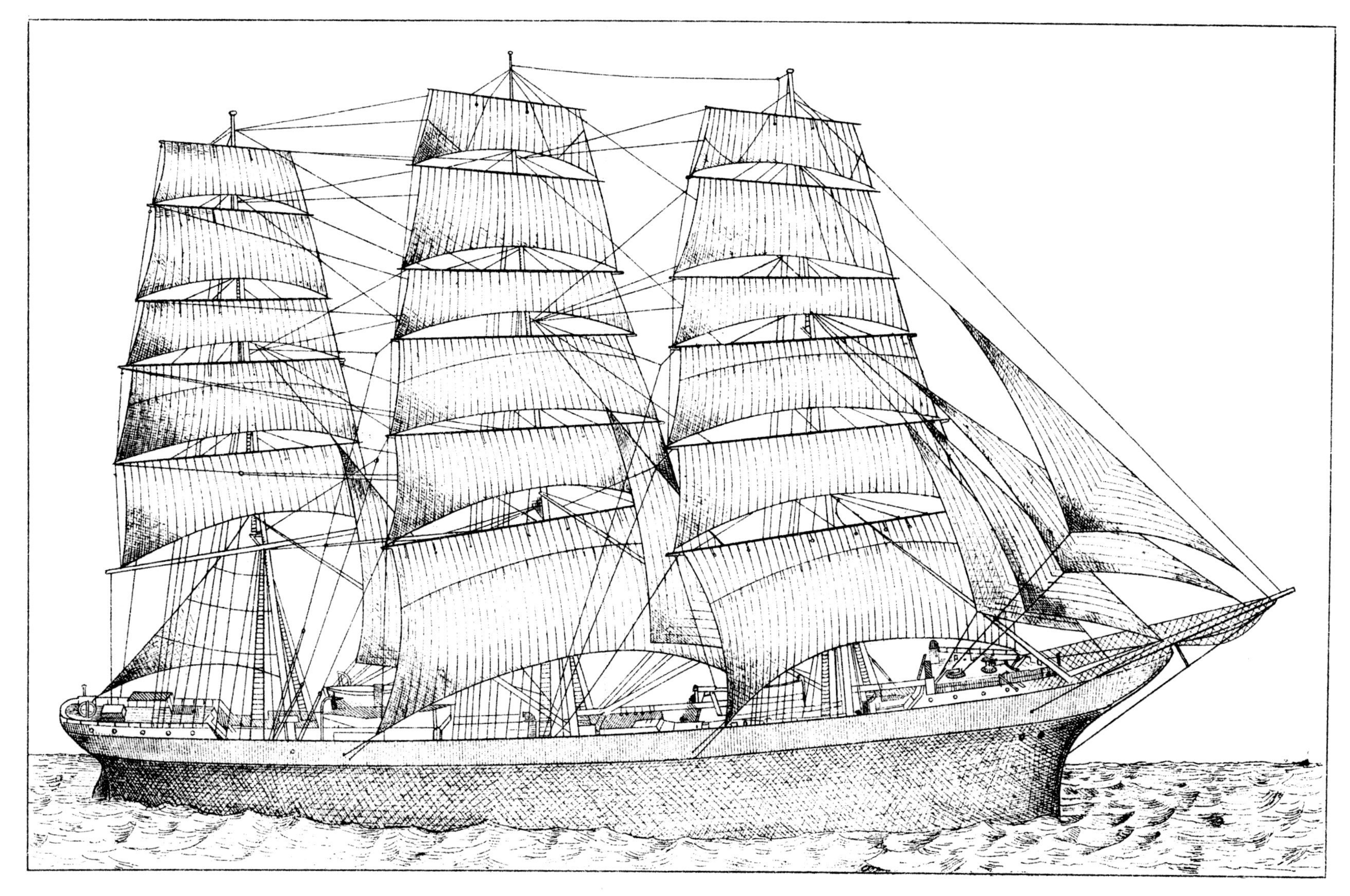

Rough Sketch made by a Master of a captured Ship.

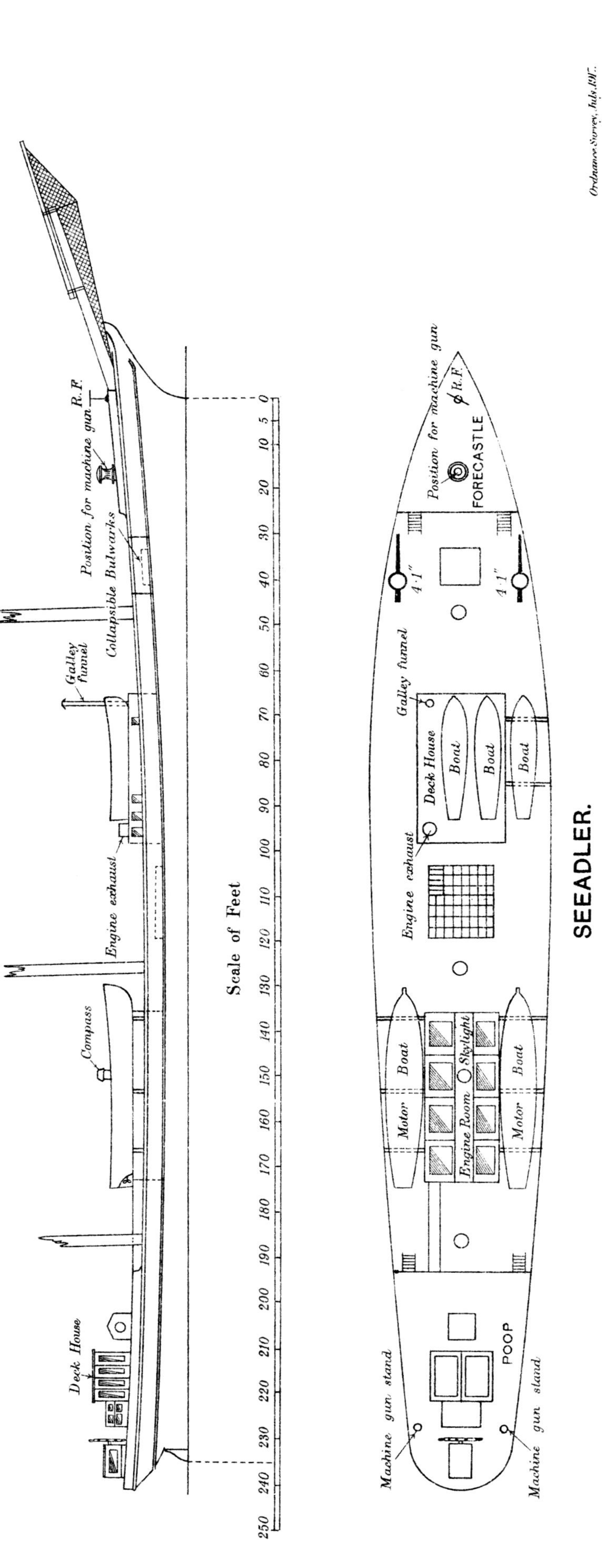

Ordnance Survey, July 1917.
Position for machine gun
R. F.
Collapsible Bulwarks
Galley funnel
Engine exhaust
Compass
Deck House
Scale of Feet
0 5 10 20 30 40 50 60 70 80 90 100 110 120 130 140 150 160 170 180 190 200 210 220 230 240 250
Position for machine gun
R. F.
FORECASTLE
4·1"
4·1"
Galley funnel
Deck House
Boat
Boat
Boat
Engine exhaust
Motor Boat
Engine Room Skylight
Motor Boat
POOP
Machine gun stand
Machine gun stand
SEEADLER.

TENDERS TO GUNNERY SCHOOL.

DRACHE.

HAY.

DELPHIN.

FUCHS.

ARMED TRAWLERS.

KARL FEUERLOB.

JOH. WESTER.
BAHRENFELD.

NORDSEE.
Leader of North Sea Patrols.

MINESWEEPING TRAWLERS.
Ashore on Jutland Coast September, 1917.

Ordnance Survey, October, 1917.

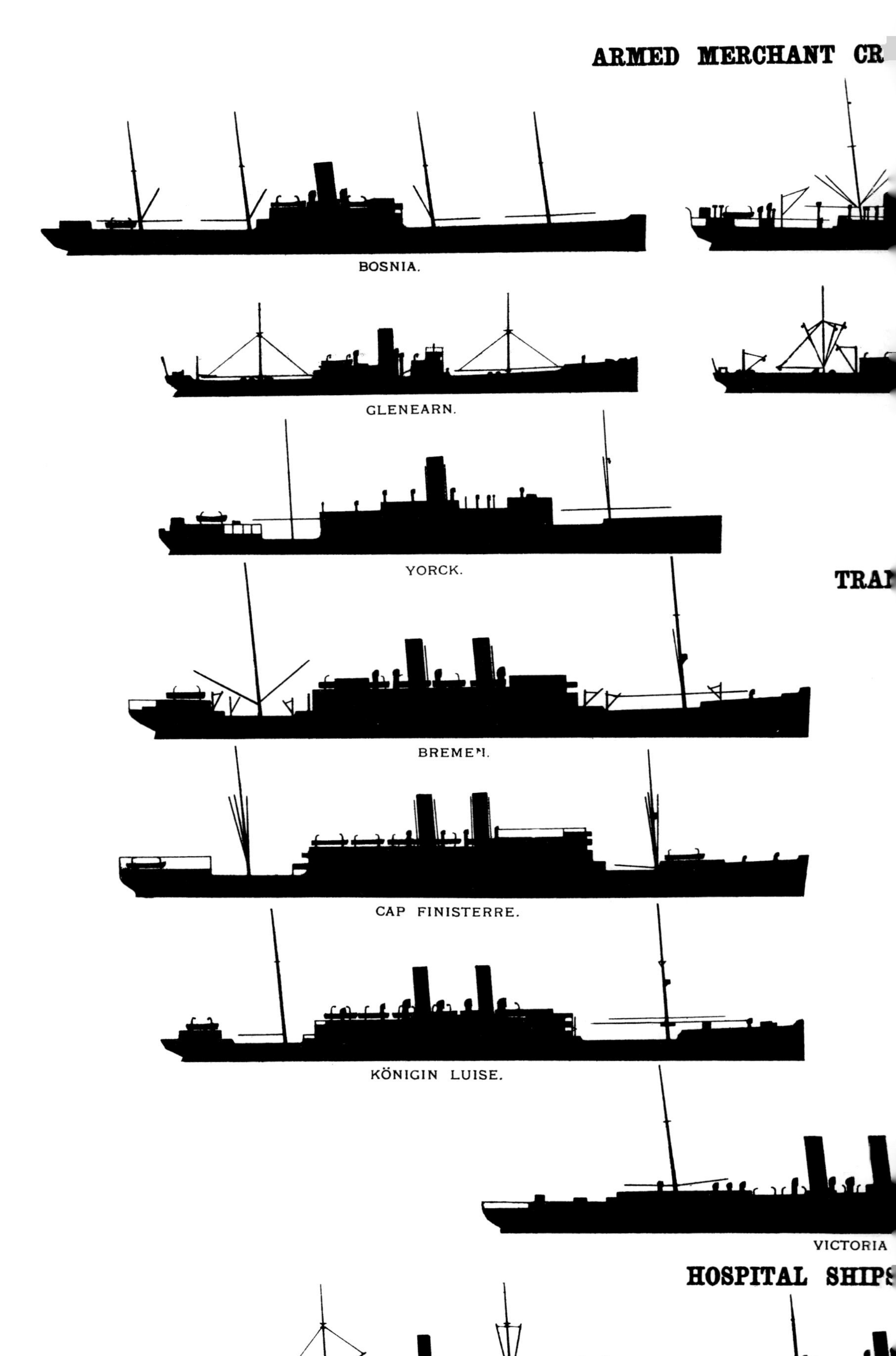

SILHO
ARMED MERCHANT CR
BOSNIA.
GLENEARN.
YORCK.
TRA
BREMEN.
CAP FINISTERRE.
KÖNIGIN LUISE.
VICTORIA
HOSPITAL SHIPS
ROSSALL.
GROSSHERZO

TTES.

Plate 108.
C.B. 1182.
Part III.
July, 1917.

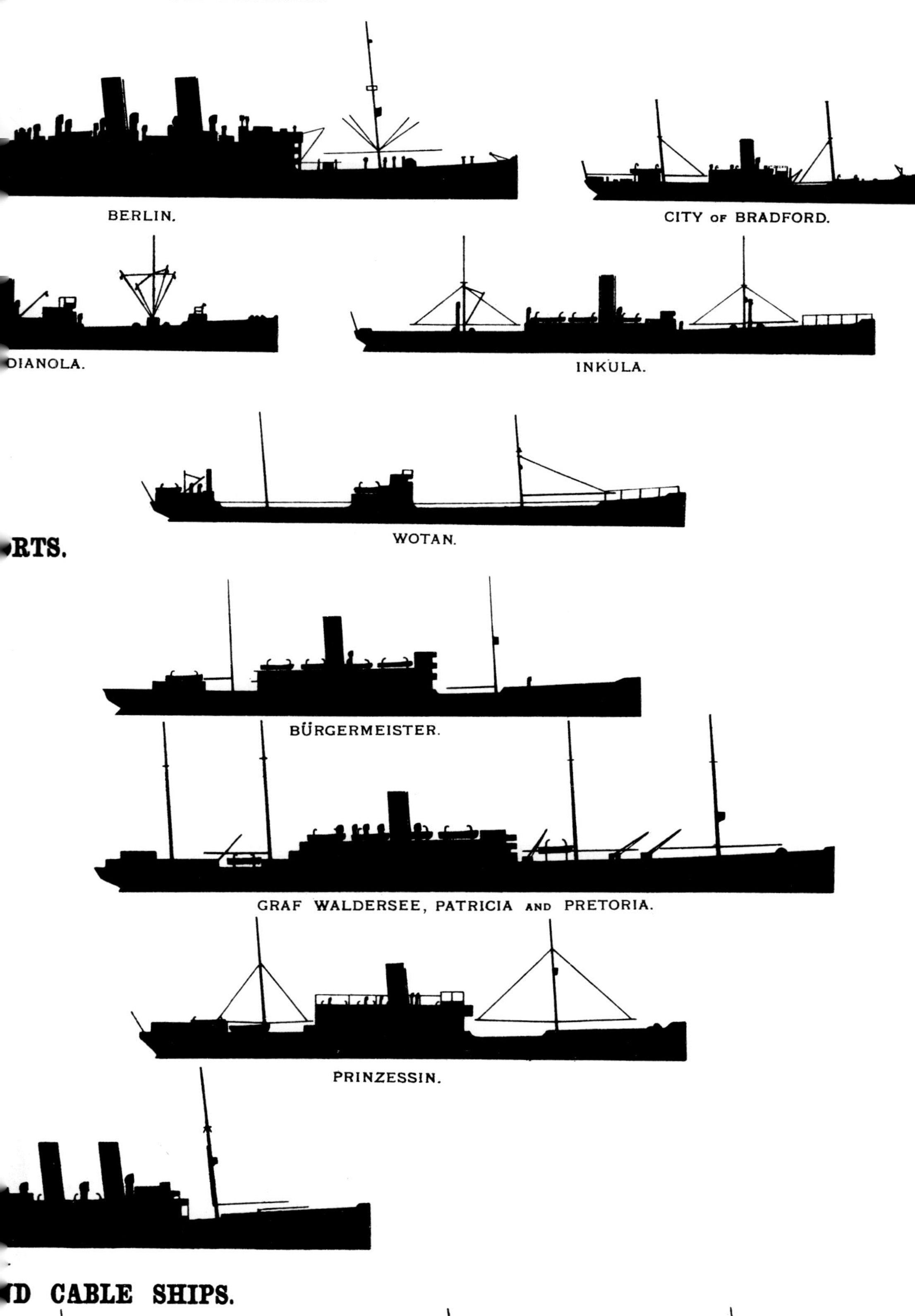

Ordnance Survey, July, 1917.

APPENDIX

Table of Naval Ordnance

Fleet Strength

TABLE OF NAVAL ORDNANCE.

The data in this table is mainly derived from Krupp's published tables, and should therefore be accepted with reserve. For fuller details concerning ammunition, *see* the table of ammunition on Part IV., Section 2.

Guns.					Ammunition.					Ballistics.			Remarks.
Calibre and Model.	Exact Calibre.	Total Weight.	Length.		Charge.		Projectile.			Muzzle Velocity.	Muzzle Energy.	Muzzle Penetration, W.I.	
			Total.	Bore.	Weight.	Powder.	Nature.	Total Weight.	Burster.				
	Ins.	Tons.	Ft. Ins.	Calrs.	Lbs.			Lbs.	Lbs.	f.s.	Ft.-Tons	Ins.	
15-in. (38·1 cm.), L/45	15	75 (?)	—	—	531·3(?)	—	—	1,675(?)	—	2,780(?)	—	—	—
12-in. (30·5 cm.) Q.F., L/50	12	52·2(?)	52 6	50	273·4	See	Table of Ammunition.			2,800(?)	—	—	In "*Thüringen*" and "*Kaiser*" classes and "*Derfflinger*."
11-in. (28 cm.) B.L., L/35	11·02	43·0	32 1	32·0	317 212	P.P. C/85 P.P. C/85	Steel A.P. shell C.I. Common	529 531	Nil 15·4	2,247	18,520	28·0	—
11-in. (28-cm.) B.L., L/40	11·02	43·3	36 6	37·0	115 108	R.P. C/00 R.P. C/98	Steel A.P. shell C.I. Common	529 531	Nil 15·4	2,313	19,630	28·9	—
11-in. (28 cm.) Q.F., L/40	11·02	38·2	36 8	37		See	Table of Ammunition			2,790	—	—	—
11-in. (28 cm.) Q.F., L/45	11·02	37	41 9	43						2,821	—	—	In "*Nassau*" class and "*Von der Tann.*"
11-in. (28 cm.) Q.F., L/50	11·02	—	—	—						—	—	—	In "*Moltke*" and "*Seydlitz.*"
9·4-in. (24 cm.) B.L., L/35	9·45	21·6	27 7	32·0	—	—	H.E. Common	474	7·49	2,263	16,830	28·9	—
9·4-in. (24 cm.) Q.F., L/40	9·45	25·2	31 4	36·9		See	Table of Ammunition			2,739	16,090	31·1	—
8·2-in. (21 cm.) Q.F., L/40	8·24	15·7	27 6	37		See	Table of Ammunition			2,526	—	—	—
8·2-in. (21 cm.) Q.F., L/40	8·24	19·9	30 11	42						—	—	—	In "*Blucher*" only.
6·7-in. (17 cm.) Q.F. L/40	6·69	7·8	22 3	37·7		See	Table of Ammunition			2,726	7,365	24·8	—
5·9 in. (15 cm.) Q.F. L/35	5·87	4·4	17 2	32·0	15·8 16·3	R.P. C/00 R.P. C/98	Capped A.P. shot C.I. Common Shrapnel Star shell	88 88 —	— 2·8 —	2,231	3,038	15·4	—
5·9-in. (15 cm.) Q.F. L/40	5·87	4·7	19 6	37	—	See	Table of Ammunition also Star shell	88 88	— —	2,624	4,207	19·7	In "*Nassau*," &c.
5·9-in. (15 cm.) Q.F. L/45	5·87	—	—	—		See	Table of Ammunition also Star shell			—	—	—	—
5·9-in. (10·5 cm.) Q.F. L/50	—	—	—	—						—	—	—	—
5·1-in. (13 cm.) Q.F.	—	—	—	—	—	—	—	—	—	—	—	—	—
4·1-in. (10·5 cm.) Q.F. L/35	4·13	1·1	11 10	32·3		See	Table of Ammunition also Star shell			2,001	1,055	—	—
4·1-in. (10·5 cm.) Q.F. L/40	4·13	1·7	13 9	37·2						2,296	1,390	—	—
4·1-in. (10·5 cm.) Q.F. L/45	4·13	—	—	—						—	—	—	In "*Breslau*" class.
4·1-in. (10 cm.) Q.F. L/50	—	—	—	—						—	—	—	Automatic B.M.
3·5-in. 22-pr. (8·8 cm.) Q.F. L/30.	3·46	·5	8 7	27·1	—	—	H.E. Common	22	3·13	1,936	—	1·97	New T.B.D. and submarines gun.
3·5-in. 22-pr. (8·8 cm.) Q.F. L/45.	3·46	—	—	—		See	Table of Ammunition also Star shell			—	—	—	Semi-automatic gun.
3·5-in. 15-pr. (8·8 cm.) Q.F. L/30.	3·46	—	—	—						—	—	—	T.B. and older T.B.D. gun.
3·5-in. 15-pr. (8·8 cm.) Q.F. L/35.	3·46	1·34	10 1	32·1						2,461	—	—	—
		Lbs.									Ft.-Tons	Ins.	
7-pr. (6 cm.) Bts. B.L. L/21	2·36	238	4 1	18·6	0·88	Grob. K.P.	C.I. Ring Shell	6·6	0·15	1,168	62	—	Boat and field guns.
7-pr. (6 cm.) Q.F. L/21	2·36	—	4 1	18·6	0·24	W.P. C/89	C.I. Ring Shell	6·6	0·15	—	—	—	
4-pr. (5·2 cm.) Q.F. L/55	2·05	860	9 5	52	—	—	C.I. Common	3·8	—	2,789	205	—	—
									Oz.				
4-pr. (5 cm.) Q.F. L/40	1·97	529	6 6	36·6	0·86	R. P. C/00	C.I. Common	3·8	0·9	2,165	124	—	—
4-pr. (5 cm.) Q.F. L/55	1·97	—	—	—	—	—	—	3·8	—	—	—	—	—
					Oz.								
1-pr. (3·7 cm.) Maxim automatic.	1·46	416	6 2	—	1·5	—	Common	1·0	0·4	1,800	22	—	—
·31-in. (8 mm.) Maxim automatic.	0·311	—	—	—	—	—	—	—	—	—	—	—	Rifle calibre M.G.

O 2016—3 C

Part III.
Section 1.

Fleet Strength.

FLEET STRENGTH.

Before the war the full strength of the German Fleet had been fixed at—

41 battleships, 20 large cruisers, 40 small cruisers } by the 1912 Amendment *Law* to the Fleet Law of 1900.

144 destroyers, 72 submarines } by the Argument to the 1912 Amendment *Bill* to the Fleet Law of 1900.

The number of gunboats and special ships was not fixed.
This strength was to be attained as follows :—

Pre-War Programme of Construction.

Until the fleet had attained the established strength, the programme of construction was to consist of (1) the additional vessels required to bring it up to full strength, and (2) the vessels built to replace those which have reached the legal limit. After it had been attained, the programme, in theory, was to become one of replacement only.

As regards (1)—The following additional ships had still to be laid down to complete the fleet to established strength :—

1 battleship "U" in 1916, 1 small cruiser "P" in 1917 } included in the programme below.

1 battleship "V", 2 small cruisers "Q" and "R" } at a date to be determined later, and therefore not included in the programme below.

As regards (2)—According to Clause 2 of the Fleet Law, **battleships, large and small cruisers,** except in case of loss, were to be replaced after 20 years, this period being reckoned from the granting of the first instalment of the ship to be replaced to the granting of the first instalment for the ship which is to replace it. In elucidation, it was stated in the Amendment Bill of 1908, that the (actual) "replacement of a ship does not take place when the first instalment of the substitute ship is granted, but when the substitute ship is completed and commissioned."

The life of **destroyers**† **and submarines**‡ was taken to be 12 years, but was not legally fixed.

Excluding gunboats and other special ships, the programme of construction followed from 1897–98 to 1913–14, and that to be followed in accordance with an Appendix to the 1912 Fleet Law Amendment Bill from 1914–15 to 1917–18 inclusive, was as follows :—

Financial Year.	Battle-ships.	Large Cruisers.	Small Cruisers.	Torpedo Boats (*i.e. Destroyers*).
1897–98	1	—	—	—
1898–99	2	1	2	6
1899–1900	3	—	2	6
1900–01	2	1	2	6
1901–02	2	1	3	6
1902–03	2	1	3	6
1903–04	2	1	2	6
1904–05	2	1	3	6
1905–06	2	1	3	6
1906–07	2	1	2	12
1907–08	2	1	2	12
1908–09	3	1	2	12
1909–10	3	1	2	12
1910–11	3	1	2	12
1911–12	3	1	2	12
1912–13	1	1	2	12
1913–14	2	1	2	12
1914–15	1	1	2	12
1915–16	1	1	2	12
1916–17	2*	1	2	12
1917–18	1	1	2*	12

* Including one additional ship.

No special programme of construction had been drawn up for the period after 1917–18.

The exact number of submarines laid down each year is not known, but 29 submarines in all were completed between 1906 (U. 1) and August 1914.

A tabular statement of battleships, large and small cruisers built, building, or to be built under the pre-war programme, either in replacement, or as additional, *i.e.*, to bring the fleet up to full strength, is given on next page.

† Memorandum accompanying Navy Estimates, 1906.
‡ Argument to the 1912 Amendment *Bill* to the Fleet Law of 1900.

Note 1.—In regard to the columns headed "Construction," where a letter follows a ship's name, or appears alone, the ship is additional to those in the first column which were reckoned towards the Established Strength in Appendix A. to the Fleet Law of 1900; where a ship's name only is given, or is preceded by the word *Ersatz* the ship is in replacement of the ship opposite it in the first column.

Note 2.—Ships in italics had been removed from the effective list before outbreak of war.

Note 3.—Dates in italics are dates of launch.

BATTLESHIPS

Ships reckoned towards the Established Strength in Appendix A. to the Fleet Law of 1900. (*See* Notes 2 and 3 above.)	Year of granting First Instalment.	Year First Instalment was to be granted.	Construction from 1901 inclusive onwards. (*See* Note 1 above.)
		1901	Braunschweig (H).
		1901	Elsass (J).
		1902	Preussen (K).
		1902	Hessen (L).
		1903	Lothringen (M).
		1903	Deutschland (N).
		1904	Pommern (O).
		1904	Hannover (P).
		1905	Schleswig-Holstein (Q).
		1905	Schlesien (R).
Bayern	—	1906	Nassau.
Sachsen	—	1906	Westfalen.
Württemberg	—	1907	Rheinland.
Baden	—	1907	Posen.
Oldenburg	1881	1908	Ostfriesland.
Siegfried	1887	1908	Helgoland.
Beowulf	1889	1908	Thüringen.
Frithjof	1889	1909	Oldenburg.
Hildebrand	1890	1909	Kaiser.
Heimdall	1891	1909	Friedrich der Grosse.
Hagen	1891	1910	Kaiserin.
Aegir	1892	1910	König Albert.
Odin	1892	1910	Prinzregent Luitpold.
Kurfürst Friedrich Wilhelm.§	1889	1911	Grosser Kurfürst.
Weissenburg§	1889	1911	Markgraf.
	—	1911	König (S).
Brandenburg	1889	1912	Kronprinz.
Wörth	1889	1913	Baden.
	—	1913	Bayern (T).
Kaiser Friedrich III.	1894	1914	Sachsen.
Kaiser Wilhelm II.	1896	1915	Württemberg.
Kaiser Wilhelm der Grosse.	1897	1916	*Ersatz* Kaiser Wilhelm der Grosse.
	—	1916	(U).
Kaiser Barbarossa	1898	1917	*Ersatz* Kaiser Barbarossa
		*	(V).
Kaiser Karl der Grosse	1898		
Wittelsbach (C)	1899		
Wettin (D)	1899		
Zähringen (E)	1899		
Mecklenburg (F)	1900		
Schwaben (G)	1900		

LARGE CRUISERS.

Ships reckoned towards the Established Strength in Appendix A. to the Fleet Law of 1900. (*See* Notes 2 and 3 above.)	Year of granting First Instalment, *or of launch.*	Year First Instalment was to be granted.	Construction from 1901 inclusive onwards. (*See* Note 1 above.)
König Wilhelm	*1868*	1901	Friedrich Carl.
Kaiser	*1874*	1902	Roon.
Deutschland	*1874*	1903	Yorck.
		1904	Gneisenau (C).
		1905	Scharnhorst (D).
		1906	Blücher (E).
		1907	von der Tann (F).
		1908	Moltke (G).
		1909	Goeben (H).
		1910	Seydlitz (J).
		1911	Derfflinger (K).
Kaiserin Augusta	1888	1912	Lützow.
Hertha	1895	1913	Hindenburg.
Victoria Louise	1895	1914	Manteuffel.
Freya	1895	1915	Mackensen.
Vineta	1896	1916	*Ersatz* Vineta.
Hansa	1896	1917	*Ersatz* Hansa.
Fürst Bismarck	1895		
Prinz Heinrich	1898		
Prinz Adalbert (B)	1900		

SMALL CRUISERS.

Ships reckoned towards the Established Strength in Appendix A. to the Fleet Law of 1900. (*See* Notes 2 and 3 above).	Year of granting First Instalment, *or of launch.*	Year First Instalment was to be granted.	Construction from 1901 inclusive onwards. (*See* Note 1 above.)
		1901	Frauenlob (G).
		1901	Arcona (H).
		1901	Undine (J).
		1902	Hamburg (K).
		1902	Bremen (L).
Zieten	*1876*	1902	Berlin.
		1903	München (M).
Merkur (ex Arcona)	*1885*	1903	Lübeck.
		1904	Leipzig (N).
Alexandrine	*1885*	1904	Danzig.
Meteor	*1890*	1904	Königsberg (*old*).
		1905	Stuttgart (O).
Wacht	*1887*	1905	Stettin.
Blitz	*1882*	1905	Nürnberg (*old*).
Pfeil	*1882*	1906	Emden (*old*).
Comet	*1892*	1906	Dresden.
Greif	*1886*	1907	Kolberg.
Jagd	*1888*	1907	Mainz.
Schwalbe	*1887*	1908	Cöln.
Sperber	*1888*	1908	Augsburg.
Bussard	*1890*	1909	Magdeburg.
Falke	*1891*	1909	Breslau.
Cormoran	*1892*	1910	Stralsund.
Condor	*1892*	1910	Strassburg.
Seeadler	*1892*	1911	Karlsruhe (*old*).
Geier	1893	1911	Rostock.
Irene	1886	1912	Regensburg.
Prinzess Wilhelm	1885	1912	Graudenz.
Gefion	1890	1913	Wiesbaden.
Hela	1893	1913	Frankfurt.
Gazelle	1896	1914	Königsberg (*new*).
Niobe (B)	1898	1914	Karlsruhe (*new*).
Nymphe (A)	1898	1915	Emden (*new*).
Thetis (C)	1899	1915	Nürnberg (*new*).
Ariadne (D)	1899	1916	*Ersatz* Ariadne.
Amazone (F)	1900	1916	*Ersatz* Amazone.
Medusa (E)	1900	1917	*Ersatz* Medusa.
		1917	(P).
		†	(Q).
		†	(R).

* The date of laying down this additional ship, sanctioned by the 1912 Amendment to the Fleet Law, was not fixed at outbreak of war.

† The date of laying down these two additional ships, sanctioned by the 1912 Amendment to the Fleet Law, was not fixed at outbreak of war.

§ Sold to Turkey in 1910.

Part III.
Section 1.

CONSTRUCTION AND LOSSES DURING THE WAR.

Fleet Strength.

It may be of interest to compare the figures given at the top of page 18, which were to be attained at some date subsequent to 1920, with the actual known strength of the German fleet in June 1917.

The figures for June 1917, excluding vessels outside the age limit, are as follows :—

Thirty-six battleships, including three vessels of the *Kaiser Friedrich* class, which are of no practical fighting value.

Seven large cruisers, viz., five battle cruisers and two cruisers, one of the latter being negligible as a modern fighting unit.

Thirty-two small cruisers.

About 150 destroyers.

About 160 submarines of all classes, including five of no fighting value, which serve for instructional purposes.

It will thus be seen that the number of battleships and large and small cruisers still falls considerably short of the intended standard; this is, of course, partly due to the losses sustained during the war, particularly in the case of cruisers, as is shown below.

On the other hand, the number of destroyers is slightly, and the number of submarines very much, in excess of the prescribed strength; great activity having been displayed in the construction of submarines throughout the war, and in the construction of destroyers during the first two years of it.

All the battleships, battle cruisers, and light cruisers authorised under the pre-war programme of construction, have been laid down on or before their appointed dates. In the case of light cruisers the dates have been considerably anticipated, and in addition, a number of smaller light cruisers of the *Brummer* type, not contemplated in the pre-war programme, have been laid down. These last-named vessels, though essentially light cruisers in general design, speed, and armament, are specially adapted for minelaying, and therefore in this book are designated minelaying cruisers.

The rate of construction for capital ships and light cruisers has been, almost without exception, slower than the peace rate, although not greatly so.

The following are (June 1917) the figures of losses and newly completed ships since August 1914 :—

——	Lost.	Added.	Increase or Decrease.
Battleships	1	4*	+ 3
Battle cruisers	1	2†	+ 1
Cruisers	6	—	− 6
Light cruisers	17	15‡	− 2

* Two *Königs*, two *Bayerns*. † *Lützow* and *Hindenburg*. ‡ Includes four *Brummers*.

EXPENDITURE ON NEW NAVAL CONSTRUCTION AND ARMAMENTS.

Financial Year.	Amounts *voted* for New Construction, including Armaments.
	£
1890–91	1,844,712
1891–92	1,734,900
1892–93	1,412,935
1893–94	1,363,945
1894–95	972,930
1895–96	1,062,535
1896–97	1,252,340
1897–98	2,454,400
1898–99	2,565,600
1899–1900	2,832,750
1900–01	3,401,907
1901–02	4,653,423
1902–03	4,662,769
1903–04	4,388,748
1904–05	4,275,489
1905–06	4,720,206
1906–07	5,167,319
1907–08	5,910,959
1908–09	7,795,499
1909–10	10,177,062
1910–11	11,392,856
1911–12	11,710,859
1912–13	11,491,187*
1913–14	11,010,883†
1914–15	10,316,264‡

* Includes 97,847*l*. for airships and experiments with airships.
† Includes 291,096*l*. for airships and experiments with airships.
‡ Amount proposed, includes 431,405*l*. for airships and experiments with airships.

C 3

INDEX TO VESSELS

Page references are to the new volume numbers, not to the original parts; illustration references are in *italics*.